To Barry,

I don't know if this book will help you how works the Internet, but I'm sure, it will help you to fall asleep after a few pages every evening of the year!

With my best regards

Maurice

Broadband Local Loops for High-Speed Internet Access

For a listing of recent titles in the *Artech House Telecommunications Library,* turn to the back of this book.

Broadband Local Loops for High-Speed Internet Access

Maurice Gagnaire

This work has been published with the help of the French Ministère de la Culture–Centre national du livre.

Artech House
Boston • London
www.artechhouse.com

Library of Congress Cataloging-in-Publication Data
Gagnaire, Maurice.
    Broadband local loops for high-speed Internet access/Maurice Gagnaire.
        p. cm.—(Artech House telecommunications library)
    Includes bibliographical references and index.
    ISBN 1-58053-089-3 (alk. paper)
    1. Local area networks (Computer networks). 2. Broadband communication systems.
    3. Internetworking (Telecommunication). 4. Internet. I. Title. II. Series.

    TK5105.7.G32 2003
    004.6'8—dc21   2003049578

British Library Cataloguing in Publication Data
    Gagnaire, Maurice
        Broadband local loops for high-speed Internet access.—
        (Artech House telecommunications library)
        1. Broadband communication systems 2. Local area networks (Computer networks)
        3. Internetworking
        I. Title
        621.3'82

    ISBN 1-58053-089-3

Cover design by Gary Ragaglia

© 2003 ARTECH HOUSE, INC.
685 Canton Street
Norwood, MA 02062

Translated from the original French work *Boucles d'accès hauts débits. Dégroupage et techniques HDSL, ASDL, G-Lite, VDSL et VoDSL,* © Dunod, Paris 2001

International Standard Book Number: 1-58053-089-3
Library of Congress Catalog Card Number: 2003049578

10 9 8 7 6 5 4 3 2 1

To my spouse Brigitte and to my parents Jean-Claude and Marie-Louise

Contents

Foreword

It is a pleasure to see the book *Broadband Local Loops for High-Speed Internet Access* by Professor Maurice Gagnaire become available to the public.

It is widely accepted that communications technology—especially the services offered through the Internet—has already become an inseparable component of our business infrastructure and daily life. The progress in broadband communications, enabling multimedia, videoconferencing, and mobility, is an important facet of this dynamic technology. Current economic conditions, with challenges to further reduce the operational expenses associated with the network infrastructure and cost-effective means of access to it, have had a temporarily adverse effect on the communications industry; nonetheless, traffic continues to increase. There still seem to be many exciting opportunities for innovations and developments in this field.

This book addresses a number of important contemporary topics in telecommunications. It is a good source of reference for professionals in the telecommunications industry, as well as for students who wish to acquire a knowledge of networking and telecommunications with a view toward the standards prevailing in the industry. Topics related to network access and infrastructure, wireline and wireless, telephony and the Internet, and optical networking have all been addressed. While certain chapters of this book familiarize readers with current telecommunications practices and their associated terminology, other chapters cover the foundations of specific technologies, such as digital subscriber lines. Furthermore, the book provides several examples of current or forthcoming communications systems. This book also provides many illustrations that make the discussions vivid and more meaningful.

The book, which is comprised of 12 chapters, covers developments in the field of telecommunications. No doubt, it has taken a significant time and effort by the author to write and interact with the publication staff to prepare this manuscript. Professor Gagnaire should be commended for this accomplishment.

Dr. Bijan Jabbari
George Mason University
Fairfax, Virginia
June 2003

Preface

Motivation

This book was originally motivated by a course entitled "Boucle Locale Large (BLL) Bande" (or broadband access systems) that I established for the final-year students (the equivalent of graduate students in the North American education system) of the *École Nationale Superieure des Telecommunications* (ENST) in 1995. The course aims to give students an up-to-date overview of the various techniques used in the local loop for the provision of enhanced services over the Internet. Accordingly, a large part of the program is dedicated to the description, analysis, and comparison of the technical aspects of the various technologies for the local loop. The course considers both existing systems used today in carriers' networks and prospective systems. The great majority of my students will work for industry, service providers, and telecommunications carriers; only a few are furthering their academic carrier to prepare a Ph.D. This is why the BLL course also covers the regulatory and economic factors that one must understand to have a complete and comprehensive overview of the domain. In July 2001, I published with Dunod Editor a book in French essentially dedicated to x-digital subscriber line (xDSL) access systems and entitled *Boucles d'Accès Haut Debit* (or *Broadband Local Loop*). The spectrum covered by the English version of this book is much wider. Thus, as a few chapters of the French edition have been translated and updated, new chapters or sections describing cable networks, power line communications, wireless, and optical access systems have been added. Chapter 3 is dedicated to unbundling compares the various approaches adopted in European countries and America for handling the open local loop market. In recent years, I have had the opportunity to collaborate with major equipment manufacturers and with telecommunications carriers to investigate specific aspects of broadband access systems. The definition of new *medium access control* (MAC) protocols and new traffic management tools for optical and wireless access systems are two examples of such investigations. First, recent advances in digital communications are at the basis of the performance of new generation modems that enable higher throughputs in access. Second, new end-to-end protocol architectures are necessary to enable the provision of enhanced services through packet-switching networks [*Internet protocol* (IP), *asynchronous transfer mode* (ATM), and frame relay networks]. Throughout this book, I attempt to consider for each technique (wired or wireless) both these factors without forgetting the level of standardization and the level of implementation by manufacturers. In terms of service, the protocol architecture within core packet-switching networks and within customer premises equipment do not differ considerably from one access system to another.

Intended Audience and Objectives

This book is intended for both an academic and a professional audience. Although it requires a basic background in digital communications and in communication protocols, most of the chapters of the book are understandable by nonspecialists of the domain. Engineers from industry and graduate students in computer science, digital communications, and networking will find in this book a practical approach that associates different fields of competence that are in general considered separately. Typically, discussions of digital communications outline the main constraints at the physical layer and describe the network protocols more or less independently from the medium characteristics. Due to space limitations, the book does not include problems or exercises and thus cannot be considered, strictly speaking, as a textbook for graduate students. Nevertheless, certain chapters, like those dealing with *asymmetrical DSL* (ADSL) modems (Chapter 6), or wireless access systems (Chapter 11), are sufficiently detailed by means of commented figures and equations to facilitate the comprehension of theoretical concepts. As well as rather academic developments (for instance, the description of the principle of multicarrier modulation), the book contains very practical discussions. Consider, for example, Chapter 3, dedicated to unbundling, or Section 11.5, dedicated to radio cell planning. Technicians and engineers working for a carrier should benefit from such considerations. The description in Chapter 9 of the provision of new services like voice and video over DSL in an end-to-end perspective also interests service providers. Globally, this book aims to serve as a basic reference enabling a comprehensive and up-to-date overview of new broadband access techniques.

Acknowledgments

First, I express my gratitude to Professor Bijan Jabbari from George Mason University in Fairfax, Virginia, who kindly agreed to write the foreword of this book. I am grateful for the strong support of Marc Oberlé, the administrator of the Groupement des Écoles des Telecommunications[1] (GET); of Marc Peyrade, head of ENST; and of Michel Riguidel, head of the computer science and network department of ENST. Several people have contributed to the detection of editorial and technical errors in the successive versions of my manuscript. I want to express my thanks to Saso Stojanovski, a research engineer at Nortel Networks, who attentively read several chapters of the French version of this book. I also thank my colleague Professor Philippe Godlewski for our numerous and fruitful discussions about error-correcting codes and modulation techniques. I am grateful to my colleague Laurent Decreusefond for his comments about the water-filling mechanism used in discrete multitone modulators. I have benefited from the experience of my colleague Claude Rigault; his very good knowledge of North American public switched telephone networks has been very useful, notably for the translation of many acronyms and technical terms. The discussions I had with my Ph.D. students Josue Kuri and Mohammed Koubaa also helped me to enrich certain sections of the book. My colleague Mary O'Connell has kindly agreed to review the annex sections of this book. I have closely collaborated from the very beginning of the project until the preparation of the final manuscript with Julie Lancashire and Tiina Ruonamaa from Artech House. Their permanent encouragement helped me to accomplish this work without an unreasonable delay. I also want to express my thanks to the multiple anonymous

---

1. GET is a subsidiary of the French Ministry of Industry. It is made up of the seven major graduate schools of France in the field of information technology.

reviewers of my English manuscript who worked under the supervision of Julie and Tiina. Without their valuable and meticulous work, this book could not have been finalized. Last, but not least, I want to express my deep thanks to my wife Brigitte. She has kindly accepted that a great part of my time out of the office has been dedicated to the writing of this book.

Introduction

Telecommunications networks may be divided into three main areas: access networks, *metropolitan area networks* (MANs), and core networks:

- Access networks, frequently considered the "last mile" of public networks, link end users to *central offices* (COs). The main objective of an access network is to concentrate upstream traffic generated by the end users to reduce the size of the access switches. Although they are traditionally built on copper wires, access networks may rely on other types of mediums such as radio access, optical fibers, power line communications, and coaxial cables. The term *local loop* refers traditionally to access networks based on twisted copper pairs, although it is today extended to alternative transmission media.

- MANs are used to federate upstream traffic coming from access networks before forwarding it to core networks. In their current configuration, the great majority of MANs are made of *synchronous optical network/synchronous digital hierarchy* (SONET/SDH) dual counter-rotating optical fiber rings. Since the end of the 1980s, SONET/SDH technology has been widely deployed by the carriers in the metropolitan area and in the core to guarantee networks' survivability in case of a fiber cut or a node failure. SONET/SDH is not used in the access network.

- Core networks are in charge of the long-haul transmission associated with general heterogeneous networking technologies that can be either connection-oriented[1] or connectionless-oriented.[2] *Public switched telephone networks* (PSTNs) on one side and ATM[3] or frame relay networks on the other side are typical examples of connection-oriented networks.

The former are based on circuit switching,[4] whereas the latter are based on packet switching.[5] The Internet, which is comprised of several autonomous IP networks,[6] is connectionless-oriented. *Internet service providers* (ISPs) manage databases connected to IP routers by means of digital leased lines. The various nodes within core networks are linked together by means of optical fiber links.

In a context of deregulation of the telecommunications market, competition between carriers is less in core and MANs than in access networks. Indeed, the huge increase in transport network capacity observed during the last 10 years makes the price of the transported bit tend to zero. Up until now, traditional telephony based on PSTN has been the major source of income for telecommunications carriers. Several factors should contribute to an important reorganization of the market in the next few years. Indeed, the recent deregulation of the local loop, the emergence of voice over packet-switching networks and *computer telephony integration* (CTI) favor the arrival of new competitors. While the number of telephony service providers is increasing, hybrid services combining voice, data, and video [e.g., interactive games and video distribution via *electronic program guides* (EPGs)] will be a new source of income. Technically and economically speaking, the provision of such services requires control of the last mile. This control enables a carrier to install the modems of its choice at each end of the access network. It facilitates a direct contact with the final customer (enterprises and residential users). This is why the local loop may be considered today to be a strategic segment of telecommunications market. Advances in the fields of transmission techniques (i.e., digital communication, signal processing, and electronics) and protocol architectures are at the origin of the new broadband access technologies described in this book. This book aims to give a comparative state of the art and a perspective view of these broadband access technologies.

---

1. Communication between distant users via a connection-oriented network needs a preliminary connection establishment between the calling party and the called party. This connection setup is used for resource reservation (bandwidth and buffer capacity) along a determined route through the network.

2. Communication between distant users via a connectionless-oriented network does not impose preliminary resource reservation. Data packets transferred between the two parties may then follow different routes through the network.

3. Asynchronous transfer mode.

4. Circuit switching assumes a deterministic bandwidth reservation on the basis of the peak bit rate of individual connections.

5. Packet switching assumes a statistical bandwidth reservation proper to each connection.

6. Autonomous IP networks are managed by independent commercial entities.

Traditional modems (also called voiceband modems[7]) enable full-duplex data rates limited to a few tens of kilobits per second. The fact that 800 million copper wires are already installed in the last mile all over the world has strongly motivated the development of a new generation of modems known as xDSL modems in the 1990s, with the prefix "x" referring to the numerous versions of DSL modems that have been developed, such as *high bit rate DSL* (HDSL), *asymmetrical DSL* (ADSL), and *very high-speed DSL* (VDSL). In parallel to xDSL, several other technologies have also been considered for the local loop. These technologies are based on various types of physical mediums such as optical fibers, coaxial cables, energy cables, and radio access. For our purposes, we consider four main aspects in the choice of technology for the local loop:

- *Industrial aspects:* New kinds of modems have to be designed and developed to enable higher bit rates (of a few megabits per second) in the local loop. Depending on the adopted transmission medium, these high bit rates require more or less complex electronic devices. Some manufacturers believe that in the next few years, the market of new generation modems, whatever the considered medium, will be a mass market, with each family being a potential consumer. The mass production of these modems will need the design of dedicated *very large scale integration* (VLSI) chips.

- *Technical aspects:* Two approaches are possible for the new generation local loop, one using existing infrastructures or one installing a fully new infrastructure. In the former case, one tries to use more efficiently twisted copper pairs or coaxial cables. The technical contribution in this case is focused in the end equipment installed at the customer premises and at the CO. In the latter case, a new physical medium must be set up and new end equipment must be developed. New generation modems must be "intelligent" in the sense that they have to integrate multiservice, multiprotocol applications. As with ATM, ideally, the hardware and software of these modems should be flexible to be adaptable to new services that do not exist at the moment. Reliability and simplicity are also two important qualities that should satisfy new technologies for the local loop. In the case of end-equipment failure or of a problem in the infrastructure (a cable cut, for instance), the reconfiguration of a system should be rapid (automatic) and cost-effective. Both these properties prohibit any visit of a representative of the service provider or of the access provider in case of problem at the customer premises.[8] Simplicity

---

7. The meaning of the term voiceband is explained in Chapter 2.

8. For instance, end users may replace a faulty modem with a new modem purchased in the shop of their choice.

is another desirable quality in such technology. Again, a solution that does not require a visit from a representative of the service provider or the access provider to the customer to active the system is economically advantageous. Ideally, solutions that do not need any cabling at the end user and that are plug-and-play–oriented are commercially less risky. By nature, wireless access systems are not sensitive to cable cuts. Nevertheless, high bit rates are more difficult to achieve on radio systems than on wired systems.

- *Financial aspects:* This aspect concerns access providers and service providers. For both of them, it is important to choose the right technology among the various possible alternatives. They need to make a trade-off between the expected return on investment duration and the expected revenue. It is clear that, on this point, the situation of incumbent carriers [also called *incumbent local exchange carriers* (LECs) (ILECs)] is totally different from the situation of competitive carriers [also called *competitive local exchange carriers* (CLECs)]. The former already own a copper or a coaxial infrastructure that they have exploited for several years. CLECs have two opportunities; either they build their own infrastructure, or they rent it from ILECs. The choice of infrastructure conditions both the type of services that can be offered and the life duration of these services.

- *Regulatory aspects:* As mentioned earlier, the main source of revenue in the telecommunications market is in innovative hybrid services. These services imply a direct contact with the end consumer. The more a carrier or a service provider has control of the last mile, the more it is autonomous in terms of service offers, tariffs, and customer satisfaction. The regulatory context of each country plays a major role. The problem for regulation bodies is finding an equilibrium point between the previous monopolistic situation of ILECs and fair competition among ILECs and CLECs. The term *unbundling* refers to this regulatory aspect.

- *Standardization aspects:* The flexibility mentioned above passes by standardized interfaces both at the network side and at the customer premises side. End equipment must be compatible between different vendors; this is particularly true when both ends of the access system need to be equipped with the same modem (as in the case of xDSL).

This book is organized into 12 chapters. Chapter 1 describes the two approaches currently adopted for access to the Internet, direct or indirect and discusses the recent innovations that aim to provide end-to-end quality of service in IP and ATM networks. Chapter 2 describes the various uses of copper wires in access networks, such as digital leased lines and basic access to narrowband *integrated services digital network* (ISDN). We outline the characteristics and limits of traditional voiceband modems, up to the V.90 and V.92 standards. Chapter 3 is divided into two independent parts: The first half explains the principle and technical constraints of unbundling; The other half briefly describes three specific approaches for the local loop—satellite access systems, cable networks, and power line communications. Chapter 4 analyzes the impact of various disruptive effects proper to the local loop on Shannon's capacity of twisted copper pairs. Chapter 5 describes HDSL access systems and related technologies such as IDSL, SDSL, HDSL2, and SHDSL. Chapter 6 covers the description of ADSL modems, detailing the various elements of the block diagram of an ADSL modem and the ADSL line initialization procedure. Chapter 7 first presents the G.lite technology that can be seen as an enlightened version of ADSL imposed by current implementation constraints. The remainder of Chapter 7 deals with VDSL technology, an enhanced version of ADSL. Chapter 8 covers two independent topics: *DSL access multiplexers* (DSLAMs) and home networking. A DSLAM corresponds to the set of xDSL modems that has to be installed at the CO in association to each xDSL modem installed at the customer premises. Home networking refers to the multiple private network configurations enabling the connection of several end terminals on a same user's modem. Chapter 9, which is dedicated to the protocol aspects of xDSL access systems with a focus on broadband access servers and the L2TP protocol, presents and discusses various protocol stacks applicable to end-to-end connections. Chapter 10 describes how packetized voice networks may interoperate with traditional PSTN networks thanks to signaling protocols such as H.323, SIP, Megaco, and Sigtran. Chapter 10 also considers video distribution over ADSL access systems. Chapter 11 deals with radio access systems, successively presenting narrowband (DECT, PHS, PACS) and broadband (LMDS, IEEE 802.16) wireless access systems. Chapter 12 deals with two partially independent topics. The first half of Chapter 12 covers the various options for the optical access networks known as *ATM over passive optical network* (APON) and *broadband passive optical network* (BPON) systems. APONs and BPONs have the support of the ITU-T. The second half of Chapter 12 focuses on the emerging concept of *Ethernet-in-the-first-mile* (EFM) access systems. EFM access networks have the support of the IEEE.

1

Telecommunications Network Evolution

1.1 Introduction

This chapter presents both the main characteristics and the evolution of public telecommunications networks. Section 1.2 gives a historical perspective on the Internet and introduces the concepts of direct and indirect access to the Internet. Section 1.3 deals with *quality of service* (QoS) provisioning in data networks. Both the cases of ATM and IP networks are investigated. Section 1.4 focuses on the concept of *virtual private networks* (VPNs) based on the IP protocol (IP-VPNs). Section 1.5 concludes the chapter.

1.2 The Internet and Its Access Modes

1.2.1 A Historical Perspective

The Internet found its origins in the ARPAnet network instituted at the end of the 1970s. The Arpanet's name is derived from the *Advanced Research Projects Agency* (ARPA), a federal American agency in charge of promoting financially the development of a new generation of data networks based on packet switching. The ARPAnet network was used to interconnect several research centers within the United States. The famous *Transmission Control Protocol* (TCP)/IP (TCP/IP) stack was developed by researchers working under the coordination of ARPA [1, 2]. The ARPAnet network was rapidly extended geographically to interconnect international research centers located in different parts of the world. The Internet, as it is known today, dates from the beginning of the 1980s. It can be seen as a concatenation of independent packet-switching networks transporting datagrams[1] with a common format, currently the IPv4 packet format.

At the beginning of the 1980s, the U.S. *National Science Foundation* (NSF) was responsible for the ARPAnet. In the remainder of this book, I use the generic term Internet to refer to this international IP-based data network. The Internet users' community was extended to academic institutions (including researchers, engineers, professors, and students) equipped with UNIX computers.[2] In 1985, the NSF promoted a new high-speed backbone for the Internet called NSFnet, itself connected to the ancient ARPAnet [1].

Whereas only a few hundred computers were connected to the Internet at the beginning of the 1980s, Vinton G. Cerf, one of the fathers of the Internet, estimates that about a billion users should be connected to this worldwide network in 2007 [3]. Today, the Internet is accessible to almost any member of the community who owns a *personal computer* (PC) based on one of the three most current operating systems (Linux, MS-DOS Windows, and MAC-OS). The success of the Internet is at the basis of the new economy, whose many facets include e-commerce, e-banking, and e-learning. Geographically speaking, the Internet is now spread among approximately 200 countries. It is not a unique universal network but rather a set of autonomous networks managed by different administrative entities (i.e., official national bodies, private companies, and governments). Today, the main innovations in terms of TCP/IP evolutions and implementations are carried out under the supervision of the *Internet Architecture Board* (IAB) and the *Internet Engineering Task Force* (IETF).

Different players with different levels of responsibility cooperate to provide new services to end users over the Internet. They are described as follows:

- *ISPs:* ISPs represent the commercial entities that provide services to the end users accessing to the Internet. These services cover the multitude of Web servers available over the world and audio and video servers using streaming multicast technology, among others.

- *Managers of private internet networks:* Private internet networks[3] correspond to *local area networks* (LANs)[4] under the responsibility of one enterprise or one private organization. In this context, the adjective "local" must not be considered in terms of range. Indeed, an enterprise can manage its distant LANs as if they were colocated at the same place[5]

---

1. The word datagram is used interchangeably with the words IP packet.

2. The TCP/IP protocol stack is related to the UNIX operating system.

3. By convention, we sometimes use in the remainder of this book the term internet, with a lowercase "i" to distinguish each autonomous domain using the IP protocol from the global Internet.

4. A LAN runs the IP protocol.

thanks to point-to-point leased lines. This service, known as VPN, is today a very important source of income for the carriers. Two kinds of *network service providers* (NSPs) may provide data transport services to enterprises for the setting of their VPN: traditional telecommunications operators (telcos) or internet transit providers. The former are in general incumbent operators of connection-oriented networks such as PSTN, ATM networks, or frame relay networks. Internet transit providers are described more specifically later in this section.

- *Regional internet carriers:* In each country, several private operators called regional internet carriers may operate their own IP network on which are connected in general the LANs of several ISPs. These regional internet carriers are in charge of the routing policy and of the *operation and maintenance* (OAM) of their IP routers interconnected by means of optical fiber links. In many cases, regional internet carriers own only the IP routers and lease the fiber links to traditional long-haul carriers. Regional internet carriers offer to ISPs a connectionless transport service from an ingress access node of their network to the site of the ISP itself. Let us note that today, regional internet carriers and ISPs correspond in many cases to the same commercial entity. As the adjective "local" refers to a virtual locality in the case of VPNs, the adjective "regional" used for regional internet carriers is also relative. Indeed, such carriers may in fact manage IP networks larger than a simple region, covering, for instance, a whole country or in some cases, several countries. In many cases, the connection between an end user and a service provider needs to cross the infrastructure of several regional Internet carriers (or ISPs). One uses the term of "peering" to refer to the cooperation that exists in general between regional internet carriers to facilitate traffic transfers through their respective infrastructures. A regional internet carrier (or ISP) is responsible for the allocation of IP addresses to end users needing to connect to the Internet.

- *Internet transit providers:* Internet transit providers are internet operators working at the international level to interconnect regional internet networks. The NSFnet mentioned above is the most important internet transit provider in the world. In Europe, the equivalent of the NSFnet is the *European Backbone* (EBONE).

- *Network access provider (NAP):* NAPs are in charge of interconnecting *customer premises equipment* (CPE) with the *point of presence* (POP) of packet-switching networks on which are connected the databases of the service providers. Currently, the infrastructure used by NAPs is largely based on frame relay (ATM being more expensive to manage). NAPs

---

5. Such networks are called *intranets*.

are in charge of filtering and concentrating traffic at the level of the CO. Thus, downstream IP packets coming from the various servers are dispatched toward the concerned end users. Upstream requests sent by the end users are grouped according to the address of the requested server.

In most cases, incumbent carriers are in charge of the local loop and of the filtering and concentration equipment at the CO. With deregulation in the local loop, new carriers called CLECs may also offer their services as NAPs.

1.2.2 The Two Types of Access to the Internet

Two kinds of access to the Internet are considered: direct and indirect. Direct access is used by enterprises that can afford to install their own IP router at their premises. Indirect access concerns residential users who own in many cases a single PC at home. In that case, the ISP provides a connection either temporarily or permanently to the Internet.

1.2.2.1 Indirect Access

A residential user accesses the Internet by means of an ISP. In the most current configuration, such access is carried out by means of an analog voiceband modem—also called a *dial-up modem*[6]—to connect a PC to a subscriber line. In this context, access to the Internet means access to the network of a regional internet carrier by way of a connection through a PSTN. Once the user has typed on his or her keyboard the *Uniform Resource Locator* (URL) of the Web server to which he or she wants to be connected, a phone connection is automatically set up by the user's voiceband modem to connect his or her PC to a *network access server* (NAS). NASs, which are controlled by ISPs, carry out three types of functionalities. The first one is to control the identity of residential users before allowing them to be connected to the distant Web server. The second one is to record eventually the duration of the connection of end users to a Web server. In most cases, this second functionality is not activated, as most of the ISPs offer nonmetered access.[7] The third functionality of a NAS is to allocate a temporary IP address to the end users. In general, a NAS is connected to one of the switches of a PSTN by means of a digital link (a narrowband ISDN link or a leased line[8]). If the identity of the end user belongs to the list of the customers of the ISP, the welcome page of the Web server is transmitted to the

---

6. Chapter 3 details the principle and limits of analog dial-up modems.

7. The term always on access is used in that case.

8. Chapter 3 details the characteristics and limits of narrowband ISDN links.

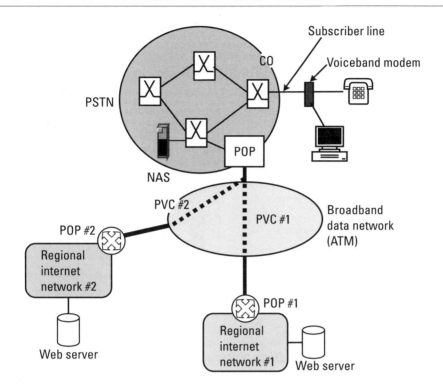

Figure 1.1 Indirect access to the Internet via an ATM backbone.

user, who may, in turn, send data to the server. The control of the end-to-end connection is managed by means of the TCP/IP stack.

In many cases, regional internet carriers rent semipermanent ATM virtual connections [called *permanent virtual channels* (PVCs)] to the traditional telcos to improve the efficiency (and thus the cost) of their own network. As shown in Figure 1.1, these PVCs (PVC #1 and PVC #2) are used to link the POPs of the regional internet carriers with the PSTN. The protocol stack used between ingress and egress POPs is *ATM adaptation layer 5* (AAL5) over ATM.

1.2.2.2   Direct Access

An IP router called an *access router* is in this case installed at the customer premises. As shown in Figure 1.2, this private router is directly connected to the POP of a regional internet carrier by means of a digital leased line. As in Figure 1.1, the regional internet carrier may rent an ATM PVC between its POP and the private router of the enterprise for efficiency issues. A NAS with the same functionalities as those mentioned for indirect access is connected to the POP of the regional internet carrier.

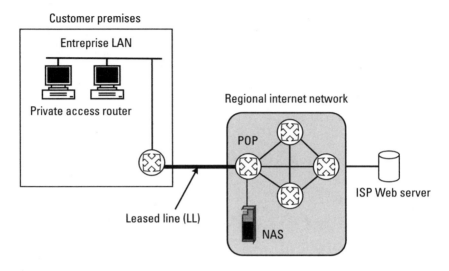

Figure 1.2   An example of a network configuration with direct access to the Internet.

1.3   QoS Provisioning in Data Networks

1.3.1   QoS in ATM Networks

A description of the ATM technique is out of the scope of this book. Nevertheless, we can simply recall that the ATM cell format (5 bytes of overhead for 48 bytes of payload) is well suited to real-time applications that, in general, generate small packets and require fast delivery. One of the ATM drawbacks that has been widely discussed is the cost in terms of protocol efficiency of the cell tax represented by this 5 bytes overhead in addition to the overheads inherent to the AAL layers. Unlike IP networks, ATM is connection-oriented and based on bidirectional virtual channels. At connection setup, end users sends their traffic descriptors (required bandwidth, residual error bit rate, admissible mean end-to-end delay, and admissible jitter) to their ATM switch. These descriptors are then analyzed by a *call-admission control* (CAC) procedure within the ATM network. ATM offers six different classes of service called *ATM transfer capabilities* (ATCs). In public networks, there are very few *switched virtual circuit* (SVC) capabilities, with most of the ATM connections being based on PVCs. Each ATC is defined by a combination of traffic and QoS parameters [4]. The ATCs have been specified with a different taxonomy by the ITU-T and the ATM Forum. These differences are described as follows.

- *Deterministic bit rate* (DBR) and *constant bit rate* (CBR): The DBR transfer capability specified by the ITU-T is used for CBR traffic with real-time constraints between a source and a destination. Its equivalent at the ATM Forum is the CBR transfer capability. DBR and CBR are typically used for *pulse-coded modulation* (PCM) voice and fixed bit rate video.

- *Statistical bit rate* (SBR) and *variable bit rate* (VBR): The ITU-T SBR and its ATM Forum equivalent called VBR concern VBR traffic with or without real-time constraints between a source and a destination. Packetized voice or video with adaptive coding is two typical examples of applications of these transfer capabilities.

- *Available bit rate* (ABR) and *guaranteed frame rate* (GFR): The ATM Forum has specified successively two transfer capabilities called ABR and GFR for TCP/IP traffic. Unlike voice and video traffic for which a minimum bandwidth must be reserved at call setup, asynchronous data traffic does not require such a reservation. TCP/IP traffic (file transfers, typically) is qualified as elastic traffic in the sense that a TCP/IP session may use transiently all the available capacity of a physical link. The ABR transfer capability is based on a dynamic control of the through-put of TCP/IP sources according to the degree of congestion within the ATM network. This control mechanism based on a closed-loop scheme provides explicit network feedback to the traffic sources. In practice, the ABR service is complex to implement and not always efficient in long-haul networks because of the inertia of the control. This is the reason why GFR has been developed. The GFR transfer capability is frame-oriented and requires minimal interaction between terminals and the network nodes. Unlike ABR, GFR guarantees a *minimum cell rate* (MCR) to elastic connections with a fair share of the available band-width between the active GFR connections. A GFR connection is associated to each TCP/IP session.

- *Unspecified bit rate* (UBR): The UBR transfer capability corresponds to the lowest priority traffic. Neither bandwidth reservation nor QoS monitoring is carried out for UBR traffic. Electronic mail (e-mail) is a typical example of traffic that can be supported by UBR.

1.3.2   QoS in IP Networks

Internet traffic volume increased significantly at the beginning of the 1990s because of the success of the World Wide Web and of e-mail. Today, e-commerce and IP telephony is accelerating this increase. In the IPv4 packet header, a specific field of 8 bits called *type of service* (TOS) enables the associa-tion of a given class of service to a packet. Various techniques presented in the remainder of this section exploit the value of the TOS field to adapt the behavior of a router to the class of service of each packet. Originally, the TOS field was divided into five subfields [see Figure 1.3(a)]. The first 3 bits indicate the level of priority of the datagram ("0" is the lowest priority; "7" is the highest one). In most current IP routers, the IP software limits traffic differentiation to the two highest levels of priority (urgent and nonurgent). The 3 following bits, "D," "T," and "R," indicate whether the packet requires a short transit delay, a high throughput, and a high resiliency. At the end of 1990, the IETF, which is in charge of IP protocol evolutions and standardization, redefined the TOS field under the name of the DiffServ codepoint as it is described in Figure 1.3(b). Section 3.2.2 comments on the various fields of the DiffServ codepoint.

In their traditional configuration, IP networks—unlike ATM networks—are unable to provide a guaranteed QoS to the end users. This QoS can be metered in terms of mean and maximum transfer delay of the packets, of residual *bit error rate* (BER), and of guaranteed bit rate. The key factor that has contributed to the success of the Internet compared to ATM networks is the simplicity of the TCP/IP stack. This section describes the principles of the mechanisms that have been successively considered for the provision of QoS in IP networks.

1.3.2.1 Integrated Services (IntServ) Approach

The IntServ approach has been the first attempt to introduce QoS in IP networks. It is based on the use of an end-to-end resource *Reservation Protocol* (RSVP). This protocol increases considerably the amount of processing to carry out in each IP router. In ATM, forwarding tables' parameters determined at connection setup are kept in the switches for the whole duration of a call. Routing tables in the routers are refreshed periodically by RSVP without an explicit demand from the end user. An IntServ router is able to detect a "flow," or a succession of IP packets belonging to a same logical session between two end equipments. The parameters associated to this logical session are kept in the IntServ routers only as long as the session is considered as alive. To control the activity of the logical session, RSVP packets are periodically exchanged between IntServ routers. Such a behavior is designated as "soft-state" protocol. It appears that a frequent reactivation of end-to-end paths by RVSP messages generates prohibitive traffic in the case of long-haul networks (as is the case with the regional

(a)   (b)

Figure 1.3 IPv4 (a) TOS byte and (b) DiffServ codepoint.

internet networks of the NSPs). Another drawback of the IntServ approach is that all the routers of an IP network have to be updated with RSVP signaling. This is why today, this approach is only considered in the case of limited-size IP networks like private IP networks.

1.3.2.2 Differentiated Services (DiffServ) Approach

The DiffServ approach is much simpler to implement than IntServ because it does not require any signaling channel [5]. Instead of trying to provide firm guaranteed QoS, DiffServ simply aims to divide IP traffic into different categories. Each category corresponds to a class of service called a *forwarding equivalent class* (FEC). The concept of the "DiffServ cloud" has been defined to proceed to this traffic differentiation. A DiffServ cloud is a set of IP routers able to distinguish IP packets according to their associated FEC. Two types of routers are used in a DiffServ cloud: internal routers and edge routers. An edge router carries out five operations: classifying, metering, shaping, dropping, and policing. Edge routers, which receive IP packets, proceed to traffic classification. Within DiffServ clouds, internal routers have the capacity to apply different buffering strategies and service disciplines to IP packets according to their assigned FEC. Whereas IntServ is based on the equivalent of virtual circuits (that is, on a determined path for a given packet flow), DiffServ internal routers serve packets on a hop-by-hop basis. In other terms, an internal DiffServ router does not have to memorize a set of information proper to each flow. Unlike IntServ, DiffServ does not require the updating of all the routers of an internet network. Within an edge router, IP packets are marked according to their FEC thanks to the DiffServ codepoint [see Figure 1.3(b)]. Only 6 bits among the 8 of the initial TOS field are used to define a subfield called the *differentiated services codepoint* (DSCP). The DSCP field enables the definition of up to 64 services. In practice, this does not mean that 64 different buffer management strategies and services disciplines have to be implemented in all the internal routers of a DiffServ cloud. Only a portion of these 64 services may be implemented in the hardware and software of IP routers. Once an IP flow is classified, an edge router controls its burstiness (the size of the bursts) in addition to its average rate. These two control operations are referred to as metering. Traffic shaping and traffic dropping aim to smooth the flows admitted within the DiffServ cloud to control packet loss in internal routers according to their FEC. Three modes of per-hop behaviors are considered by the DiffServ approach. They are described as follows:

- *Expedited forwarding (EF):* EF traffic is characterized by a single DSCP value (101110). This traffic is subject to an admission control and to a shaping operation in the edge routers of a DiffServ cloud. Within the DiffServ cloud, EF packets are always served with a higher priority than

AF packets. Within an internal router, EF packets are all inserted in a common *first-in, first-out* (FIFO) queue. Aggregated EF traffic benefits from a guaranteed minimum throughput at the output of the internal routers. To benefit also from a lower loss and from a lower end-to-end delay than *assured forwarding* (AF) traffic, the maximum bit rate of aggregated EF traffic at the input of an internal router must never exceed a minimum fixed service rate.

- *AF:* Four levels of AF have been specified in the *Request for Comment* (RFC) 2474 of IETF. The AF packets are inserted into dedicated FIFO queues within the internal routers. Three drop precedence values are possible within each AF class for a total of 12 DSCP values. Packets with a high drop precedence value are dropped with higher probability than packets with smaller drop precedence value. For each AF level (AF1, AF2, AF3, AF4), drop precedence is implemented by means of the *random early discard* (RED) buffer-management algorithm [6]. The RED algorithm is based on the use of two thresholds in the buffers of the routers. A low threshold Min(k) and a high threshold Max(k) are associated to the each of the four AF-k classes. Figure 1.4 illustrates the principle of the RED algorithm for a given AF-k queue.

If the AF-k buffer occupancy is smaller than Min(k) when a new AF-k packet arrives, this packet is accepted into the queue. If the buffer occupancy is greater than Max(k), the packet is discarded. If the buffer occupancy is between Min(k) and Max(k), the packet is discarded with a probability $P(k)$. As it is described on Figure 1.4, $P(k)$ increases linearly with buffer occupancy at the arrival of the new packet.

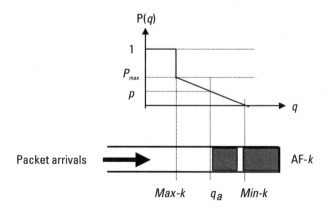

Figure 1.4 Principle of the RED buffer management for AF traffic.

- *Default forwarding (DF):* The traffic considered by this type of per-hop behavior is typically "best effort." It accesses the output port only if all the other traffic categories are inactive.

In summary, a DiffServ router carries out four operations. An IPv4 packet arriving on one of the inputs of an edge router is first classified on the basis of its TOS field. According to this TOS value and to the buffer occupancy at the instant of arrival of the packet, the drop precedence (the packet loss probability) of the packet is computed. Packet dropping is carried out in the last stage of the router (case of internal routers). In the case of edge routers, input traffic is shaped before being inserted into the DiffServ cloud.

1.3.2.3 Multiprotocol Label Switching (MPLS) Approach

At the end of the 1990s, several techniques were investigated to improve the speed and scalability of the Internet, including IP switching (Ipsilon/Nokia), tag switching (Cisco), and aggregate route-based IP switching (IBM) [7–10]. The basic idea of these techniques is to replace the traditional forwarding operation of IP routers with a simpler layer 2 forwarding. Facing the lack of compatibility between these different solutions, the MPLS approach has been promoted within IETF as a recognized standard. The ultimate objective of MPLS is to enable multiservice provisioning in IP networks. As in DiffServ, one defines MPLS clouds[9] made of internal routers [*label-switched routers* (LSRs)] and of edge routers [*label edge routers* (LERs)]. All the routers of the cloud are updated with the implementation of the MPLS control plane. The MPLS control plane is used to emulate connection-oriented virtual circuits within IP networks thanks to a label mechanism. Such virtual circuits are called *label-switched paths* (LSPs). On the basis of its header, each IP packet arriving at an ingress edge router is classified according to its FEC. An FEC corresponds to a set of IP packets with the same destination address that are served by a same policy within the LSRs. Once a packet is classified, the FEC is coded within the label. Figure 1.5 illustrates the format of an MPLS label with the following fields:

- On the basis of its FEC, a 20-bit label is assigned to each packet. Like an ATM *virtual circuit identifier* (VCI), a label is recalculated hop-by-hop in each intermediate LSR. This technique is called *label swapping.*
- A class of service field (field Exp) is coded on 3 bits enabling up to eight service classes.

---

9. An MPLS cloud is a subset of IP routers within an IP network on which MPLS functionalities have been implemented.

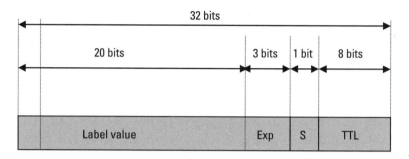

Figure 1.5 Format of an MPLS label.

- The single bit "S" is inherent to the *label-stacking* mechanism. The label-stacking mechanism has been introduced to provide scalability of the forwarding tables within the LSRs. This mechanism is quite similar to the concept of virtual path in ATM, but it is extended to several hierarchical levels. In other terms, several LSPs may be aggregated by adding to all the packets belonging to these individual connections a new common label. Similarly, several LSPs of second level may be aggregated by means of a third level of label and so on. Progressively, when a packet leaves an MPLS cloud, its labels are successively destacked. The bit "S" is set to "1" for the last label of the stack. An egress LER withdraws this last label from each packet.

- The last 4 bits of the MPLS label format correspond to the *time-to-live* (TTL) field. The value of the field TTL is set in an ingress LER. This value is decremented at each hop within the cloud. A packet with TTL = 0 may be deleted by an LSR.

- Different types of techniques may be used at layer two to proceed to label swapping, including frame relay, SONET/SDH, Ethernet, and ATM. In the case of IP over Ethernet, a label, or a list of labels, is inserted between layer 3 and layer 2 headers (see Figure 1.6).

In the majority of current MPLS implementations, label swapping is carried out by ATM switches. In that case, the value of the first label of the stack (the most recent label located at the top of the list) is inserted in the VCI/VPI field of the first cell of the ATM connection associated to the LSP. Figure 1.7 shows how MPLS labels are adapted to ATM formatting [case of ATM cells at the *user network interface* (UNI)]. The second label of the stack is transported in the ATM cell payload. The following cells of the same ATM connection (same

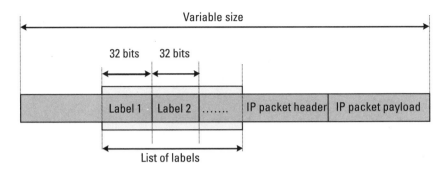

Figure 1.6 Label stacking in case of layer 2 switches.

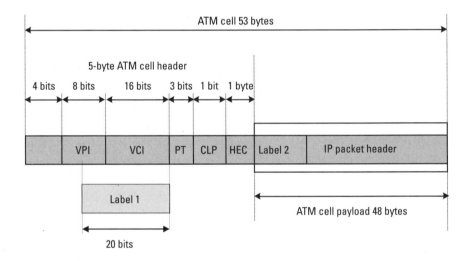

Figure 1.7 Label stacking in the case of ATM switching.

LSP) only transport the 20 bits of the first label in the VCI/VPI field. If ATM switches are not able to recognize MPLS labels in the ATM cells' header, the labels' encapsulation is carried out in the same way as in frame relay or in Ethernet (tunneling).

A *Label Distribution Protocol* (LDP) has been specified for LSP establishment under the initiative of ingress LERs. Figure 1.8 describes an example of an MPLS cloud with two unidirectional LSPs enabling the creation of a bidirectional channel between ingress LER *A* and egress LER *B*. If an LER *A* is not able to manage traffic engineering, it determines, thanks to its routing protocol, the first LSR to reach for the packets of a given FEC *X*. This first LSR determines in its turn the next hop for this FEC *X* and so on as far as the egress LER *B*. When

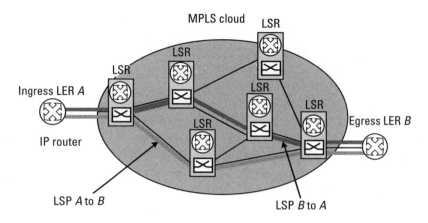

Figure 1.8 Example of unidirectional LSPs between two LERs.

the egress LER *B* is informed of the FEC *X*, it assigns a label, which becomes the first label of the stack (with the oldest one located at the bottom of the list). Hop by hop, an LDP message propagates back from *B* to *A* via the same intermediate LSRs. At each hop, the switch routers update their control plane by adding to their label-swapping tables the new association (number of the input port, input label) and (number of the output port, output label). Once it has received the LDP message, the ingress LER *A* knows that the LSP has been set up. It is important to note that unlike ATM but like frame relay, MPLS relies on unidirectional LSP. In other terms, the same procedure must be activated in order to set up the reverse LSP from *B* to *A*. The LDP protocol has been standardized within IETF under the reference RFC 3036. In order to be protected, LDP messages are transported under the control of a TCP session between the ingress LER and the egress LER.

Traffic engineering (TE) in MPLS refers to intelligent routing. Intelligent routing refers itself to the possibility of setting up LSP routes not only on the basis of a shortest path algorithm in terms of hops but also to prevent the overload of the most used physical links [11, 12]. Two protocols are candidates for MPLS traffic engineering:

- *TE-RSVP:* This protocol exploits extensions of the RSVP protocol (mentioned in Section 3.2.1 for IntServ) to set up and release LSPs. LSP establishment messages are exchanged in using the *User Datagram Protocol* (UDP) transport protocol.[10] TE-RSVP is qualified as a *soft-state*

---

10. UDP is a connectionless end-to-end protocol that is used over IP in the case of urgent messages or real-time traffic like voice traffic.

protocol. This means that an established LSP is released if it is not used over a given period of time. Periodically, signaling messages are exchanged between the routers of the cloud to determine if the connection is still alive. The amount of such "keep-alive" messages increases with the amount of RSVP active sessions.

- *Constraint-based routing (CR) LDP (CR-LDP):* During a first phase, the LSRs discover their respective neighbors using the UDP transport protocol on top of signaling messages. Label assignment to the various concerned LSRs is carried out under the control of the TCP protocol. Unlike TE-RSVP, CR-LDP is a *hard-state* protocol for which signaling information is exchanged only at connection setup and release (corresponding to TCP connection setup and release) and not during the life duration of the LSP.

In the next years, MPLS clouds should be surrounded either by classical IP networks or by DiffServ clouds. In the latter case, which should be the most frequent in the future, the 6 bits of the DSCP [see Figure 1.3(b)] transported in IP packets headers are converted by ingress LERs either into a code word transported in the 20-bit "label" field of the MPLS label or into the 3 bits of the "Exp" field of the MPLS label. In the two cases, this conversion is carried out either statically by the network manager, or dynamically thanks to LDP signaling. Figure 1.9 summarizes the functional modules implemented within an ingress LER.

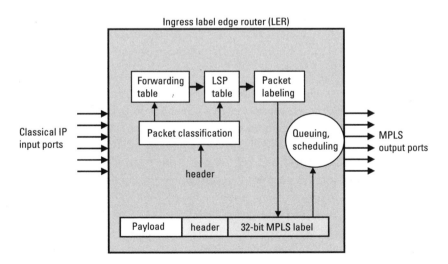

Figure 1.9 Functional modules of an ingress LER.

Various service disciplines are considered to manage hierarchical scheduling of the different flows that want access simultaneously to the same output port. Most of these disciplines were developed initially for the ATM environment. *Weighted fair queuing* (WFQ) and *class-based weighted fair queuing* (CB-WFQ) are among the most popular service disciplines that have been adapted to the IP context [3]. Different queuing strategies are also used to distinguish the priorities of the different flows. RED[11] and *weighted RED* (WRED) are two more popular buffer management strategies inspired by the ATM environment [6]. Different criteria enable one to determine the relative merits of these techniques. A technique may be more or less efficient to guarantee certain performance requirements. An efficient service discipline results in a higher network utilization. Protection is a second criteria; a service discipline must meet the performance requirements of packets generated by well-behaving users even in the presence of ill-behaving users or best effort traffic. A service discipline must be simple to facilitate its high-speed implementation.

1.4 IP-VPNs

Although they are not directly related to Internet access, IP-VPNs are one of the new major services offered by internet network providers. A VPN may be used for essentially three types of service:

- The interconnection of distant LANs of the same enterprise to define an intranet;
- The access of nomadic users to an intranet, with these nomadic users being members of the enterprise in charge of this intranet;
- The controlled access of certain persons to an intranet, although they do not work for the enterprise that manages this intranet.

In other terms, a VPN may be seen as a component of a corporate network. It includes additional services such as firewalls to prevent illegal access to the enterprise network resources. A VPN may rely on two types of tunnels: layer 2 tunnels and layer 3 tunnels. In the case of layer 2 tunnels, which are the most frequent, data flows transit over a private backbone network, meaning semipermanent virtual circuits that are only accessible to the members of same enterprise. This privacy is easily manageable at layer 2 in frame relay or in ATM networks. Layer 3 tunnels correspond to VPNs based on the IP technology. Whereas layer 2 VPNs are several years old, layer 3 VPNs—that is, IP-VPNs—are newly emerging technologies. They rely on specific protocols

---

11. Section 2.2.2 describes the RED buffer management strategy.

such as IPsec and MPLS. From a carriers' perspective, IP equipment and IP network management costs are lower than in the case of ATM. This is why, frame relay VPNs, which for several years have been the most popular, should be replaced in the coming years by IP-VPNs. Figure 1.10 illustrates an example of a layer 2 VPN using a frame relay network. Such a VPN is today point-to-point (static), the two remote sites of an enterprise being connected to the central site by means of semipermanent PVCs. Each access router is connected to the nearest POP of the NSP operating the frame relay network by means of a *leased line* (LL). Note that in Figure 1.10, for survivability purposes, two access routers are used at the central site, each of these routers being connected to two different POPs. The main drawback of FR-VPNs is their lack of scalability in terms of access rates. Indeed, frame relay does not authorize access rates greater than 1.55 Mbps in North America or than 2 Mbps in Europe. Such a limitation prevents any use of FR-VPN for video services.

VPNs are based either on point-to-point tunnels or on any-to-any tunnels. In the former case, tunnels must be preestablished before data transfers, whereas in the latter case, tunnels are set up dynamically. The IPsec protocol in charge of confidentiality is currently statically configured. Some carriers have announced for 2003 a dynamically configurable version of IPsec. We have seen how MPLS enables QoS guarantees in IP networks. The main expected benefit of IP-VPNs should be then the fact they are any-to-any–oriented and that they should facilitate multiservice integration (data, voice, and video) for the enterprises.

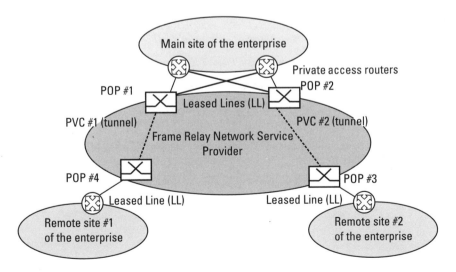

Figure 1.10 An example of a frame relay VPN.

1.5   Conclusion and Perspectives

This chapter presents the main characteristics of IP and ATM networks. As a first step, it defines the basic taxonomy associated to Internet access. This taxonomy is adopted in the following chapters of this book. The chapter then presents the two possible types of access to the Internet, direct and indirect. In addition, we have considered the various approaches enabling the provision of end-to-end QoS guarantees in ATM and in IP networks. From this perspective, the principles of the IntServ, DiffServ, and MPLS techniques have been discussed. This chapter ends with an introduction to the concept of IP-VPN, which appears to be a promising service for enterprises thanks to the IPsec protocol and to MPLS.

References

[1]   Huitema, C., *Routing in the Internet,* Upper Saddle River, NJ: Prentice Hall, 2000.

[2]   Comer, D., *Internetworking with TCP/IP, Volume 1: Principles, Protocols, and Architecture,* 4th ed., Upper Saddle River, NJ: Prentice Hall, 2000.

[3]   Zhang, H., *Scheduling Algorithms for Integrated Services Networks,* http://cs.cmu.edu/~hzhang.

[4]   *ATM Forum Traffic Management Specification 4.1,* ATM Forum af-tm-0121.000, March 1999.

[5]   Weiss, W., "QoS with Differentiated Services," *Bell Labs Technical Journal,* October–December 1998, pp. 48–62.

[6]   Floyd, S., and V. Jacobson, "RED Gateways for Congestion Avoidance," *IEEE/ACM Trans. on Networking,* Vol. 1, No. 4, August 1993.

[7]   IETF, RFC 1953, *Ipsilon Flow Management Protocol for IPv4 version 1.0.*

[8]   IETF, RFC 1954, *Transmission of Flow Labeled IPv4 on ATM Data Links, version 1.0.*

[9]   IETF, RFC 2105, *Tag Switching Architecture: Overview.*

[10]   Ahmed, H. N., et al., "IP Switching for Scalable IP Services," *Proc. of IEEE,* Vol. 85, No. 12, December 1997.

[11]   Awduche, D., et al., "Requirements for Traffic Engineering over MPLS," *IETF RFC 2702,* September 1999.

[12]   Awduche, D., "MPLS and Traffic Engineering in IP Networks," *IEEE Communications Magazine,* December 1999.

Selected Bibliography

Harrington, D., "Link Local Addressing and Name Resolution in IPv6," *IETF Draft*, draft-ietf-ipngwg-linkname-01.txt, January 1997.

Rekhter, Y., et al., "An IPv6 Provider-Based Unicast Address Format," *IETF Draft*, draft-ietf-ipngwg-ipv6-arch-00.txt, March 1997.

2

Existing Infrastructure and Dial-Up Modems

2.1 Introduction

The expression "local loop" has historically been used for designating the very last miles of telephone carriers' networks. The various nodes of a PSTN could be classified into three main hierarchical levels. Starting from the *customer premises equipment* (CPE), these nodes are the *end office* (EO), the CO, and the *intermediate switching offices* (ISOs). The EO [sometimes called concentrator or local exchange (LEX)] is used to concentrate traffic to and from the end users. The CO serves as a toll office to identify the calling party and the called party, to set up, to release, and to bill telephone communications. An EO is connected to a CO by means of copper lines called toll-connecting trunks. When the calling party and the called party are not connected to the same CO, the path linking these two parties must transit through one or several ISOs. Physical links between COs and ISOs or between ISOs are called intertoll trunks. Different hierarchical levels of ISOs are used in carriers' networks. For instance, traffic coming from various COs is routed by a first-level ISO to a distant second-level ISO. A second-level ISO may play the role of telephone gateway between two different countries.[1] The expression local loop for telephony refers to the portion of the copper wire infrastructure going from the CPE to the EO along which a metallic continuity is observed. The end user's telephone set is powered directly from the EO by means of a continuous 48V. The two wires of twisted copper pair, the input impedance of the telephone set, and the input impedance

---

1. In that case, the ISO is called an international switching center.

21

of the line interface circuit at the EO may thus be seen as a current loop; the following describes the origin of the expression "local loop."

The meaning of local loop has been progressively extended these last years with the emergence of new access network technologies and new end-to-end services. Today, this expression is used in a generic way to refer to the various kinds of techniques that can be used to link end users to their CO. These techniques are based on the transmission of various kinds of signals over different types of mediums: electrical voltages over twisted copper pairs or coaxial cables, optical pulses over optical fibers, or radio waves using free-space propagation. Today, copper pairs are used in the local loop for three types of service:

- Access to the PSTN;

- LLs;

- Basic access to narrowband ISDN.

Data transmission over the PSTN began with modem technology. Section 2.2 gives a historical overview of dial-up modems. We briefly enumerate the principles and limitations of the successive generations of such modems, which are also called voiceband modems or analog modems. The most recent V.90 and V.92 dial-up modems are detailed. Section 2.3 aims to describe the typical configuration of the access network of the PSTN. Section 2.4 outlines the main characteristics of LLs and of basic access to narrowband ISDN, which both use copper wires in the access network. Section 2.5 discusses the necessity of a standardized interface between EOs and COs. We then describe the main characteristics of the V5 interface specified within ETSI. This interface manages signaling channels associated to individual calls between EOs and COs. The equivalent interface in North America is known as the GR-303/TR-008 interface. Section 2.6 introduces the concepts of *digital loop carrier* (DLC) and of *carrier serving area* (CSA). Section 2.7 discusses the impact of dial-up modems on the PSTN.

2.2 Dial-Up Modems

Since 1980, PSTN users have begun to use their subscriber lines to access data network, and more particularly the Internet. For that purpose, a voiceband modem must be installed at the CPE. The term *voiceband* refers to the fact that the on-line signal must be included in the [0, 4 kHz] band. The reason for this constraint is simple: With analog modems, the subscriber line is dedicated exclusively either to the telephone set or the modem. When an end user switches his or her line from the telephone set to the modem, nothing indicates to the other end of the line (that is, to the line interface circuit of the EO) that voice traffic is being replaced by IP traffic. In other terms, the signals associated with modulated IP packets must be assimilated to a voice signal by the EO to be sampled and digitized in a similar way. Knowing the presence of a [0, 4 kHz] low-pass filter at each input of the EO (the role of this filter is explained in Section 2.3), it is mandatory for a voiceband modem to guarantee a signal bandwidth limited to 4 kHz. Voiceband modems are also known as dial-up modems because they assume indirect access to the Internet (to an IP network) through a PSTN network. In other words, a TCP/IP session between an end user's PC and a distant Web server needs the use of a telephone circuit through PSTN. This characteristic is particularly compulsive for the end users who have to pay for a telephone call each time they access the Internet. In North America, this cost remains acceptable because PSTN access carriers apply flat-rate billing. The situation is much less favorable in Europe where phone calls are paid pro rata of their duration. One knows that, on average, Web connections are much longer than telephone communications. Section 2.7 details indirect access to the Internet and outlines its consequences on PSTN dimensioning. Several steps related to advances in electronics and signal processing have yielded several generations of voiceband modems (see Table 2.1).

The oldest versions of dial-up modems such as V.21, V.22, and V.22bis are based on the use of frequency or phase modulation also known as *frequency shift keying* (FSK) and *phase shift keying* (PSK). They also use *frequency division duplexing* (FDD) to separate in the frequency domain upstream and downstream flows.

From V.32 to V.34bis modems, *quadrature amplitude modulation* (QAM) has been adopted. This modulation technique is a mix of amplitude modulation

Table 2.1
Successive Generations of Voiceband Modems

Generation	Date	Bit Rate	Technique
V.21	1978	300 bps	FSK+FDD
V.22	1980	1,200 bps	PSK+FDD
V.22bis	1981	2,400 bps	PSK+FDD
V.32	1985	9,600 bps	QAM+EC
V.33	1988	14.400 bps	QAM+EC
V.34	1990	28.800 bps	QAM+EC
V.34bis	1995	33.600 bps	QAM+EC
V.90	1998	33.6 Kbps (U); 56 Kbps (D)	QAM (U) +PAM (D)
V.44	Pending	N/A	Compression
V.92	Pending	48.0 Kbps (U), 56 Kbps (D)	PAM

and of phase modulation. To enable full-duplex transmission in the same frequency band, these modems use *echo cancellation* (EC). Bits to be transmitted are grouped into words of k bits. An alphabet associates to each possible word obtained by the combinations of these k bits a given amplitude and a given phase to the on-line signal at a fundamental frequency f_0. Since 1985, EC has allowed a sharp increase in the achievable throughputs. The following generations starting from V.32 use EC to transmit in the same frequency spectrum upstream and downstream data. The principles of QAM and of EC are described in the chapter dedicated to ADSL. Figure 2.1 illustrates the basic principle of a voiceband modem. The signal generated by a voiceband modem must be smoothly timed to prevent high-frequency harmonics in the spectral domain.

Sections 2.2.1 and 2.2.2 first briefly recall the principle of V.34bis modems, which are still very popular, before describing the more recent V.90 and V.92 generations.

2.2.1 V.34bis Modems

Numerous impairments degrade the quality of transmission over copper wires in the local loop: attenuation distortion, impulse noise, crosstalk, delay distortion, jitter, and impedance distortion. Attenuation distortion and delay distortion are two of the main drawbacks of transmission media.

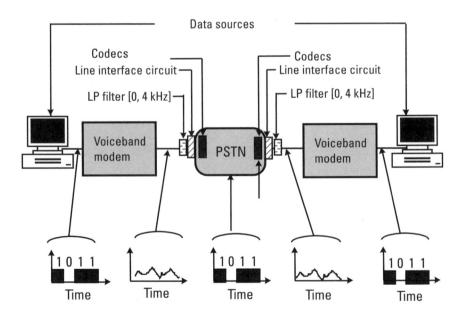

Figure 2.1 Voiceband modem principle.

Electrical signals are subject to attenuation during their propagation over a copper wire, this attenuation being in most cases strongly dependent on the frequency of the signal. On a subscriber line, the higher frequencies are in average more attenuated than the lower ones. The more complex the dependency between attenuation and frequency, the more difficult amplitude equalization at the receiver. The oldest modem generations (from V.22 to V.33) use a fixed linear equalization in order to compensate attenuation distortion introduced by the medium. As will be underlined in Chapter 4, which is dedicated to the electrical characteristics of copper wires, nonlinearities of the medium may strongly differ from one subscriber line to another. In other terms, adaptive equalization is much better suited to the local loop environment than fixed equalization.[2] Since V.34 modems, adaptive equalizers, also called *decision feedback equalizers* (DFEs), have been used. Such equalizers require an initialization phase to record the medium characteristics in the frequency range at which the modem is supposed to operate. A DFE is a digital filter whose coefficients are determined at the end of the initialization phase.

Delay distortion is another important disruptive effect on transmission media. Because of copper pairs' characteristics, the various frequency spectral components of a signal do not propagate at the same speed.[3] The wider the spectrum of the transmitted signal, the more numerous the spectral components and the more noticeable the impact of delay distortion. Delay distortion is at the origin of *intersymbol interference* (ISI).

Figure 2.2 illustrates the principle of ISI in the case of baseband transmission. A transmitter A sends data with a clock H_e to a distant receiver B. Due to ISI, the high-frequency components of the signal propagates faster than the low-frequency ones. The receiver uses a *phase-locked loop* (PLL) to determine the best clock H_r at which it has to sample the received signal. If distance d between A and B is high enough (or if the bit duration is low enough), ISI induces an overlap between the first bits of the high-frequency train with the last bits of the low-frequency train. In baseband transmission, the value of the bits is deduced from the amplitude of the received signal. Thus, successive erroneous bits may be introduced at receiver B. In general, delay distortion (also known as *phase distortion*) is reduced by means of power control.

Let us consider the case of two end users A and B exchanging e-mail by means of their PCs. Each PC is connected to the PSTN by means of a V.34bis voiceband dial-up modem (see Figure 2.3). We first notice from Figure 2.3 that within PSTN networks, upstream and downstream traffic are clearly separated thanks to a special electronic system called hybrid (or 2 wires to 4 wires adapter)

---

2. This remark is even more true in the case of the wireless environment, which will be considered in the chapter dedicated to *wireless-in-the-loop* (WITL) access systems.

3. Propagation speed of the spectral components issued from the Fourier transform of a signal is known as the *phase velocity*.

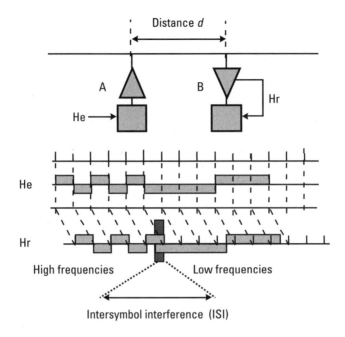

Figure 2.2 Principle of ISI.

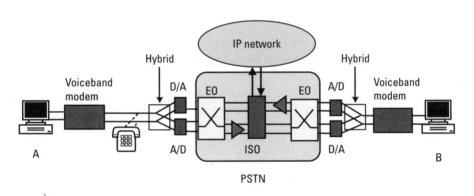

Figure 2.3 Home-to-home Internet connection by means of voiceband modems.

implemented at the end of each subscriber line at the EO side. The data sent by *A* are converted from *analog to digital* (A/D) when they arrive at the EO. Within PSTN, an ISO extracts these data from E1/T1 frames in order to send them to the POP of an ISP. These data are reinjected further in the PSTN network at the level of another ISO. Before being transmitted on user *B*'s subscriber line, data are again converted from *digital to analog* (D/A). The same kind of operations are carried out for the traffic going from *B* to *A*. Figure 2.4 shows the details of the A/D operation carried by the EO.

As mentioned above, bits generated by a PC are modulated by means of analog QAM modulation. When it arrives at the EO, the analog signal obtained at the output of a modem is filtered by the [0, 4 kHz] lowpass filter, sampled, and quantized. The Fourier transform of the sampled signal is represented in Figure 2.4. One observes that the condition for which the band [0, F_{max}] does not interfere with the bands centered around the sampling frequency F_e is $F_e - F_{max} > F_{max}$, which corresponds to the Nyquist criteria. The PCM encoder associates with each sample a code word that is an approximation of the real value of the sample. Section 2.4.1 defines the quantizing noise introduced by this approximation. Figure 2.5 illustrates the D/A operation carried out the EO.

Successive bits arriving on E1/T1 frames and corresponding to *time slots* (TSs) proper to the considered connection are converted into discrete voltage amplitudes every 125 μs. The distribution of these amplitudes corresponds to the companding techniques of the A-law or μ-law described in Section 2.4.1. Figure 2.6 describes the shape of the *pulse amplitude–modulated* (PAM) signal obtained after decoding.

A lowpass filter is used to smooth the shape of this signal. Thanks to this shaping operation, the signal sent to the called party becomes almost similar to

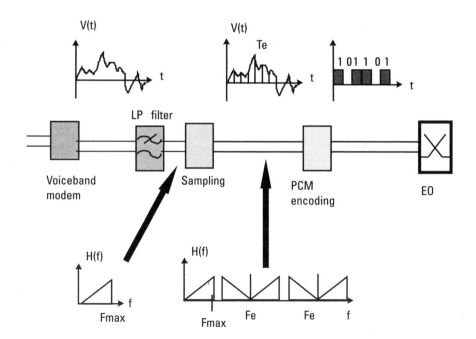

Figure 2.4 Analog-to-digital (A/D) conversion.

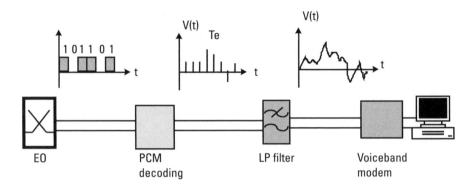

Figure 2.5 D/A conversion.

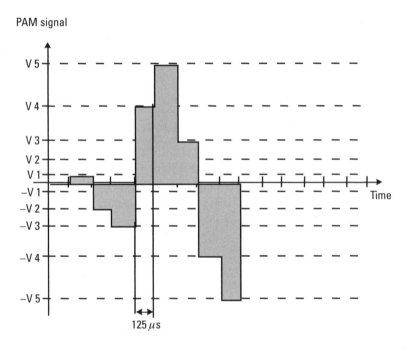

Figure 2.6 PAM signal obtained after PCM decoding.

the signal sent by the calling party. When it arrives at its destination, the signal is finally QAM-demodulated by the receiver's V.34bis modem.

Claude Shannon outlined about 50 years ago that any channel of transmission with *additive white Gaussian noise* (AWGN) and a bandwidth limited to W could be characterized by a maximum data rate capacity C_{max} expressed in bits per second (each of these terms is detailed in Chapter 4, dedicated to the electrical characteristics of copper wires). If S and N stand for the average power of the signal and the average power of the noise, respectively, at the output of the receiver's modem, one has:

$$C_{max} = W \cdot \log_2\left(1 + S/N\right) \qquad (2.1)$$

Even in optimal conditions (no noise due to propagation over the twisted pair), the quantizing noise introduced by A/D conversion is between 38 and 39 dB. Knowing that $W = 4$ kHz, we obtain in this case a maximum capacity of 34 Kbps. This is the reason why one may estimate that V.34bis voiceband modems enabling 33.6 Kbps are very efficient.

2.2.2 V.90 Modems

V.90 modems were standardized within ITU-T in 1998 to achieve higher downstream data rates than V.34bis modems. The basic idea of V.90 modems is to suppress quantizing noise for downstream traffic to increase S/N and then to increase C_{max} in a client-server environment. In order to prevent A/D conversion between the server (for instance, a Web server) and the client (the end user), one assumes that the former is connected to PSTN by means of a digital link. Strictly speaking, the server is indirectly connected to PSTN via the POP of an ISP (see Figure 1.1). In our description of V.90 modems, we then assimilate the POP to the server. This POP is in general connected to an ISO by means of a LL. Rather than interpreting the received signal from the POP by means of a QAM demodulator, the end user's V.90 modem tries to recover the codewords associated to the discrete levels hidden in the smoothed shape of the PAM signal. For that purpose, the V.90 modem must have knowledge of the companding law used at the EO. It must also synchronize its sampling clock with the sampling clock of the codec used at the EO. On the basis of these two requirements, a V.90 modem behaves like a PCM decoder. Each connection by means of a V.90 modem requires the subscriber line to have an initialization phase. Because of the various noise sources on the medium, a trade-off has to be found between the achievable downstream data rate and the target BER. It has been shown experimentally that 56 Kbps was the very best achievable bit rate by means of PAM modulation between the EO and the end user's PC. In the upstream direction, it is not possible to use the same PAM modulation as in the downstream direction. Indeed, due to the 8-kHz clock of the modem, the signal sent by the end user should be truncated by the [0, 4 kHz] filter at the EO. This is the reason why V.90 modems operate in the upstream direction as V.34bis modems—that is at most at 33.6 Kbps.

Before the complete V.90 specification was approved, several manufacturers proposed modems inspired by the V.90 technique. Today, most of the ISPs have implemented the 56-Kbps achievable transmission rate in the ports of their

POP. In the United States, the *Federal Communications Commission* (FCC) has decided to limit the downstream data rate from V.90 compatible servers to 54 Kbps.[4] In practice, it seems that this bit rate is achievable only if the end user is quite close from his or her end office (less than 2 km). In case of a too-high detected BER, a V.90 modem switches automatically in the V.34bis mode. In terms of BER, one shows that in the case of a binary modulation[5] with AWGN, the average error bit probability is given by:

$$P_b = Q\left(\frac{d_{min}}{\sqrt{2 \cdot N_0}}\right) = Q\left(\frac{2 \cdot E_b}{N_0}\right) \qquad (2.2)$$

where E_b stands for the energy of a bit, $\sigma^2 = N_0/2$ stands for the variance of the Gaussian noise, and d_{min} stands for the minimal distance between the two code-words. The Q function is known as the Gaussian error function [1] and is given by:

$$Q(x) = \frac{1}{\sqrt{2\pi}} \int_x^\infty e^{-t^2/2} dt \qquad (2.3)$$

with x positive. A generalization of (2.3) to the case of generalized PAM with M signals is possible. Such a M-ary PAM is quite similar to the case of V.90 modems for downstream traffic. It has been shown that BERs of 10^{-7} were achievable in the case of V.90 transmission at 54 Kbps.

2.2.3 V.92 Modems

The V.92 standard is under definition within ITU-T. It refers to an extension of the V.90 standard with the objective of providing higher bit rates to upstream traffic. Whereas V.90 modems operate as V.34bis modems at 33.6 Kbps for upstream traffic, V.92s offer up to 48 Kbps in this same direction. For that purpose, PCM is also used for upstream transmission without enlarging the bandwidth of the upstream signal over 4 kHz. Thanks to the advances in the speed of electronics, the line initialization phase of a V.92 modem is sensibly reduced in comparison to V.90 modems. For that purpose, a new compression technique known as V.44 has been developed. Rather than restarting this initialization phase at each connection, the subscriber line parameters are stored in the modem after the first connection. A V.92 modem also enables one to receive or

---

4. This limitation is due to a limit imposed on the maximum permissible transmit power levels.

5. Binary (or antipodal) modulation is the simplest PAM modulation, voltage pulses +V and –V being associated to "1" and "0," respectively.

to initiate a phone call without interrupting an Internet connection. A mechanism referred to as *called waiting* developed by V.92-modem manufacturers, ISPs, and telephone companies fixes the limit of time during which the TCP/IP connection or the phone connection may remain active (up to a few tens of minutes). This technique is referred as *modem-on hold*. Currently, several ISPs and modem vendors provide V.92 access line in North America, Europe, and Asia.

2.3 Access to the Telephone System

2.3.1 Typical Configuration of PSTN Access Networks

Each PSTN user is directly connected to a line interface circuit of the nearest EO by means of a dedicated copper pair. The analog voice signal of each user is transported on the local loop from the CPE to the EO. Before the 1970s, traffic trunking was carried out by means of coaxial cables and *frequency division multiplexing* (FDM). To optimize these cables' capacity, [0, 4 kHz] lowpass filters have been used at the end of subscriber lines on the EO side. Since the 1970s, analog voice signals obtained at the output of the [0, 4 kHz] filters have been progressively digitized at the EO by means of voice coders.[6] Chapter 6 shows that the presence of these filters is at the origin of serious disruptive effects. Figure 2.7 describes the current configuration of an access network going from the CO to a set of CPEs. A point-to-point electrical continuity exists between each CPE and an input port of the EO. The physical junction between a twisted pair in the local loop and an input port of the EO is carried out by means of a distribution frame. Similarly, a distribution frame called a *main distribution frame* (MDF) is used at the CO at the end of the toll-connecting trunks. In European urban areas, the distance between a CPE and an EO is on average 2 km with a minimum of 500m and a maximum of 5 km. In rural areas, this distance may reach around 10 km. Chapter 3 describes how such distances may be reached over copper wires.

Typically, an access network is comprised of a few tens of thousands of subscriber lines. Depending on the population density, the amount of subscribers connected indirectly to a same CO may vary from 10,000 to 100,000. Within the same access network area, any end user may be connected to any other user belonging to the same area by means of a single CO.

2.3.2 PSTN Network Dimensioning

Because of the statistical activity of the access lines, ISOs proceed to a progressive concentration of the digital traffic they receive from the COs. A capacity corresponding to a voice channel in the core (or transport) network is called a

---

6. Voice coders are also called *codecs* for coders-decoders.

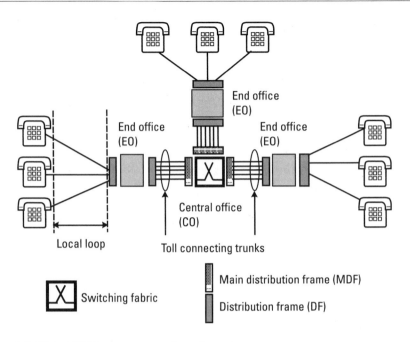

Figure 2.7 Digital PSTN access network configuration.

circuit. In the core network, COs and ISOs are connected to each other by means of a meshed network of cable trunks. Each trunk corresponds either to an optical fiber, to a coaxial cable, or to a set of copper wires. The capacity of each trunk is expressed as a multiple of synchronous T1 or E1 links at 1.544 Mbps and 2.048 Mbps, respectively. T1 or E1 links correspond to 24 and 30 voice channels (or circuits), respectively. Each cable trunk has the capacity to transport simultaneously up to N voice channels. The choice of the value of N strongly depends on the expected load on the telephone system and on the blocking probability that the operator intends to offer to its customers. In practice, the blocking probability of a call is calculated at the busiest hour. Many statistical studies of telephone systems have been carried out since the 1960s. These studies outline, for instance, that on average, phone calls are about 3 minutes long and that their arrival rate corresponds on average to a Poisson process. It can be shown that the probability B that a new call is rejected on a trunk of capacity N is given by the well known Erlang-B formula:

$$B = \frac{\dfrac{\rho^N}{N!}}{\displaystyle\sum_{i=0}^{N} \dfrac{\rho^i}{i!}} \qquad (2.4)$$

In this formula, ρ stands for the traffic intensity submitted to the trunk. This intensity is expressed in Erlangs. An Erlang is the measuring unit of telephone traffic intensity. It corresponds to the amount of calls multiplied by their mean duration relatively to the most loaded hour in the telephone network. As an example, a telephone line permanently busy during the busiest hour corresponds to a traffic intensity of 1 Erlang. By extension, a cable binder of N lines serves a traffic intensity of k Erlang if, on average, one has during the busiest hour k busy lines among these N lines. It is generally admitted that traffic intensity on a subscriber line is between 0.03 and 0.1 Erlang, whereas it is between 0.3 and 0.8 Erlang on a circuit. It is possible to determine for different values of B and ρ the number N of circuits to be used on a trunk. For a given probability B, N increases much more slowly than ρ. This is why carriers have an interest in setting in their core network high-capacity cable trunks rather than numerous low-capacity cable trunks representing globally the same capacity. We shall come back later to the importance of the Erlang formula relative to the increase in the Internet traffic.

Subscriber lines corresponding to the local loop are progressively grouped in cable binders of increasing capacity, from a few tens of pairs close to the end users to a few thousands of pairs close to the EOs. The location of the points at which pairs are grouped in the upstream direction or separated in the downstream direction are called distribution points. Distribution points are in general incorporated into cabinets at the street level. A CO is made of three main units: a switching fabric, a control unit, and set of line interface circuits. Like a CO, an EO is made of a switching fabric, a control unit, and a set of line interface circuits. On the basis of signaling information (dialed number), the EO control unit commands the setup of connections from an input line interface circuit to an output line interface circuit. A part of the switching function is carried out by EOs, which analyze some of the digits of the dialed numbers. Setup, release, and billing of individual connections is under the responsibility of the CO. Interoperability between the switching fabric of a CO with the switching fabric of one of its EOs is in most cases proprietary (i.e., vendor-dependent). Meanwhile, two versions of signaling interfaces between EOs and COs have been standardized by ETSI. Section 2.5 describes these interfaces.

2.4 Leased Lines and Basic Access to Narrowband ISDN

2.4.1 Codecs and PCM

Let us detail the operations carried out by voice coder/decoders (or codecs) implemented at the input of the EOs (before the line interface circuits). After it has been filtered, an analog voice signal is digitized by means of PCM to be transported in digital PSTN. The analog signal obtained at the output of a

[0, 4 kHz] filter is sampled 8,000 times per second according to the Nyquist criteria. This criteria specifies at which frequency an analog signal must be sampled before being quantized and encoded in order to be rebuilt at destination. This frequency f_e must be at least two times greater than the upper bound f_{max} of the spectrum of the analog signal (f_{max} = 4 kHz and f_e = 8 kHz). When telephone networks were digitized in the 1970s, national carrier members of CCITT[7] did not reach an agreement on an international standard. This lack of consensus explains the various versions of the *plesiochronous digital hierarchy* (PDH) used today in telephone systems in Japan, Europe, and North America. These different versions are outlined in Section 2.4.2.

Each sample in volts of the analog signal is quantized on the basis of 256 discrete levels. These levels are symmetrically distributed around the average tension V_{mean} of the voice signal. An 8-bit word is associated to each of the 256 discrete levels. In Europe, a voice channel corresponds in digital PSTN to a 64-Kbps channel. In North America, 1 bit among the 8 bits of each word is used for signaling purposes. This implies a voice channel rate equal to 56 Kbps with an associated signaling rate equal to 8 Kbps. In the quantizing process, the real value of a sample is rounded off to the nearest discrete level. If a uniform quantizing rule was used—that is if the 256 discrete levels were regularly spread out over the possible amplitudes V_{min} and V_{max} of a voice signal—the quantizing noise should be much more penalizing for the samples with small amplitude than for those with high amplitude. This is why CCITT has recommended nonuniform quantizing in order to achieve a constant signal-to-quantizing noise ratio. The main idea of nonuniform quantizing consists in using more codewords (that is, a shorter distance between discrete levels) for low-amplitude samples than for high-amplitude samples. For that purpose, two techniques called compressing and expanding are used at encoding and at decoding, respectively. The global compressing and expanding process is known as the companding technique. Two different nonuniform quantizing rules known as the A-law and the μ-law have been standardized within ITU-T. The former has been adopted in Europe under the recommendation G.732. The latter is used in North America and Japan under the recommendation G.733.

2.4.2 Digital Leased Lines

LLs have been for several decades the most efficient way for enterprises to interconnect their distant LANs. A LL may be either analog or digital. It corresponds to predetermined data rates permanently reserved between the endpoints to be connected. Today, LLs are also used to interconnect IP routers with the switching nodes of PSTN or to interconnect base stations within a radio-mobile

---

7. CCITT is the ancient name of the ITU-T based in Geneva, Switzerland.

network to their *mobile-services switching center* (MSC). In general, an enterprise, a radio-mobile operator or an ISP, leases lines from long-distance carriers for a duration of at least several months.

Digital leased lines aim to deliver end-to-end data rate multiples of 64 Kbps up to 1,920 Kbps (in the case of Europe). Other comparable data rates in multiples of 56 Kbps are achieved in North America and in Japan. The provision of digital LLs by a carrier requires the setting of a new copper infrastructure specifically dedicated to that service. Three kinds of equipment are generally used for that purpose: two or three preconditioned copper pairs (the number of pairs depends on the expected range of the leased line), digital multiplexers, and digital cross-connects. Chapter 4 outlines that, according to the data rates and to the spectrum transported on each pair, digital LLs cannot reuse standard subscriber lines. For instance, it has been shown that coexistence in the same binder of digital LLs and of copper pairs used for Internet access by means of voiceband modems is not possible. Indeed, interference due to high frequencies transported on a digital LL strongly degrades the *signal-to-noise ratio* (SNR) for Internet access. This is why digital LLs have been so expensive during these last 30 years. Let us note that this remark was valid at the beginning of the 1990s; today it is not true any longer thanks to the advances in electronics and signal processing.

In general, large enterprises lease from a long-distance carrier four-wire LL. Such a service assumes the installation of digital multiplexers within the buildings of the enterprise. In Europe, digital LLs enable E1 access rate or its multiple in the PDH hierarchy. The 2.048-Mbps capacity proper to E1 is a gross bit rate that includes overheads proper to the E1 frames format. The effective capacity offered to the enterprise is the equivalent of 30 voice channels—that is, 1,920 bps. In North America, T1 digital LLs with a gross bit rate of 1.544 Mbps are the equivalent of 24 voice channels. *Small and medium enterprises* (SMEs) use two-wire LL. Both T1 and E1 LLs assume baseband transmission. The word "baseband" means that the on-line signal is characterized by discrete voltage levels with stable states of which the duration is a multiple or submultiple of the bit duration. The choice of the line codes used to convert the information bits into successive voltage levels strongly influences the performance of the system. In Europe, the line code used for E1 LL is the *high-density bipolar code of order 3* (HDB3). This code uses three voltage levels: +3V, 0V, and −3V. For any type of transmission medium (copper pair, coaxial cable, optical fiber, radio propagation), baseband transmission requires a minimum power spectral density of the signal in the very low frequencies.[8] In addition, in synchronous transmission as is the case with E1/T1 LLs, it is necessary to have a minimum of transitions per time unit in the on-line signal to facilitate clock

---

8. This constraint is also referred to as the minimization of the dc component on the medium.

recognition at the receiver. Figure 2.8 illustrates the principle of the HDB3 line code.

As can be seen from Figure 2.8, a "0" bit corresponds to the absence of voltage on the line. A "1" bit corresponds to a pulse of voltage with a value that alternates between +3V and −3V between successive "1"s. To prevent long suites of null voltage in the case of successive "0"s of information, HDB3 line code imposes a fourth bit to zero to be modulated by means of a 3-V pulse with the same polarity as the voltage pulse associated to the last transmitted "1" bit. This violation of polarity is recognized at the receiver in order to avoid confusing between a "0" and a "1." Figure 2.9 compares the power density spectrums of the *nonreturn-to-zero* (NRZ) and of the HDB3 line codes. One notices that the rapidity of HDB3 is equal to 1, which means that 1 bps of information rate corresponds to 1 Hz on the medium. Knowing that with E1 LLs, a bit duration T is 125 μs divided by 256, the necessary bandwidth on the medium is 2.048 MHz. In North America, another code called *alternate mark inversion* (AMI) is used for T1 LLs. This line code is very similar to HDB3 but without the violation polarity in the case of long suites of zeros. It is also characterized by a rapidity of 1. The required bandwidth for a T1 LL is then 1.544 MHz.

2.4.3   Constraints for the Provision of E1/T1 LLs

We mention above that E1/T1 leased lines require two preconditioned unidirectional copper pairs between a CPE and a digital cross-connect, a packet switch, or an IP router. At high frequencies such 2.048 MHz or 1.544 MHz, attenuation of the copper pair is such that *repeaters/regenerators* (RRs) have to be installed along each pair. Distance between successive RRs depends on the data

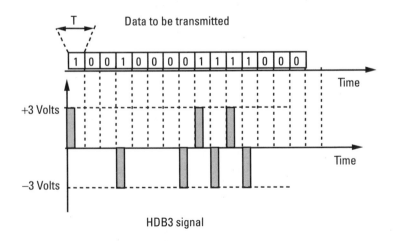

Figure 2.8 Principle of the HDB3 line code used by E1 LLs.

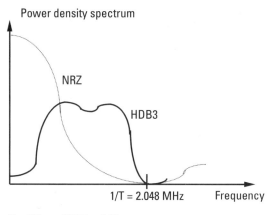

$T = 125 \ \mu s / 32*8 = 0.49 \ \mu s$

Figure 2.9 Power density spectrum of the HDB3 line code.

rate (E1 or T1) and on the copper wire diameter. In the case of T1 LLs, this distance is 900m with 0.4-mm diameter copper pairs, whereas it is 1,800m with 0.6-mm diameter copper pairs. The setting of RRs is very expensive because such equipment requires remote powering. In terms of maintenance, carriers have to control regularly the good state of their powering equipment along their LLs. Figure 2.10 illustrates the configuration of a E1/T1 LL.

Until the 1970s, many carriers installed regularly spaced induction coils (also called load coils) along their subscriber lines. Indeed, it is possible to reduce

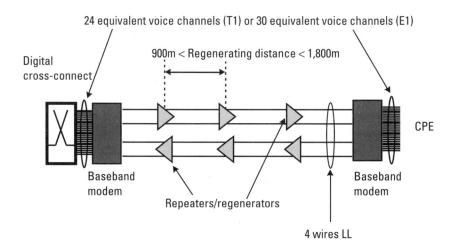

Figure 2.10 Configuration of a E1/T1 LL.

the attenuation of a copper wire in the low frequencies (under 100 kHz) by increasing the imaginary impedance of the line. The principle of this technique is described into more details in Chapter 4. In this way, carriers are able to optimize the amount of telephone lines connected to their COs. Such loaded subscriber lines are in fact unusable for the provision of E1/T1 LLs. It has been shown that load coils may induce cutting frequencies in the E1/T1 bands. It is also important for a carrier to check that a subscriber line does not include passive derivations. Chapter 4 describes why such derivations called bridge taps may also induce the equivalent of cutting frequencies on the medium. The knowledge of such bridge taps requires from the carriers a rigorous management of the chronological evolution of their copper wires in the access network. Let us recall here that copper wires, once they are installed, may be used for several decades. Depending on the fluctuations of the customers' location, carriers add or remove bridge taps along the line. Additional problems may be induced by the fact that successive extensions of the subscriber lines may use inhomogeneous copper wires. Such a disparity in terms of impedance results in general in parasite echoes on the line. For all these reasons, the reader may understand why traditional LLs are so expensive for the end users and only affordable for professionals and not for residential users.

2.4.4   Basic Access to Narrowband ISDN

One of the objectives of narrowband ISDN at the end of the 1980s has been to provide affordable digital pipes for SMEs and residential users. Another objective of narrowband ISDN was to enable simultaneous voice and data services via a unique user-network interface and a single twisted pair. In this context, voice codecs are located at the CPEs and not within the EOs. Compared to LLs, which assume a semipermanent end-to-end resource reservation in the public network, narrowband ISDN enables resource reservation in real time, connection by connection. Before the emergence of xDSL modems at the beginning of the 1990s, basic access to narrowband ISDN at 160 Kbps (144 Kbps for user data plus 16 Kbps for signaling) has required the setting of dedicated copper pairs. As was the case for E1/T1 LLs, it was not possible to reuse existing analog subscriber lines for the provision of such access.

A basic ISDN access (see Figure 2.11) needs the installation of two baseband modems, one at the customer premises called the *network termination of type 1* (NT1) and another one at the switch called *line termination* (LT). ISDN subscriber lines are used in full-duplex thanks to echo cancellation (echo cancellation is detailed in Chapter 5). Such a technique does not authorize the use of RRs along the line. Due to the 160-Kbps gross bit rate of a basic access, the maximum achievable distance between the end users and the switch is 3,600m. An optional piece of equipment called *network termination of type 2* (NT2) may

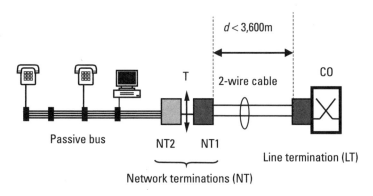

Figure 2.11 Basic ISDN access configuration.

be used by the end user to benefit a home network. Based on a passive bus, this LAN allows ISDN access to be shared between several pieces of terminal equipment. The NT1 and LT are provided by the carrier, whereas passive bus and NT2 are provided either by the carrier or by an independent company. The role of NT2 is to distribute incoming data to the corresponding destinations in the home network and to share the 144 Kbps useful capacity between the various terminal equipments. Chapter 8 consider adaptations of NT2 to the context of xDSL access. The NT1 and LT baseband modems use PAM based on the 2B1Q line code. Figure 2.12 illustrates the principle of this code. Every two

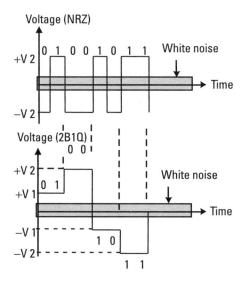

Figure 2.12 Principle of the 2B1Q line code.

successive bits, one associates a voltage pulse with four possible discrete levels: +2.5V, +0.83V, −0.83V, and −2.5V. The modulation speed of the 2B1Q is 80 kBauds (a Baud is the equivalent of a hertz), which corresponds to a 80-kHz frequency.

Basic ISDN access has been widely deployed in a few countries like Germany in Europe. In other countries like France, the deployment of ISDN has remained rather limited, probably due to the cost of installation of the dedicated lines between the NT1 and LT. As will be outlined in the following chapters, the problem of the cost of the access line is today solved thanks to HDSL technologies.

2.5 V5 Interface and Unbundling

Section 3.2 outlines the possible incompatibility between a CO and its EOs in terms of signaling interface if both of them are not provided by the same vendor. The terms unbundling and DSLAM have already been used in the general introduction of this book. Chapters 3 and 8 are dedicated largely to the explanation of these two terms, respectively. Unbundling refers to the deregulation of the local loop in order to facilitate fair competition between ILECs and CLECs. A DSLAM corresponds to the set of xDSL modems that have to be installed at the EO (with a traffic concentration). The problem of incompatibility between EOs and COs has an impact on the feasibility of unbundling. Let us consider, for instance, the case of a CLEC that only wants to offer ADSL services. If colocation of the CLEC DSLAM with the ILEC DSLAM is not possible, this CLEC has to install "in parallel" to the ILEC's EO its own EO in another building. Because the CLEC is not interested in PSTN services provision, a connection must be set between its EO and the CO of the ILEC. In such a scenario, unbundling is feasible only if the EO of the CLEC and the CO of the ILEC come from the same vendor (i.e., they have the same control unit). This is why the problem of the standardization of the signaling interface between EOs and COs is important in context of deregulation of the local loop.

Two versions of such an interface called V5 have been standardized within ETSI as "a European standard for the interface between public networks and local exchanges (LEXs)"[9] [2, 3]. The V5 interface has been designed to support both PSTN and ISDN lines on metallic E1 links. It is proper to the European market and it is not compatible with T1 links, which are used in North-America. Two types of signaling paths are considered within the V5 specification: *general signaling* paths and *service-related signaling* paths. The former, which deal with setup, release, and protection of connections, are called *control paths, bearer channel connections* (BCCs), and *protection paths,* respectively. The

---

9. Let us recall that the expression "local exchange" is synonymous with the expression "CO."

latter refers to the services offered by PSTN and ISDN. One can divide ISDN signaling into three types: "P" for packet data, "F" for frame data, and "S" for D-channel signaling. Two variants of the V5 interface have been defined:

- *V5.1. interface:* This interface does not provide traffic concentration, which means that line interface circuits at the input of the CO and line interface circuits at the output of the EO are coupled two by two. The V5.1 interface consists in a single E1 link with fixed-time slot assignment. It can support 30 PSTN ports, 15 basic access to narrowband ISDN ports, or a mix of the two service types. Within an E1 frame, up to three TSs can be used for the transport of signaling information: TS-15, TS-16, and TS-31. A single general signaling path is considered by the V5.1 interface: the control path on TS-16.

- *V5.2. interface:* This interface provides traffic concentration on the basis of the Erlang formula mentioned above. It consists of up to 16 E1, links which support both concentration and dynamic slot assignment. In comparison to the V5.1 interface, the V5.2 interface considers two additional general signaling paths—the BCCs and the protection paths. These additional general signaling paths share the same time slot as the control path (that is, TS-16). In case of malfunctioning time slots, the V5.2 interface allows the use of back-up time slots. Up to three back-up time slots may be used in V5.2. In case of protection switching, all of the signaling paths within the faulty time slot are redirected to the same protection time slot simultaneously.

Let us recall that the GR-303/TR-008 interface used in North America is the equivalent of the European V5 interface.

2.6 Digital Loop Carrier and Carrier Serving Area

In the middle of the 1970s, several carriers decided to introduce point-to-point digital links called DLCs in their access networks in order to serve more rapidly the increasing demand in subscriber lines for PSTN services. As it is shown in Figure 2.13, the basic idea of DLC consists of reducing the length of twisted pairs to install in the access network between the CPEs and the EOs. For that purpose, an intermediate piece of equipment called a *remote terminal* (RT) digitizes upstream analog signals and transports them on a unique link to the EO. The DLC technique is also known as the *pair-gain technique.*

In the United States, the pair-gain technique has been applied to about 30% of the subscriber lines. A DLC may be compared to a cable trunk as it is

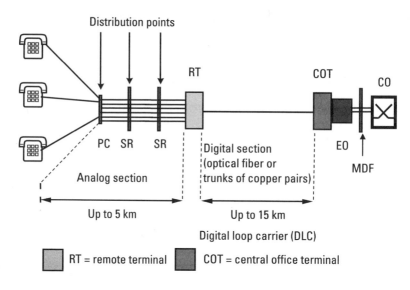

Figure 2.13 Principle of DLC.

described in Section 3.2 in the context of core networks. The point-to-point DLC link corresponds either to an optical fiber or to a cable binder made of several copper wires. A DLC cable enables the transport of $k*24$ voice channels (in North America). This capacity is equivalent to k parallel T1 links from the RT to another piece of equipment called the *CO terminal* (COT), which is located close to the EO. In addition to the fact that it allows a rapid extension of access networks' capacity, the DLC technique allows the range of the access network to be extended. We see from Figure 2.13 that thanks to DLC, the distance between an end user and his or her EO is extended from 5 km to about 20 km.

In the United States, major carriers defined in coordination with the FCC at the end of the 1970s a list of constraints that the new installed subscriber lines had to satisfy in order to facilitate their future use for digital transmission. In this context, we mean by "digital transmission" transmissions with on-line spectrums larger than the [0, 4 kHz] band. This list of constraints is known as the CSA. For instance, a CSA-compatible line must be at most 12,000 feet—that is, 3,600m if it is made with a single-section cable of 0.6 mm. Similarly, a CSA line must be at most 9,000 feet, or 2,700m if it is made with a single section cable of 0.4 mm. The CSA recommendations also specify that newly installed copper pairs must not include load coils nor bridge taps between the RT and the CPE. Today, some carriers use DSL modems at each end of a copper pair (at the RT and at the CPE) in order to provide two telephone lines over a single cable.

2.7 Dial-Up Modems and Erlang's Formula

Indirect access to the Internet by means of a dial-up modem concerns essentially residential users and SMEs. In this context, an end user accesses a server via his or her subscriber line and a telephone call through the PSTN network. We have underlined the strong increase in Internet traffic at the beginning of this book. In the context of Internet access by means of dial-up modems, such an evolution is a serious problem for telephone operators. Indeed, the mean duration, for instance, of a Web-browsing connection is much longer than the mean duration of a phone call. The former may last several tens of minutes, up to an hour, whereas the latter is about 3–4 minutes long. By consequence, some telephone operators begin to receive complaints from their customers because of the lack of availability of their telephone lines. This situation is more critical in the countries that apply flat-rate billing rather than those that apply duration-based billing of the telephone calls.[10] Cable trunks' dimensioning between COs and ISOs relies on the assumption of a 0.1-Erlang load per subscriber line (see Section 3.2). This load may reach up to 0.25 Erlang in the case of professional access. Section 1.3.3 discusses the fact that such statistics are unsuitable to describe Internet traffic behavior. Thus, the Erlang-B formula presented above is not applicable to Internet traffic. Two solutions to this problem are then possible:

- With the extension of IP networks, it is possible to estimate that in the future, POPs of NSPs will be installed closer to the end users. Thus, Internet traffic should be extracted much earlier from PSTN to be forwarded toward a packet-switching network. Such an evolution should then reduce the probability that a PSTN trunk is overloaded by Internet traffic. The problem is that call admission control in PSTN is circuit switch–oriented. This means that a new phone call is accepted only if the capacity inherent to this phone call is available on each of the physical links between the calling party and the called party. The fact the CO is still loaded by Internet traffic prevents from any improvement in terms of blocking probability for voice services.

- Another solution consists of replacing dial-up modems with new generation modems. Whereas a single modem is used at the customer premises in the case of dial-up access, these new generation modems assume the installation of a modem at each end of the subscriber line. The first role of the modem used at the EO is to separate analog voice traffic from data (Internet) traffic. This separation carried out at the input distribution frame of the EOs requires the installation of

---

10. In the next years, flat-rate billing should be generalized because of the cost of detailed billing required in the case of duration-based billing.

dedicated filters at the end of each subscriber line. This option is the basis of ADSL access.

2.8 Conclusion and Perspective

This chapter provides an overview of the various generations of dial-up modems, from V.22 to V.92. The V.44 standard mentioned in Table 2.1 is about to be adopted within ITU-T. It is not strictly speaking a new modem generation but a new compression technique designed to be implemented in various types of voiceband modems. The benefits of V.44 in terms of compression rate may reach 70%. Applied to V.90 or to V.92 modems, the V.44 compression technique should yield the ultimate version of dial-up modems. This chapter also outlines the main characteristics of access networks and of local loops at the periphery of PSTN networks. In addition, two utilization domains of twisted copper pairs are discussed: digital LLs and basic access to narrowband ISDN. We also outline the necessity of a standardized interface between EOs and COs in order to facilitate unbundling and determine that the V5 interface proposed within ETSI is a step in this direction. In addition, the chapter introduces the concepts of DLC and of CSA. Finally, this chapter ends by outlining the negative impact of dial-up modems on PSTN network availability, noting that one of the objectives of xDSL modems is to be a solution to this problem.

References

[1]   Proakis, J. G., *Digital Communications,* 3rd ed., New York: McGraw-Hill, 1995.

[2]   *Digital Central Offices Interfaces: V.5.1 and V.5.2,* Recommendation ITU-T Q.512, February 1995.

[3]   ETSI Draft EN/SPS-03047-1, *V Interfaces at the Digital Service Node (SN)*; ETSI recommendation, May 1997.

Selected Bibliography

ANSI T1.601-1992, *Integrated Services Digital Network (ISDN) Basic Access Interface for Use on Metallic Loops for Applications on the Network Side of the NT (Layer 1 Specification),* 1992.

Roberts, J., "Traffic Theory and the Internet," *IEEE Communications Magazine,* January 2001, pp. 94–99.

3

Unbundling Characteristics and Alternative Technologies

3.1 Introduction

Telecommunications services may be decomposed into three sectors: long-haul transport, local transport, and service provisioning. In Western Europe, deregulation policy was first set by the European Union authorities for the long-haul market in January 1998. Thanks to this new regulation context known as interconnection, a competitive operator may offer alternative telephone services by reusing a part of the PSTN infrastructure of a traditional telephony carrier. The concept of interconnection is also used when a competitive carrier wants to offer IP transport services (or IP services directly) via the PSTN of a traditional carrier. The book's introduction explains why access to the local loop is today a necessity for the carriers.

Compared to MANs or to core networks, implementation and operation of access networks is very expensive. The reason for this cost disparity is related to the lack of economy of scale inherent in the various equipment involved in access networks. For instance, a twisted pair is dedicated to a single user in the local loop whereas a backbone optical fiber is shared by several hundreds or thousands of users in the metro and in the core. In terms of civil engineering, laying a new twisted pair represents the same cost as laying a new optical fiber. In other terms, the fact that traditional carriers (ILECs) own a dense copper pair infrastructure in the last mile is a key advantage. Chapter 2 detailed the multiple uses of copper wires installed in the local loop: PSTN access, ISDN access, LLs, and analog Internet access. In addition, it described the principle and limitations of voiceband modems currently used for Internet access over subscriber

45

lines. Thanks to advances in the field of high-speed electronics, in VLSI circuits, and in signal processing, since the end of the 1980s we have known that voice-band modems underutilize the real capacity of copper wires in the local loop. In addition, about 800 million subscriber lines are today operated by telecommunications carriers.

For all these reasons, we can understand why competition between various carriers in the local loop is a hot topic. In most cases, ILECs were formerly national PTT (Poste, Télégraphe, Téléphone is the traditional name for an ILEC in Europe) operators benefiting from a monopolistic position in their respective markets. During the last decade, both in North America and in western Europe, national regulation bodies have favored competition between ILECs and CLECs. The term *unbundling* refers to the obligation for ILECs to rent or to sell their copper infrastructure to CLECs that would intend to provide high-speed access to the Internet to the end users. As an example, the French telecommunications regulation authority or *Autorité de Régulation des Télécommunications* (ART) has decided to apply unbundling to the national carrier France Telecom since the first of January 2001. Finland and Germany have been among the very first European countries to adopt unbundling. In the United States, the FCC has imposed various forms of unbundling to the traditional regional carriers known as the *regional Bell operating companies* (RBOCs) born from the AT&T company since the second half of the 1980s. Additional rules have been imposed ever since.

Section 3.2 describes the interconnection characteristics. Section 3.3 presents the various possible alternatives for unbundling. Two technical difficulties are inherent to the application of unbundling: the colocation problem and the "churn" problem. Colocation refers to the necessity to install CLECs and ILEC equipment in a same building. The "churn" problem refers to the operations that must be carried out on the MDF of the CO of the ILEC each time an end user changes its access provider. Some of the CLECs have decided to circumvent these difficulties by adopting new technologies for the local loop. Most of these new technologies will be described in detail throughout the various chapters of this book. A few of the alternatives to the copper wires do not merit dedicated chapters either because of their lack of maturity or because of their less innovating nature. For example, Sections 3.6–3.8 are dedicated to the description of satellite networks, *hybrid fiber coaxial* (HFC) networks, and *power line communication* (PLC) networks.

3.2   The Interconnection Concept

The various countries belonging to the European Union adopted the principle of interconnection in January 1998. Within each of these countries, the national regulation bodies have decided on the rules on which interconnection is based (i.e., the way the equipment of a CLEC or of a service provider can be connected to the nodes of an ILEC). Two types of interconnection are considered: telephony services provisioning and Internet access provisioning.

3.2.1 The Interconnection Principle

Let us consider in a first step the case of interconnection for telephony services provisioning. As it is described in Figure 3.1, interconnection may occur at different levels of a PSTN. The POP of a CLEC (i.e., a telephone switch managed by this CLEC) may be connected to the different switching nodes of the ILEC. At the lowest level, such a POP may be connected to the output[1] distribution frame of a CO[2] of the ILEC (solution 1). We see on the same figure that two other interconnection alternatives are possible. Solutions 2 and 3 correspond to the connection of a POP to the output distribution frame of an ISO[3] of level 1 or 2, respectively.

In Figure 3.1, the vertical solid lines at the ingress and egress of the different PSTN nodes (EO, CO, and ISOs) represent their corresponding distribution frames. The definition and the role of a distribution frame are given in Section 3.3.1. The higher the level of the interconnection, the lighter the infrastructure required by the CLEC. For instance, solution 1 requires that the CLEC install its own ISOs of level 1 and of level 2. Solution 3 only requires that this carrier install a few ISOs of level 2. Let us consider now the case of

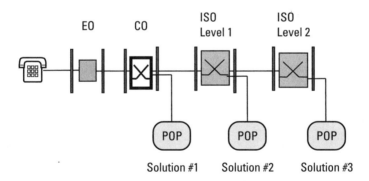

Figure 3.1 The principle of interconnection.

---

1. In the following, one considers the terms *ingress* and *egress* under the assumption of upstream traffic transmitted from an end user toward the heart of a PSTN network.

2. This distribution frame is also called *digital distribution frame* (DDF).

3. ISOs are also called *digital trunk exchanges* (DTXs).

interconnection for Internet access provisioning. The POP mentioned in Figure 3.1 corresponds then in general to an IP router of a NSP. The servers of various ISPs are accessible via the IP network of this NSP. Protocol conversion must be carried out at the interconnection point between the POP and its attachment ISO. Again, the higher the level of the interconnection, the lighter the IP infrastructure required by the CLEC. The flip side of the coin for the competitive telephony service provider or for the NSP is that, the higher the level of the interconnection, the more expensive this interconnection. Indeed, the competitive telephony services provider or the NSP use the switching and transmission resources of the ILEC. We mentioned above that thanks to the increasing capacity of the transmission links, the cost of transmission becomes negligible. This is not the case for the cost of switching. ILECs bill the renting of their switching nodes on a per-call basis in proportion to the connections' duration, just like traditional voice calls. In practice, a new billing principle has been defined by the European authorities for interconnection. This billing principle is known as the *long-term mean incremental cost* (LTMIC). The LTMIC is based on the investment that the ILEC is supposed to realize in its local loop in order to facilitate interconnection of its infrastructure with CLECs. The LTMIC calculation makes the assumption that the most advanced technologies available on the market are used by the ILEC at the considered date. One may notice a form of subtlety in the European recommendations about interconnection. For instance, the recommendations mention that the price lists of the ILECs must be affordable for the CLECs but not too low. Indeed, too low tariffs could dissuade CLECs from investing in new technologies for the local loop.

3.2.2   The Colocation Problem

Colocation refers to the possibility offered by an ILEC to a CLEC to install a POP in the same building as the access switch. For instance, colocation means in solution 1 mentioned in Figure 3.1 that the ISO of the CLEC is located in the same building as the CO of the ILEC. If collocation is not possible, the CLEC has to build or to rent its own distant premises for the installation of its POP. To be physically connected to the access node of the ILEC, the CLEC has to use a digital leased line (E1 or T1) between its premises and the ILEC's access node. Section 3.2.4 outlines the fact that LLs are very expensive. An E1 or a T1 LL requires two preconditioned copper pairs with eventually RRs. Let us remark that the higher the interconnection level, the larger the amount of potential subscribers accessible by the CLEC. In other terms, at a given distance between the POP and the switching access node, the cost of the LL decreases with the level of the interconnection.

3.2.3 Drawbacks of Interconnection

In terms of operational cost, the three alternatives mentioned in Figure 3.1 assume that the ILEC invoices CLECs in terms of the duration and of the distance of the phone calls established through its PSTN infrastructure. The exact rules on which this billing is based are under the jurisdiction of the national regulation bodies. We mentioned above two types of interconnection, one for telephone services provisioning, the other one for Internet access provisioning. Chapter 1 presented the various network configurations inherent to Internet access by means of voiceband modems. Interconnection explicitly refers to indirect access to the Internet. Enriched telephony services provisioning by a CLEC totally depends on the potentialities of the switches, and more particularly on the ILEC's CO. Chapter 2 underlined the limited capacity of traditional voiceband modems. The xDSL modem generation appeared on the market in the 1990s. Such modems enable higher bit rates (and subsequently innovative services) than voiceband modems, essentially because of their much wider bandwidth. A constraint inherent to this larger bandwidth is the need for a modem at the line termination (i.e., at the level of the MDF of the EO). It is clear that interconnection does not authorize the installation of line termination xDSL modems by the CLECs. A solution could entail renting the use of these modems from the ILEC. Meanwhile, the potential services offered by the CLEC should depend in that case on the good will of the ILEC.

3.3 The Unbundling Concept

In urban areas with a high population density, EOs may be colocated with COs. New generation xDSL modems assume a length of the local loops under 5 km. This condition is satisfied only in the case of dense urban areas. This is why, in the remainder of this chapter, we shall only refer to situations where the CO and its attached EOs are colocated.

3.3.1 Distribution Frame's Utility

At the input of an EO, all the subscriber lines arriving via different binder cables are connected to a MDF. An MDF provides a convenient way to connect subscriber lines with the line circuits physically associated to the input ports of the EO. A line circuit associated to each subscriber line supplies a battery feed to energize the transmitter in the telephone set. The direct current generated by the line circuit reaches the telephone set via the customer loop to carry signaling and conversation signals. The second role of a line circuit is to convert the two-wire subscriber line into a four-wire circuit in order to enable A/D and D/A conversion by a coder/decoder (also called *codec*). Figure 3.2 illustrates a typical digital switching system. One assumes in Figure 3.2 that the EO (i.e., the set of A/D and D/A converters and of traffic concentrators represented by the multiplexers/demultiplexers) are colocated with the CO (the switching fabric). As shown in Figure 3.2, subscriber lines and line circuits arrive on the vertical side and on the horizontal side of the MDF, respectively. A jumper (a movable copper pair) is used to link a subscriber line to a line circuit via the MDF. Instead of disconnecting a copper pair at the level of a line circuit in the case of desubscribing an end user, the MDF allows one to just disconnect the jumper associated with this end user at the MDF. In Figure 3.2, this disconnection is illustrated in the case of user B. Line circuits are grouped on line interface cards. Once individual voice signals have been digitized, the associated data are concentrated by means of a multiplexer before being forwarded through a switching fabric. In the reverse direction, data getting out from the switching fabric are demultiplexed before being converted from digital to analog and finally transmitted on a subscriber line. Between the multiplexer/demultiplexer (also called mux/demux) and the switching fabric, digital information is transported over a very short distance in E1 or T1 frames (see Chapter 2) over toll-connecting trunks.

Figure 3.3 describes the various network elements involved in an access network. Most of the terms mentioned in Figure 3.3 have already been defined in the previous chapters. In addition to the MDF, another distribution frame is used to connect E1/T1 copper pairs getting out from the switching fabric to the inputs of transmission equipment (a transmission modem). This second distribution frame is called a *digital distribution frame* (DDF). In many cases, the

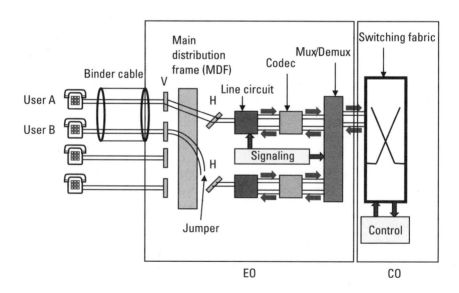

Figure 3.2 Elements of a digital switching system.

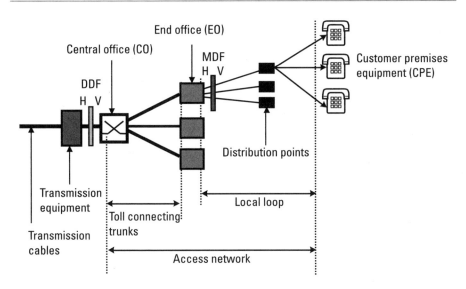

Figure 3.3 Global access network infrastructure (under low population density).

transmission equipment concentrates E1/T1 flows into higher bit rates like E2/T2 at 8.448 Mbps or 6.312 Mbps. These higher rates modulate directly a laser diode for an optical transmission over an optical fiber. This fiber links the CO to the first level ISO. Similar to an MDF, the two sides of a DDF are referenced by the letters "V" (vertical) and "H" (horizontal). In general, the MDF and the transmission equipment are placed in two dedicated rooms adjacent to the room in which the switch is installed.

3.3.2   Unbundling Definition

The unbundling concept refers to the possibility given to a CLEC to connect its POP at a lower level than a CO on the infrastructure of an ILEC. The unbundling term assumes a separation of the local loop ownership from particular service provisioning such as telephony and high-speed Internet access. The installation of xDSL modems by a CLEC at the EO is possible only in the context of unbundling. Figure 3.4 illustrates the typical network configuration in the case of unbundling. The POP of the CLEC is either an EO or an IP router, the CLEC aiming to provision telephony services or Internet access, respectively.

Two types of unbundling are possible: full and partial. Full unbundling means that the use of the global capacity of a copper pair is left under the responsibility of a CLEC. In other terms, both analog voice service and full-duplex data services are provisioned by the CLEC. Partial unbundling means

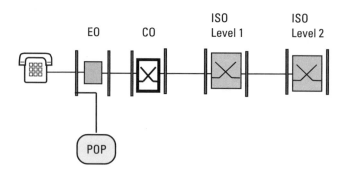

Figure 3.4 Principle of unbundling.

that the ILEC keeps telephony services provisioning, high-speed Internet access being provisioned by a CLEC. Partial unbundling may also mean that the ILEC provisions high-speed Internet access whereas a CLEC provisions telephony services. (Such a combination should remain the exception in the next years.) As outlined in the following sections, full unbundling and partial unbundling are facilitated if colocation is possible. Colocation means that some of the CLEC equipment, typically an EO, a DSLAM, or a CO, may be installed on the premises of the ILEC.

3.3.3   Partial Unbundling for Telephony Services Provisioning

Let us consider in a first example the case of a new carrier that wants to offer telephony services concurrently to those offered by the ILEC (see Figure 3.5). In such a situation, the adjective "partial" in the expression partial unbundling refers to the fact a fraction of the subscriber lines are used for telephony services provisioning by the CLEC. The CLEC installs its own CO in parallel to the CO of the ILEC. The line circuits of the CLEC's CO are connected on the same output distribution frame as the ILEC's CO line cards. Ideally, the ILEC authorizes the CLEC to install its CO in a dedicated room of the same building as its own CO. Technically speaking, colocation is very easy in most cases. Indeed, the CO buildings of traditional carriers were designed during a period (the 1970s) when the first electronic switches were requiring a lot of square meters. Thanks to the progress in electronic chip integration, the same capacity switches require today much less surface. Depending on the national regulation bodies, the unbundling rules do not impose systematically colocation. If colocation is not possible, CLECs have to rent or to build their own building (as has been mentioned in the case of interconnection). They have also to ask for a LL from their building and the CO building of the ILEC. This second option appears to be much less favorable to competition in the last mile. Let us remark

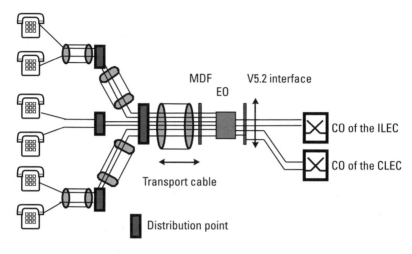

Figure 3.5 Partial unbundling for concurrent telephony services.

that unbundling enables a CLEC to offer its own enriched telephony services, with such services being implemented in the CO. The identification of the calling party by the called party is an example of enriched telephony service.

Another configuration for partial unbundling for telephony services provisioning corresponds to the case where the ILEC wishes to have the monopoly of high-speed Internet access provisioning. Figure 3.6 describes such a situation. The DSLAM installed by the ILEC is connected only to the subscriber lines that are used for both analog telephony and data transfers. Within the DSLAM, filters called *POTS splitters* isolate in the frequency domain analog voice channels and data channels.[4] The former get into the EO to be digitized, the latter are directed to a POP (an IP router or a packet switch).

Telephony services provisioning shared by the ILEC and one or several CLEC in a same access network raises a technical difficulty. Indeed, because of the historical monopolistic position of ILECs, EOs have been designed by the manufacturers in such a way that they can be mastered only by a single CO. In Figures 3.5 and 3.6, two COs are supposed to control the same EO. This problem can be solved by a software updating of the EO.

3.3.4 Partial Unbundling for Internet Access Provisioning

High-speed Internet access provisioning requires from a CLEC that it owns and manages a proper DSLAM. For this type of configuration, the ILEC is supposed

---

4. By simplification, we have not represented in Figure 3.6 POTS splitters associated with the DSLAM.

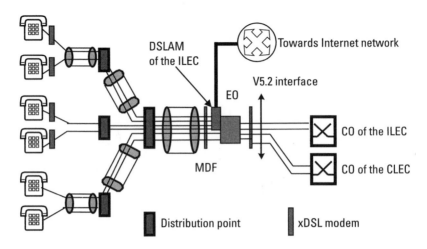

Figure 3.6 Partial unbundling for high-speed Internet access provisioning.

to retain the monopoly in telephony services. Like in the case of partial unbundling for telephony services, partial unbundling for high-speed Internet access provisioning raises the colocation problem. Two configurations are possible: Either the DSLAM of the CLEC is installed in the same building as the ILEC's EO (same configuration as in Figure 3.6), or this DSLAM has to be installed in distant premises owned by the CLEC (Figure 3.7).

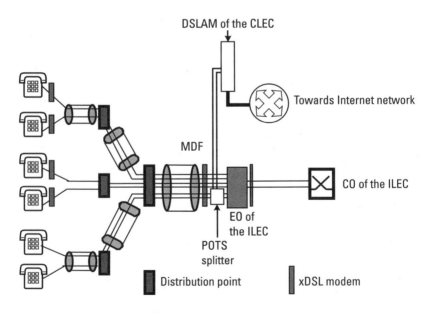

Figure 3.7 Partial unbundling for Internet access without colocation.

The main constraint in the absence of colocation is the limited acceptable distance between the end users and the DSLAM. In general, this distance is supposed to be under 5 km to enable a good performance by the xDSL modems. Let us notice that POTS splitters are in this case under the responsibility of the ILEC.

We have mentioned that in urban areas, a CO with its attached EOs and transmission equipment (see Figure 3.3) are colocated in a same building. The field of application of xDSL access corresponds in most cases to such a configuration where the length of the local loop remains acceptable for xDSL modems. To clearly separate the equipment under the responsibility of the ILEC from that under the responsibility of the CLEC, a four-room configuration may be adopted as illustrated in Figure 3.8. Subscriber lines arrive in room 1 on the vertical side of the MDF. On the horizontal side of the MDF, POTS splitters isolate the [0, 4 kHz] band proper to analog voice signals from upper frequencies inherent to data traffic. Copper pairs link POTS splitters to traffic concentrators (codecs and mux/demux) at the input of the CO of the ILEC located in room 2. At the output of the CO, copper pairs are used to forward digital voice channels toward transmission equipment located in room 3 via the DDF. In room 3, transmission equipment (optical modems) converts electrical signals associated to voice traffic into optical signals. An *optical distribution frame* (ODF) is used to link the transport optical fibers with the transceivers of the transmission

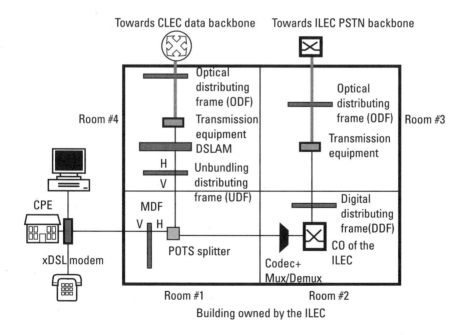

Figure 3.8 Detailed configuration of collocation in the case of partial unbundling.

equipment. Voice traffic is then forwarded to an ISO in the PSTN of the ILEC. Let us notice that POTS splitters are in this case under the responsibility of the ILEC.

High-speed access to the Internet is provisioned by the CLEC thanks to its equipment installed in room 4. The CLEC manages its own distribution frame, which it calls an *unbundling distribution frame* (UDF). Copper pairs link POTS splitters located in room 1 with the DSLAM located in room 4 via the UDF. At the output of the DSLAM, data traffic (Internet traffic) is forwarded to transmission equipment. Like in room 3, an ODF is used to link the transport optical fibers with the transceiver's transmission equipment. Let us recall that the main role of a DSLAM is to concentrate data traffic coming from the CPEs before forwarding it toward the POPs of different NSPs. If several CLECs operate on the same access network, it is necessary to assign a dedicated room adjacent to the CO of the ILEC for each of them. Rooms 1, 2, and 3 are under the control of the ILEC.

3.3.5   Full Unbundling

In the case of full unbundling, the CLEC provides both analog voice services and Internet access services to the end users. Again, two configurations are possible depending on whether colocation is possible or not. The CLEC must in this case install its DSLAM and its EO in its own building. At the level of the MDF, jumpers are used to reroute traffic coming from unbundled pairs toward the DSLAM of the CLEC. In practice, the distance between the MDF of the ILEC and the DSLAM of the CLEC must remain compatible with the maximum admissible range for the local loop. Unlike the situation described in Figure 3.7, the CLEC is in this case in charge of the POTS splitters.

Figure 3.9 illustrates the colocation configuration under full unbundling. An unbundled copper pair arriving on the MDF is redirected on the vertical side of the UDF. On the horizontal side of the UDF, these pairs are connected to POTS splitters. After POTS splitters, analog voice channels are forwarded to traffic concentrators for digitization before being switched. Frequencies above 4 kHz are transferred to the DSLAM. One has assumed in our example that the ILEC is still offering analog telephony services. Room 4 is under the control of the CLEC, whereas rooms 1, 2, and 3 are under the control of the ILEC.

3.4   The "Churn" Problem

Between 20 and 40 million subscribers lines are installed in each of the larger European countries (e.g., France, Germany, and Great Britain). In these countries, between 10,000 and 15,000 EOs and between 1,000 and 1,500 COs are

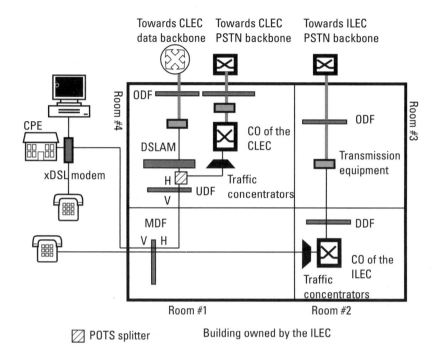

Figure 3.9 Full unbundling network configuration with collocation.

operated, in most cases by incumbent carriers. The number of POPs operated by IP network access providers is between 50 and 100. With the deregulation of the local loop, the number of CLECs should increase considerably in the next few years. Previous sections have underlined the impact of this multiplicity of operators on the colocation problem. Another consequence of the multiplicity of operators in the local loop is the necessity to move more and more frequently jumpers within the MDF of the ILEC. In the next few years, the number of customers who will change their subscription from one carrier to another or who will ask for new telephony or data services should increase considerably. Such an evolution will require an increasingly frequent change and turn on the jumpers used within MDFs. This operation of change and turn, which is called *churn,* is relatively costly because it is carried out manually in most cases. Today, the annual churning rate reaches between 15% and 50% according to the countries (France, Germany, and England). Facing such a situation, some manufacturers have developed totally automatic cross-connecting machines to carry out jumpers shifting within MDFs. These new types of machines also enable the automatic testing of the electrical characteristics of the subscriber lines once they have been reconfigured without a technician's visit to the CPE or the CO.

3.5   Colocation Status in Europe and in North America

3.5.1   European Market

Table 3.1 gives the colocation status in Europe at the end of October 2000 [1]. The second column of Table 3.1 refers to the *national regulation authorities* (NRAs) of the countries mentioned in the first column. We see from Table 3.1 that Germany and Denmark are the two first countries to have found an agreement on colocation.

In Germany, the law instituting unbundling was enacted in 1997. Unbundling became effective in 1999. About 170,000 of the subscriber lines were unbundled in May 2000. The reasons for this delay were essentially due to the lack of consensus between Deutsche Telekom and its competitors about the rent of the copper pairs. According to the national regulation bodies, 82 unbundling agreements were signed in Germany at the end of 2000. A specificity of the German market is that many of the unbundled lines are used by CLECs to provide ISDN services.[5] The reason for this specificity is due to the fact that Germany is

Table 3.1
Collocation Status in Western Europe

Country	Colocation Status (October 2000)
Austria	Sample collocation contract is available from Telecom Austria. This is based on the rules determined by the NRA following a failure to reach commercial agreement on terms.
Belgium	No rules or contractual terms available.
Denmark	A framework for colocation agreements was published by the NRA in Denmark. Based on this, Tele Denmark produced a standard agreement on collocation and this was published on April 22, 1999.
Finland	No rules or contractual terms available.
France	No rules or contractual terms available.
Germany	Rules for colocation are available from the NRA.
Ireland	No rules or contractual terms available.
Italy	Detailed terms and conditions available from Telecom Italia.
Netherlands	A sample contract is available from the incumbent.
Spain	No rules or contractual terms available.
Sweden	Terms available form the incumbent. These were developed unilaterally and were not subject to guidance from the regulator.
United Kingdom	Terms available from the incumbent

5. Both narrowband ISDN access at 128 Kbps and primary ISDN access at 2 Mbps are proposed by CLECs.

the European country where ISDN services have known the greatest success. This success is itself related to the high density of SMEs in the territory that could not afford during the last 20 years the cost of LLs. Thus, 18 million ISDN access lines were operational in Germany at the end of 1999.

Denmark and the Netherlands adopted unbundling in 1998 and in 1997, respectively. In the Netherlands, about 100% of the households are connected to cable networks. This is why the xDSL access market remains relatively modest in this country.

3.5.2 North American Market

Unbundling was decided by the FCC with the Communications Act in 1996. This document aims to enable competition between telecommunications carriers both in long-distance and local-distance markets. The RBOCs issued from the original AT&T telephone company in 1984 were not authorized to provide long-distance calls because of their monopoly over local calls. The Communications Act of 1996 suppressed this restriction to the RBOCS on the condition that they facilitate unbundling. A few years after the Communications Act, the situation of unbundling in the United States is unclear. The xDSL services offered by RBOCs through their own infrastructure are a relative success, with the amount of unbundled subscriber lines remaining rather modest (700,000 unbundled subscriber lines at the end of 1999).

Associated to the Communications Act, the FCC has determined the billing rules to be used by the ILECs for renting their copper pairs. This rule is known as the *total element long-run incremental cost* (TELRIC). The consensus obtained on that hot topic is that the copper pair rent is not homogenous over the territory. This price depends on the potential market, that is on population density, and particularly on the density of enterprises.

Some of the first CLECs on the market went bankrupt because of the high cost of deployment of their infrastructure. Pessimistic observers estimate that at least one or two CLECs should remain on each local loop market to prove that deregulation in the last mile is effective. In spite of favorable unbundling rules and economies of scale proper to the large American urban areas, it seems that the local loop's market requires stronger capitalistic resources than expected initially.

Relative to the size of each country, the DSL market size seems quite similar in European countries like France and Germany than in the United States. Under the authority of the FCC, it has taken one year in the United States to find an agreement on the terms of unbundling and colocation.

In Canada, unbundling has been facilitated by the regulation bodies but for a limited duration. Thus, an CLEC may benefit from the coppers of an

Table 3.2
Target Bit Rates for New Generation Modems

Service	Bit Rate	Interactivity
Web browsing	240 Kbps	Yes
E-mail	A few kilobits per second	No
Interactive games	Around 1 Mbps	Yes
Digital telephony	5–6 Kbps	Yes
File transfers	Around 1 Mbps	No

ILEC for, at most, 5 years. Such a limitation has been imposed to motivate the deployment of new infrastructures by the CLECs.

3.6   Alternative Technologies for the Local Loop

Chapter 2 underlined the limited performance of voiceband modems. Typically, such modems enable a few tens of kilobits per second for upstream and downstream traffic.[6] In the short term, one estimates according to the state of the technology in terms of coding and compression techniques that most current applications require the target bit rates mentioned in Table 3.2.

Multiple techniques are possible to satisfy such target data rates in the local loop. In a first approach, it is possible to classify these techniques into two categories: wired and wireless.

3.6.1   Wired Alternatives

Figure 3.10 summarizes the four main techniques based on different types of wired infrastructure: twisted copper pairs, optical fibers, coaxial cables, and energy cables. For each of these techniques, one notices that two modems are systematically involved, one at the CO side and one at the CPE side.

Three alternatives to xDSL are possible: *fiber in the loop* (FITL), HFC, and PLC. The FITL technique has been considered during the last decade as economically flawed, even if some major carriers have pursued intensive research activities in this domain. After 2000, this opinion has been widely discussed because of the limits of xDSL technologies and the recent important advances in optical communication. The HFC technique is inherent to existing cable networks. Last, the PLC technique uses the low-voltage energy cables of electricity suppliers to serve end users from *low-voltage transformers* (LVTs). One notices

---

6. In the very most favorable configuration, V.92 modems support 56 Kbps downstream and 48 Kbps upstream.

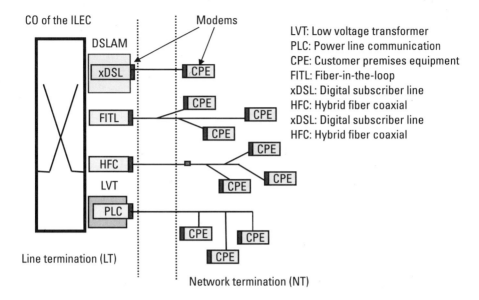

CO of the ILEC

DSLAM

Modems

LVT: Low voltage transformer
PLC: Power line communication
CPE: Customer premises equipment
FITL: Fiber-in-the-loop
xDSL: Digital subscriber line
HFC: Hybrid fiber coaxial
xDSL: Digital subscriber line
HFC: Hybrid fiber coaxial

Line termination (LT)

Network termination (NT)

Figure 3.10 Wired access technologies.

that, unlike xDSL, which is a point-to-point technique, FITL, HFC, and PLC are point-to-multipoint–oriented. Point-to-multipoint is well suited to broadcasted information (TV channels typically). The main drawback of a point-to-multipoint network configuration is the necessity of a MAC protocol to manage upstream transmission. Globally, two types of MAC protocols have been developed for such infrastructures: those that accept collisions and those that prevent collisions. Among the four techniques shown in Figure 3.10, two of them are detailed in this book: xDSL in Chapters 4–10 and FITL in Chapter 12. We describe more briefly in Sections 3.7 and 3.8 the principle of HFC and PLC access systems.

3.6.2   Wireless Alternatives

Figure 3.11 summarizes the three main techniques based on wireless access. WITL access systems, which are described in detail in Chapter 11, are today largely deployed by CLECs. Access systems based on satellite or on sounding balloons are described next.

3.6.2.1   Satellite for the Local Loop

Geostationary satellites (GEOs) placed at 36,000-km altitude cannot be considered for Internet access due to the prohibitive round-trip time to the Earth stations. In the 1990s, *low Earth orbit* (LEO) satellite constellations aimed to

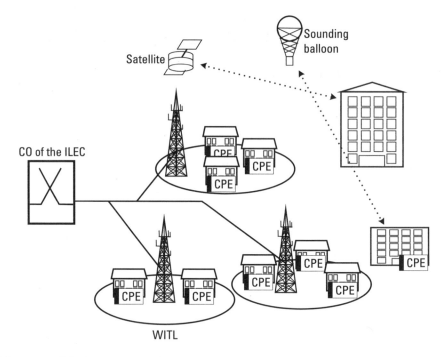

Figure 3.11 Wireless access technologies.

provide worldwide mobile telephony. Located at around 750-km altitude, LEO satellites move at such a speed that handover procedures are mandatory during phone calls. These handover procedures are complex and expensive to implement. They assume a permanent volume of signaling information exchanged between satellites and Earth stations. The gain in propagation delay obtained thanks to low orbits is spent on another side by signaling processing time. Satellite handover is in general at the origin of a SNR degradation. If this degradation remains acceptable in the case of voice traffic, it may be unacceptable in the case of Internet traffic, particularly when the two successive satellites move in opposite directions. Let us add that voice channels available on LEO satellites are very low speed, under 10 Kbps. Considering the commercial failure of the two first satellite constellations, one can conclude that Internet access by means of LEO satellite constellations is not a realistic approach for the short term. In the medium term (that is, before the end of this decade), *medium Earth orbit* (MEO) satellite constellations could be a suitable solution for wireless access to the Internet. Thanks to their altitude around 5,000 km, LEO satellites authorize both admissible propagation delays and lighter signaling traffic. Hybrid solutions combining downstream GEO satellites channels with terrestrial upstream return channels are also investigated. In such systems, downstream channels should provide a few megabits-per-second capacity to each end user. Upstream channels using xDSL connections between CPEs and Earth stations could provide several hundreds of kilobits per second to individual upstream channels [2–4].

3.6.2.2 Sounding Balloons

Experiments using sounding balloons are also considered for wireless local loop [5]. Systems known as *high-altitude aerial platforms* (HAAPs) are used as a radio platform located at a 20-km altitude. To give an idea of the range of such systems, let us say that agglomerations like Paris or London with about 10 million inhabitants could be covered by HAAP. In the United States, the Sky Station research program aims to develop sounding balloons stabilized in position by means of onboard engines powered with photo-voltaic batteries.[7] A global upstream plus downstream capacity of 7 Gbps is considered. This capacity should enable symmetrical individual capacities between 64 Kbps and 2 Mbps. The American FCC has already reserved the [47.2 GHz, 48.2 GHz] band for HAAP systems. The ITU-T approved this frequency plan at the end of the year 1997. The main advantage of HAAP systems in comparison to LEO satellite constellations in their cheaper implementation. The cost of implementation of the 66 satellites of the Iridium constellation reached $4 billion, whereas the implementation cost of the Proteus HAAP project is fixed to about $10 million.

3.7 Cable Networks

Cable networks were designed initially to distribute analog TV channels by means of a coaxial cable infrastructure. During the last 20 years, cable networks have been largely deployed in the United States and in Canada, especially in large urban areas. In comparison to radio systems for which the frequency resource is limited, cable networks can broadcast a larger amount of TV channels. In Europe, the success of cable television has been more contrasted. For instance, up to 11 million cable modem network terminations had been installed in France by the end of 2001. This figure must be compared to a population of 60 million inhabitants, with, in fact, only 3.1 million effective subscribers to cable networks. At the opposite end of the spectrum, cable networks are very popular in countries like Belgium where population density is higher.

In the 1990s, several factors such as the emergence of Internet services, advances in video coding and compression techniques, and the deregulation of the local loop motivated the development of a new generation of cable

---

7. The Sky Station program is granted by private American and European companies such as EADS, Daimler Benz, Comsat, and the Jet Propulsion Laboratory.

networks. Unlike the first generation cable networks, which were limited to TV channel broadcasting, new generation cable networks are bidirectional. This property enables companies to provide interactive services such as *video-on-demand* (VoD), digital telephony, and high-speed access to the Internet. Analog TV channels are in this case replaced by digital TV channels. This digitization—using in most cases the MPEG-2 coding and compression technique—presents two main advantages. First, an MPEG-2 digital TV channel requires a bandwidth of about 2 MHz with QPSK modulation whereas an analog TV channel (PAL, SECAM, or NTSC) needs around 8 MHz. Second, digital TV is less sensitive to noise distortion inherent in propagation than analog TV. Unlike first generation cable modems, new generation cable modems have required the development of a MAC protocol for upstream traffic. Several efforts have been carried out to standardize such MAC protocols. Three main approaches have been considered:

- The IEEE 802.14 standard: The IEEE 802.14 MAC protocol [6, 7] was standardized in 1998. It was developed thanks to the contributions of several institutions and companies such as IBM, AT&T, Columbia University, and NIST.

- The *digital video broadcasting* (DVB)-Digital Audio and Video Council (DAVIC) (DVB-DAVIC) standard: This approach has been considered by two professional consortiums in the domain of public television: the DVB and the DAVIC consortiums. The basic objective of the DVB-DAVIC standard is to provide interactive IP-oriented services to end users in addition to the distribution of analog or digital TV channels.

- The Multimedia Cable Network System (MCNS)–Data Over Cable Service Interoperability Specification (DOCSIS) (MCNS-DOCSIS) standard: The MCNS and the DOCSIS consortiums have the same objectives as DVB-DAVIC [8, 9].

At the end of 1995, the majority of U.S. cable operators and of cable modems manufacturers found a consensus on the MCNS-DOCSIS approach that became a de facto standard.[8] This first version is known as the DOCSIS 1.0 standard. During 1999, an updated version of the DOCSIS specifications was published. This new version known as DOCSIS 1.1 considers advanced mechanisms for QoS guarantee provisioning, IP multicast, and improved security [3, 4].

---

8. The IEEE 802.14 and DVB-DAVIC standards have not been supported by the cable modem vendors.

3.7.1   HFC Network Configuration

Figure 3.12 illustrates the typical configuration of a point-to-multipoint HFC network. At the head-end of the tree, a modem called a *cable modem termination system* (CMTS) supervises dynamic bandwidth allocation in the network. The role of the CMTS is to link the coaxial tree with the POP of an IP NSP. At each leaf of the tree, modems located at the customer premises—called *cable modems* (CMs)—enable the connection of users' equipment via an Ethernet interface. Three types of equipment can be connected to a CM: digital phones either for CBR-PCM telephony or for *voice-over-IP telephony* (VoIP), a TV set coupled with a set-top box enabling interactivity, and a PC. Up to 2,000 end users may be connected to an HFC network. RRs are needed on average every 4 km both for upstream and downstream traffic.[9] To extend the range of the coaxial infrastructure, it is possible to use between the CMTS and the head-end of the coaxial tree a pair of unidirectional optical fibers.[10] The optical modems used at each

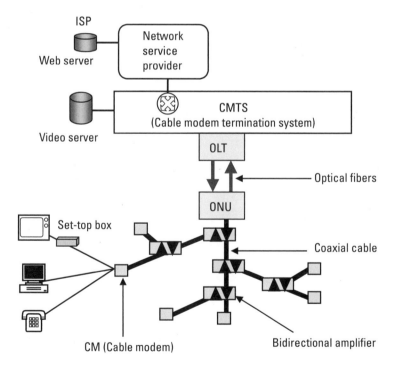

Figure 3.12 Typical HFC network configuration.

---

9. This point is one of the main drawbacks of HFC networks because of the cost of implementation and maintenance of RRs.

10. Such a configuration justifies the acronym *hybrid fiber coaxial* (HFC).

end of the fiber links are called *optical line terminations* (OLTs) and *optical network units* (ONUs).

3.7.2 Physical Layer

Two different analog TV standards have been adopted in the United States and in Europe, called NTSC and the PAL/SECAM, respectively. NTSC TV and PAL/SECAM TV channels require 6 MHz and 8 MHz of bandwidth, respectively. Two different physical layer specifications are considered in the MCNS-DOCSIS standard, called DOCSIS, and its European version, called EuroDOCSIS, respectively. Figure 3.13 describes the spectral occupancy in a MCNS-DOCSIS system. Upstream traffic is transmitted on the medium by means of N parallel carriers. These channels are located in the [5 MHz, 42 MHz] band in the United States or in the [5 MHz, 60 MHz] band in Europe. The typical bandwidth of each upstream channel is between 200 kHz and 3.2 MHz. Downstream traffic is transmitted on the medium by means of M parallel carriers located in the [50 MHz, 850 MHz] band in the United States or in the [65 MHz, 850 MHz] band in Europe. The typical bandwidth for each downstream channel corresponding to one TV channel is 6 MHz for MCNS-DOCSIS or 8 MHz for MCNS-EuroDOCSIS. The values of N and M are let at the choice of the carriers in the limit of the total capacity of the system. For instance, the number of video channels in the system may depend on the

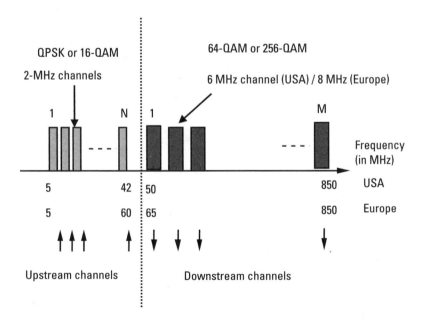

Figure 3.13 Spectral occupancy in a MCNS-DOCSIS system.

capacity a carrier intends to assign to data services. The data rates of each upstream or downstream channel depends on the adopted modulation technique.

3.7.2.1 Downstream Channels

In Figure 3.13, a certain amount M' of downstream channels may be used for analog TV channel distribution, the $M - M'$ remaining channels being used for data services. Two types of modulation are considered for downstream transmission from the CMTS to the CMs. Either 64-QAM or 256-QAM quadrature amplitude and phase modulations are adopted depending on the requested SNR. To reduce the error rate due to the various noise sources in the system, FEC and interleaving techniques are implemented in the CMTS and the CMs. The FEC technique based on Reed-Solomon encoding improves SNR, thanks to redundant bits in order detect and to correct burst of errors. Interleaving mixes the order of transmitted bits to prevent long bursts of errors. The more efficient the interleaving, the longer it takes at the sending and at the receiving nodes. Considering the real-time constraints of the applications offered on HFC networks, a delay limited to 4 ms for interleaving and deinterleaving is considered as admissible. According to the adopted FEC and interleaving techniques, useful bit rates per channel between 38 Mbps and 57 Mbps with 256-QAM and between 27 Mbps and 36 Mbps with 64-QAM are achievable.[11]

3.7.2.2 Upstream Traffic

Two types of modulation are considered for upstream channels: *quadrature phase shift keying* (QPSK) and 16-QAM. Like for downstream channels, various techniques are possible in order to improve SNR. Depending on the complexity of these techniques, useful data rates between 320 Kbps and 5.12 Mbps with QPSK and between 640 Kbps and 10.24 Mbps are achievable on each upstream channel.

3.7.3 Protocol Architecture

Figure 3.14 describes the protocol architecture of an MCNS-DOCSIS system made of two layers, a physical layer and a data link layer. Section 3.7.2 described the characteristics of the MCNS-DOCSIS physical layer. The data link layer is itself made of four sublayers described in the following sections.

3.7.3.1 MPEG-2 Transmission Convergence Sublayer

Above the physical layer, a convergence layer is used exclusively for downstream data traffic, which is systematically encapsulated into MPEG-2 frames. This

---

11. The principles of QAM modulation and of FEC and interleaving/deinterleaving techniques are detailed in Chapter 6.

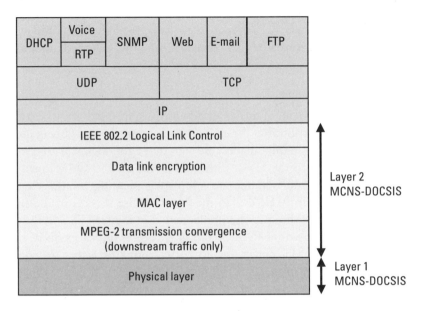

Figure 3.14 The MCNS-DOCSIS protocol stack. (*From:* [8]. © 2001 IEEE. Reprinted with permission.)

convergence sublayer also allows for the broadcast of digital TV channels in addition to analog TV channels on an HFC network. A 188-byte MPEG-2 frame is made of a 187-byte data field preceded by a synchronization byte. Within the data field, a 4-byte header specifies if the contents of the frame corresponds to an MPEG-2 video sequence or to user's data. When it is used for data traffic, the 183-byte payload of an MPEG-2 frame may contain one or more *MAC-protocol data units* (PDUs) (MAC-PDUs), the size of each MAC-PDU being variable. The same MAC-PDU may also be transported in several successive MPEG-2 frames. In that case, a pointer may be used at the beginning of the payload of the MPEG-2 frame to determine where a MAC-PDU in frame number k ends, whereas its beginning has been transmitted in a previous frame. At last, if a MAC-PDU is too short to fill in the MPEG-2 frame payload, the unused capacity is filled by means of bit stuffing. Figure 3.15 illustrates three examples of the configuration of MPEG-2 frames in the case of different size MAC-PDUs.

3.7.3.2 MAC Sublayer

Above the convergence layer, a MAC layer is used by both upstream and downstream traffic. Two main objectives are assigned to the MCNS-DOCSIS MAC protocol. First, it must share efficiently and fairly the network capacity between different active CMs. Second, it aims at guaranteeing QoS constraints inherent

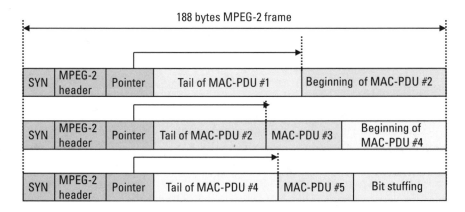

Figure 3.15   Examples of MPEG-2 frames used for MAC-PDU transport.

to real-time applications. An active CM must send a bandwidth request to the CMTS for each of its pending packets. A packet corresponds in this context to an IP packet encapsulated in a MAC-PDU. The MCNS-DOCSIS MAC protocol is designed in order to authorize variable size packets' transmission. According to the instantaneous available capacity on the system, the CMTS assigns a certain amount of minislots to active CMs in specifying on which frequency and on which time slots each CM is authorized to send its MAC-PDU. Section 3.7.4 details the principle and the characteristics of the MCNS-DOCSIS MAC protocol.

3.7.3.3   Encryption Sublayer

Above the MAC layer, the MCNS-DOCSIS standard specifies an encryption layer that is used to insure the confidentiality of users' data for both upstream and downstream traffic. The data link encryption layer is based on the *baseline privacy international* (BPI) specifications. The BPI specifications use the public key *data encryption standard* (DES) encryption algorithm.[12] Let us notice that the DES algorithm does not proceed to the modem's authentication. This means that a malicious user could connect to the HFC network an illegal modem and try to find the key with which the data of a given user have been encrypted.

3.7.3.4   Logical Link Control (LLC) Sublayer

The CMTS and the CMs are equipped with Ethernet interfaces. The IEEE 802.2 LLC protocol, which is not strictly speaking specific to MCNS-DOCSIS, is used as a bridge between a private Ethernet LAN in the CPE and the HFC

---

12. The description of the principle of public key algorithms is out of the scope of this book.

network. This bridge enables the association of sequencing numbers to MAC-PDUs. Thanks to this sequencing, a sliding window flow control may be associated with the HFC MAC protocol.

3.7.3.5   Upper Layers

Unlike the IEEE 802.14 protocol, which is ATM-oriented, MCNS-DOCSIS is IP-oriented. Several applications using either the UDP or the TCP transport protocol are considered on top of MCNS-DOCSIS layers. Thus, IP addresses' assignment by means of the DHCP protocol,[13] VoIP, or SNMP messages use the UDP transport protocol on top of the IP layer. Due to the lack of IP addresses, temporary IP addresses are assigned dynamically to each newly active CM by a DHCP server. This server may be connected at the CO or higher in the network architecture, typically at the level of an IP router of the NSP. Non-real-time Internet services such as Web browsing, e-mail, and file transfers use the TCP on top of the IP layer.

3.7.3.6   Ranging Process

Before any data transmission, the CMTS must be synchronized with a newly active CM. Indeed, clock distortion inherent to propagation may vary according to the distance between this CM and the head-end. Similarly, due to propagation disparity between the CMs and the CMTS, power ranging is necessary in order to guarantee a fixed SNR. These two operations are called clock ranging and power ranging, respectively.

3.7.4   MAC Protocol

3.7.4.1   MCNS-DOCSIS 1.0

Upstream traffic accesses the medium by means of a combination of *frequency division multiple access* (FDMA) and of *time division multiple access* (TDMA) techniques. Each upstream carrier is divided into successive time slots called minislots. The size of a minislot is equal to $2^n \times 6.25\ \mu s$ with $n = 0, 1 \ldots 7$. In [4], a minimum minislot length of 16 bytes is specified ($n = 1$), the maximum size being equal to 128 bytes ($n = 7$). The value of n depends on the adopted modulation technique. MAC-PDUs containing IP packets are transported in successive minislots. Downstream traffic accesses the medium by means of a simple *time division multiplexing* (TDM). Periodically, the CMTS reserves a certain number of minislots. A clock ranging procedure is applied between the CMTS and the CMs in order to facilitate upstream transmission scheduling into successive cycles. This clock ranging procedure aims to give a common clock reference to all the CMs of the HFC network. The composition of each

---

13. The use of DHCP is presented in Chapter 1.

upstream cycle is determined dynamically by the CMTS. For that purpose, the CMTS informs the CMs of a cycle structure by means of periodical MAP messages. Three types of minislots are considered in a cycle:

- A certain amount of minislots composing a *request contention area* are dedicated to the transmission of bandwidth requests by the CMs. Multiple CMs may send simultaneously a request in a same minislot. Collisions may then occur and can be solved by a contention resolution algorithm. The adopted collision resolution algorithm is inspired by the slotted Aloha protocol [10].

- A certain amount of minislots are reserved by the CMTS in response to bandwidth requests sent previously by the CMs. Such reserved minislots are called *transmission opportunities*. Up to 255 minislots may be requested at once by a single CM to send its variable size data packets.

- A certain amount of minislots may be dedicated to the transport of signaling messages. For instance, when a new CM is connected to the HFC network, signaling messages are exchanged between this CM and the CMTS in order to proceed to clock and power ranging. The format of a MAP message is illustrated in Figure 3.16.

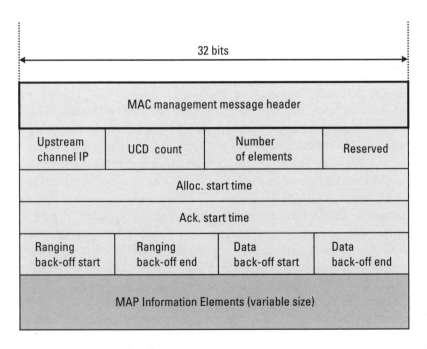

Figure 3.16 Format of a MAP message.

The meanings of the various fields of the MAP message are described as follows.

- *Upstream channel ID:* This field identifies the index (between 1 and N) of the considered upstream carrier.
- *Upstream channel descriptor (UCD) count:* This field is incremented after each new assignment of the minislots for the considered upstream carrier by the CMTS.[14] It informs then a CM of such a modification.
- *Number of elements:* This field gives the number of *information elements* (IEs) composing the variable-size portion of the MAP message.
- *Allocation start time:* This field specifies at which instant the information transported in the MAP IEs becomes effective.
- *Acknowledge time:* The value of this field enables a CM to know if the information transported in this MAP message is anterior or posterior to the last bandwidth request sent by this CM.
- *Ranging back-off start and end:* These two fields specify the initial ranging window size and the maximum ranging window size.
- *Data back-off start and end:* These two fields specify the initial size and the maximum size of the contention window by analogy with the contention window used in the slotted Aloha protocol for contention resolution.
- *MAP IEs:* Between 2 and 240 IEs may be transported in the same MAP message.

Figure 3.17 describes the format of a typical list of IEs. This list may be divided into two parts. In a first part of the list, the CMTS informs the CMs of the various minislots assignment for the next cycle. These minislots are either reserved for a given CM or open to contention. The second part of the list is used for maintenance. It contains explicit acknowledgments associated to transmissions in contention minislots carried out in the previous upstream cycle. It may also contain pending data grants if a fraction of the requests sent by the CMs cannot be satisfied in the next upstream cycle. An upstream cycle is organized in *time intervals,* each time interval corresponding to a certain amount of minislots. A cycle is organized in two portions: A first portion corresponds to the *request contention area;* the second part corresponds to the *transmission opportunities.*

Within the next cycle structure description, a *station identification* (SID) field specifies to which CM the associated *time interval* is dedicated. The *interval*

---

14. Two successive MAP messages may or may not correspond to the same cycle structure.

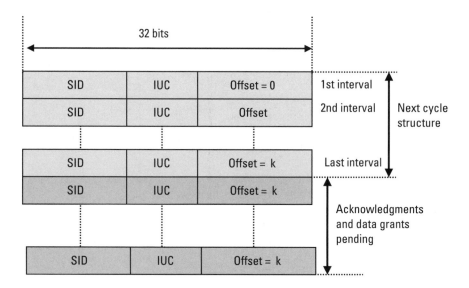

Figure 3.17 Format of IEs listed in a MAP message.

usage code (IUC) field specifies if a time interval belongs to the *request contention area* or to the to the *transmission opportunities*. The offset field indicates the index of the first minislot of the time interval within the cycle. Within the second portion of the IE list, the offset field is useless and set to the highest offset value k. Figure 3.18 illustrates an example of upstream cycle structure.

Figure 3.19 illustrates the principle of the exponential back-off resolution algorithm inspired from the slotted Aloha protocol. A CM detects a collision if the MAP message following its last transmission in the *request contention area* does not contain an explicit acknowledgment. This CM is then invited to try

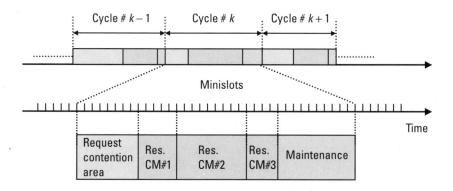

Figure 3.18 Example of cycle structure.

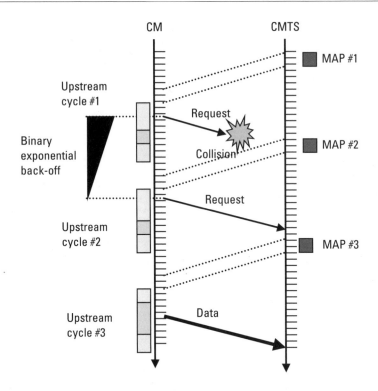

Figure 3.19 Example of collision resolution.

again to send its request during a given time window that may cover several cycles. If a collision occurs again, the size of the time window is doubled up to a maximum admissible upper bound. One has adopted in Figure 3.19 the same colors' convention as in Figure 3.18 to represent the request contention area, the transmission opportunities, and the maintenance area in the upstream cycles.

3.7.4.2   MCNS-DOCSIS 1.1

The most recent version 1.1 of the MCNS-DOCSIS specifications focus on QoS guarantees at the MAC layer. For instance, packet classification based on traffic classes is considered within the CMs and the CMTS in order to guarantee a minimum bandwidth, a maximum delay jitter, or a maximum packet loss ratio to the different flows. Classification criteria are based on the following:

- TCP/IP connections parameters such as source and destination addresses, TCP ports, and the value of the TOS field;
- LLC parameters such as MAC Ethernet addresses;
- Classification parameters inherent to the IEEE 802.1P/Q protocol.

In order to facilitate delay constraints proper to real-time applications, a polling mechanism allocates dedicated minislots to the active CMs for their requests' transmission. Nonreal-time applications use contention minislots to send their requests. Various types of schedulers are considered at the MAC layer within the CMs. The assignment of a given scheduler to a data flow depends on the burstiness of this flow (that is on its granularity) and on the delay requirements of the application associated to this flow. For instance, VoIP using constant bit rate PCM coders benefits of an *unsolicited grant service* (UGS). In that case, a periodic grant's attribution is activated after connection setup at the CMTS. These periodic grants correspond to reserved minislots in MAP messages. A variant of the UGS scheduler called *UGS with activity detection* (AD) (UGS/AD) is considered for packetized VoIP. In that case, a request is sent at the beginning of each new burst either in a contention minislot or in a polling minislot. Once it has been received by the CMTS, this request activates a periodic grants' attribution. This periodic grant's attribution is deactivated as soon as an unused reserved minislot is detected by the CMTS. Best effort traffic corresponding, for instance, to e-mail applications use contention minislots to access the medium. A CM can fragment long data packets in order to avoid penalizing smaller real-time packets in terms of access delay.

Scheduling classifying mechanisms are implemented in other point-to-multipoint access systems such as APON or EPON systems, which will be detailed in Chapter 12.

3.8 PLC

Two fields of application of PLC are considered, either for in-house networking or for outdoor networking. The former refers to a new access technique in the local loop, whereas the latter concerns home networking.

Section 3.6.1 mentions various alternatives based on wired infrastructures for the last mile of telecommunications networks. After a few unsuccessful commercial announcements of PLC products in the mid-1990s, it seems that this technology is now mature. PLC consists of offering services equivalent to HFC or to xDSL by using the medium-voltage grid of an electricity supplier. The fact that PLC does not require the setup of a new infrastructure as is the case for FITL or HFC is one of the positive aspects of PLC. As it has been described in Figure 3.10, a PLC network is based on a point-to-multipoint infrastructure between a low-voltage transformer and the electricity meters of about 50 households. PLC modems are located at each end of the network. The data rate offered to each user is about 1 Mbps. In principle, a PLC modem may be moved from one location to another one. The typical range of a PLC network from the LVT to the electricity meter is between 200m and 500m.

PLC networks' operation by a carrier requires a specific license provided by the national regulation bodies. The number of countries that have already specified the terms of such licenses remains limited. One of the drawbacks of PLC access systems compared to other wired technologies is that the network becomes unusable in case of power failure. Other systems such as xDSL, FITL, or HFC may be secured by batteries for a few hours in case of electricity failure. Several experimental commercial services have been proposed in different European countries including Austria, France, Germany, Italy, Norway, Portugal, Spain, and Switzerland. There is not yet a worldwide consensus on PLC specifications for an international standard. Nevertheless, a consortium named PLC Forum was established in March 2000. This consortium is comprised of approximately 100 members, including equipment manufacturers and electricity suppliers [11]. In North America, another consortium called the HomePlug Powerline Alliance is more interested in the standardization of PLC for in-house networking.

3.9 Conclusion and Perspectives

This chapter introduced the interconnection and unbundling concepts. Today, a majority of countries in Europe and North America have adopted through their respective regulation bodies the principle of unbundling. Two aspects tend to limit a rapid uptake in the use of unbundling: the price-list of the ILECs and the colocation rules. Competition between ILECs and CLECs may concern either telephony services, or Internet access services. On the basis of these two services, two types of unbundling have been defined: full unbundling and partial unbundling. We have presented the state of advancement of unbundling and of colocation in Europe and North America. We then presented an up-to-date overview of the various technologies for the last mile. We have classified these technologies into two groups, according to the wired (xDSL, HFC, FITL, PLC) or the wireless (WITL, LEO, MEO, GEO, HAAP) nature of their infrastructure. The satellite, HAAP, PLC, and HFC technologies have also been described in this chapter. Unlike first generation cable networks, which were used only for analog TV channel distribution, new generation cable networks aim to provide digital TV channels, high-speed Internet access, and digital telephony services. PLC techniques are a promising alternative for the last mile. Meanwhile, the offered capacity per user is, in most cases, less than the capacities offered by ADSL systems. Like HFC systems, PLC is subject to electromagnetic incompatibilities.

References

[1] Van de Vel, K., "Avoiding xDSL Provisioning Pitfalls: How to Mass Market and Roll Out xDSL Services," *Proc. of 40th European Telecommunications Congress,* Barcelona, Spain, August 22–25, 2001.

[2] Hu, Y., and V. O. K. Li, "Satellite-Based Internet: A Tutorial," *IEEE Communications Magazine,* March 2001.

[3] Seidman, L. P., "Satellites for Wideband Access," *IEEE Communications Magazine,* October 1996.

[4] Garg, V. K., and E. L. Sneed, "Digital Wireless Local Loop Systems," Lucent Technologies Inc., 1995.

[5] Carrere, G., "High-Altitude Aerial Platforms," *Proc. of the Seminar New Multiservice Access Technologies,* Paris, France, May 1999.

[6] IEEE 802.14 Working Group, *Media Access Control, IEEE Std. 802.14,* Draft 3, Revision 1, April 1998.

[7] Golmie, N., Y. Saintillan, and D. Su, "A Review of Contention Resolution Algorithms for IEEE 802.14 Networks," *NIST Publications,* 1998.

[8] Fellows, D., P. H. Ventures, and D. Jones, "DOCSIS Cable Modem Technology," *IEEE Communications Magazine,* March 2001.

[9] "Data over Cable Service Interface Specifications: Radio Frequency Interface Specification," *Interim specification SP-RFIv1.1-IO6-001215,* Cable Television Laboratories, Inc., http://www.cablemodem.com.

[10] Tannenbaum, A., *Computer Networks,* 3rd ed., Upper Saddle River, NJ: Prentice Hall, 1996.

[11] PLC Forum, http://www.atanvo.de/plcforum/.

Selected Bibliography

Gagen, P. F., and W. E. Pugh, "Hybrid Fiber-Coax Access Networks," *Bell Labs Technical Journal,* Summer 1996.

McDonald, J. C., *Fundamentals of Digital Switching,* 2nd ed., New York: Plenum Press, 1990.

4

Twisted Pair Electrical Characteristics

4.1 Introduction

A twisted pair is characterized by two parameters: the diameter of its copper wire and its twisted period. Typically, copper wire's diameter is 0.4–1 mm. In North America, this diameter is not expressed in inches but in gauge. In the United States, the "gauge" is a standardized unit called *American wire gauge* (AWG). The larger the copper wire diameter, the smaller the AWG of the cable. The main diameters used in North America are 19, 22, 24, and 26 AWG. Pairs with a diameter less or equal to 22 AWG enable local loops with a range of 5,400m for analog voice traffic. The thinnest 26-AWG pairs authorize only 4,500-m-long local loops for voice traffic. This chapter presents several experimental results referring to the most currently used diameters of 24 AWG and 26 AWG. These two diameters correspond to 0.6 and 0.4 mm, respectively. The larger the copper wire diameter and the shorter the twisted period, the better the insulation of the cable facing electromagnetic interference. The copper pairs used in the local loop are *unshielded twisted pairs* (UTPs). These pairs are used either in underground ducts or suspended at pole tops (aerial cables). Twisted pairs are also frequently used for in-home cabling for telephony and data services. Voice-grade UTP pairs are frequently marketed by group of four included in a common plastic gain. As outlined in the previous chapters, UTPs in the local loop are grouped into successive binder cables from the CPEs to the EO.

Section 4.2 describes the electrical characteristics of a homogeneous copper pair. Section 4.3 investigates the Shannon's capacity of such a transmission medium. Section 4.4 analyzes the electrical characteristics of a twisted copper pair in the local loop environment. Section 4.5 investigates the impact of electromagnetic interference between copper pairs belonging to a same binder cable in an access network.

4.2 Electrical Characteristics of a Copper Pair

As a first step, let us consider for the sake of simplicity the case of a single twisted pair in a protected electromagnetic environment. This is, for instance, the case for a pair used as the medium of a 10BaseT Ethernet. From experimental measurements, it is possible to characterize the electrical behavior of a twisted pair at low frequencies by a succession of four-port filters as described in Figure 4.1. Each of these four-port filters models the electric behavior of an infinitesimal length of cable dx. The primary constants characterizing a four-port filter are resistance R, inductance L, shunt conductance G, and shunt capacitance C per unit length dx. One defines the characteristic impedance Z_c of the cable as the theoretical complex impedance of a perfectly homogeneous cable.[1]

The expression of Z_c is given by:

$$Z_c = \sqrt{\frac{R + jL\omega}{G + jC\omega}} \tag{4.1}$$

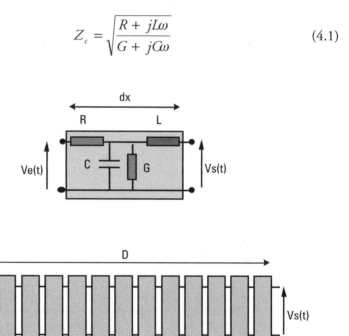

Figure 4.1 Four-port filter model of a twisted pair.

---

1. One means by homogenous cable a cable with a constant diameter and a constant twisted period on all its length.

If a voltage $V_e(t)$ is applied at the input of the cable, a tension $V_s(t)$ is detected at the other end of the cable at a distance D from the origin. The dependence between $V_e(t)$ and $V_s(t)$ is given by their respective Fourier transforms $V_e(f)$ and $V_s(f)$ and by the transfer function of the cable $H(f, D)$, where f stands for the signal frequency. One has:

$$V_s(f) = H(f, D) \cdot V_e(f) \tag{4.2}$$

Ideally, transmission efficiency on a twisted pair is optimal if this pair is terminated at each of its two ends by its characteristic impedance Z_c. It is shown that in this case, the energy of the received signal is totally absorbed by the equipment located at the end of the cable. One has:

$$V_s(f) = e^{-\gamma \cdot D} \cdot Ve(f) = e^{-(\alpha + j\beta)D} \cdot V_e(f) \tag{4.3}$$

where $\gamma = (\alpha + j\beta)$ represents the propagation constant. The expression of γ is given by the primary constants of the cable:

$$\gamma = \sqrt{(R + jL\omega) \cdot (G + jC\omega)} \tag{4.4}$$

The parameter α represents the attenuation of the cable and is expressed in decibels per kilometer (dB/km). The parameter β represents the phase distortion induced by propagation. This distortion corresponds to the disparity in velocity between high-frequency and low-frequency components of the transmitted signal. The transient frequency of the transmitted signal depends itself on the line coding and modulation technique adopted in the modems attached to each station of the network [1].

4.2.1 Attenuation Distortion

The evolution of attenuation α versus frequency f is obtained by a generalization of the expression of the propa7gation constant γ:

If $f < 10$ kHz, ωL is negligible facing R and G is close to zero. In using a power series expansion of the terms contained in the expression of γ, one gets:

$$\gamma = \alpha + j\beta \approx \sqrt{\omega RC} \cdot \sqrt{\left\{ 1 + \frac{(\omega L)^2}{2R^2} - \frac{(\omega L)^4}{8R^4} + \ldots \right\}}$$

$$\cdot \exp\left\{ j \cdot \left[\frac{\pi}{4} + \frac{\omega L}{2R} - \frac{(\omega L)^3}{6R^3} + \ldots \right] \right\} \tag{4.5}$$

Equation (4.5) gives when $f < 10$ kHz the same expression for α and β, which are in this case proportional to the square root of f:

$$\gamma = \alpha + j\beta \approx \sqrt{\pi f RC} + j\sqrt{\pi f RC} \qquad (4.6)$$

If 10 kHz $< f <$ 20 kHz, the expressions of α and β differ from each other and are not any longer proportional to the square root of f:

$$\gamma = \alpha + j\beta \approx \sqrt{\frac{\omega RC}{2}}\left(1 - \frac{\omega L}{2R}\right) + j\sqrt{\frac{\omega RC}{2}}\left(1 + \frac{\omega L}{2R}\right) \qquad (4.7)$$

If 20 kHz $< f <$ 150 kHz, the expressions of α and β are unpredictable. Indeed, at such frequencies, R and L are not any longer constant but also depend on frequency f. It is not possible to determine a generic expression of $R(\omega)$ and of $L(\omega)$, these functions appearing from experimental measurements strongly dependent of the considered environment. One has in that case:

$$\gamma = \alpha + j\beta \approx \frac{R(\omega)}{2}\sqrt{\frac{C}{L(\omega)}} + j\omega\sqrt{CL(\omega)} \qquad (4.8)$$

For $f < 150 <$ kHz, one observes experimentally a linear dependency between α and β with the square root of f. Such a dependency may be explained by the skin effect that imposes electrons to propagate at the periphery of the copper cable. On has in this case:

$$\alpha \approx k\sqrt{f} \qquad (4.9)$$

Figure 4.2 illustrates the evolution of attenuation α of a twisted pair of 3,600m (that is, 12,000 feet) long with a 0.6-mm diameter (that is, 24 AWG) versus frequency f. The solid line curve corresponds to the real attenuation measured on the cable [2]. One notices two linear portions in this curve for frequencies lower than 20 kHz and for frequencies higher than 200 kHz.[2] These experimental results are in coherence with the analytical expressions given in (4.6) and (4.9). Between 20 kHz and 200 kHz, α is not linear and cannot be described by a generic analytic expression. The dotted line curve with a negative slope $-\phi$ corresponds to an approximation of the average slope of the real attenuation α. Such an approximation may be used in order to design the

---

2. The reader may notice that the vertical axis uses a logarithmic scale, whereas the horizontal axis uses a linear scale.

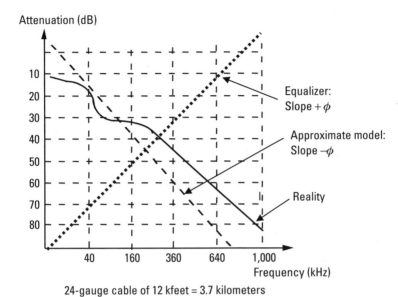

Figure 4.2 Attenuation of an isolated twisted pair. (*From:* [2]. © 1991 IEEE. Reprinted with permission.)

equalization filter to be implemented in the receivers. The transfer function of the linear equalizer is then characterized by a positive slope $+\phi$.

As will be outlined in Section 4.5, the electrical behavior of a twisted pair in the local loop may be in fact quite unpredictable. This means that the design of equalization filters in xDSL modems is more complex than a simple linear equalization. Because of electromagnetic perturbation between copper pairs contained in a same duct, the higher the bit rate, the larger the on-line spectrum, the more unpredictable the shape of attenuation α.

4.2.2 Loaded Twisted Pairs

When PSTN was analog in the 1960s, many carriers set the value of the primary constants resistance R and conductance G in order to facilitate the extension of the range of their subscriber lines. In practice, we notice that the higher the value of L, the lower the value of α. We also observe that the lower the value of C, the lower the value of α. Upgrading artificially the value of L along each wire of a copper pair is in fact easier than decreasing the value of C between these two wires. This operation can be carried out easily in inserting additional coils (a coil is an inductance L') regularly spaced along the pair. For specific values of R and C, it is possible to reduce α by increasing artificially the value of inductance L. In the [0, 4 kHz] band, the values of R and G may be chosen very low compared to the values of $L\omega$ and $C\omega$, respectively. This implies a simplification of the expression of the propagation constant γ given in (4.5) such as:

$$\alpha \approx \frac{R}{2}\sqrt{\frac{C}{L}} \qquad (4.10)$$

$$\beta \approx \omega\sqrt{L \cdot C} \qquad (4.11)$$

One observes experimentally that α can be set to 0.5 dB/km for frequencies lower than 3 kHz, thanks to additional coils regularly spaced every 1,800m. A same value of α can be obtained up to 5 kHz with load coils regularly spaced every 900m. For each of these two loaded-line configurations, one also observes that frequencies above 3 kHz or above 5 kHz, respectively, induce a strong increase in α. This is why the principle of loaded twisted pairs is only applicable to voiceband modems and not to xDSL modems. When spacing between two successive coils is lower or equal to 900m, it has been shown that the range of a telephone line could be extended up to 9 km for analog voice traffic. In the case of xDSL modems operating at frequencies greater than 4 kHz, load coils may also induce a strong reduction in the SNR of a copper pair for discrete frequencies. The impact of these discrete frequencies on the transfer function of the cable may be compared to cutoff frequencies that strongly reduce the Shannon's capacity[3] of a cable.

4.2.3 Phase Distortion

Whatever the transmission media (e.g., copper pair, optical fiber, and air interface), phase distortion due to a disparity of the velocity $V(f)$ of the on-line signal according to its transient frequency f is a frequently observed phenomenon. Its impact in terms of erroneous bits at the receiver increases with propagation distance and frequency range. After demodulation at the receiver, successive symbols in the NRZ signal[4] may overlap. The sampler implemented in a receiver in charge of translating the measured voltage levels into information bits may then induce erroneous bits after decoding. Phase distortion β mentioned in (4.3) of the expression of propagation constant γ is at the origin of ISI. Two types of delays characterize the on-line signal: the phase delay $V_\phi(f)$ and the envelope delay $V_g(f)$. One has:

---

3. We have briefly introduced the Shannon's capacity in Chapter 2.

4. An NRZ signal corresponds to a signal made of regular pulses of tension equal to +V Volts in the case of a bit equal to "1" and of −V Volts in the case of a bit equal to "0."

$$V_\phi(f) = \frac{\omega}{\beta(\omega)} \tag{4.12}$$

$$V_g(f) = \frac{dw}{d\beta(\omega)} \tag{4.13}$$

These two delays are both expressed in kilometers per second. Phase delay $V_\phi(f)$ represents the propagation speed of a pure wave centered on a unique frequency f. It typically corresponds to the propagation speed of one of the harmonics composing the Fourier transform of a more complex signal such as the on-line signal generated by an xDSL modem. Envelope delay $V_g(f)$ represents the velocity of a group of adjacent pure waves in the Fourier transform of the on-line signal. It is shown in [3] that for frequencies lower than 10 kHz, envelope delay is such that:

$$\frac{1}{V_g(f)} = \frac{d\beta(\omega)}{d\omega} \approx \sqrt{\frac{R \cdot C}{8\omega}} \left(1 + \frac{3L\omega}{2R}\right) \tag{4.14}$$

For $f > 150$ kHz, the value of the phase delay and of the envelope delay are comparable and almost independent from f.

$$\frac{1}{V_g(f)} = \frac{d\beta(\omega)}{d\omega} \approx \frac{1}{V_\phi(f)} \approx \frac{\beta(\omega)}{\omega} = \sqrt{L \cdot C} \tag{4.15}$$

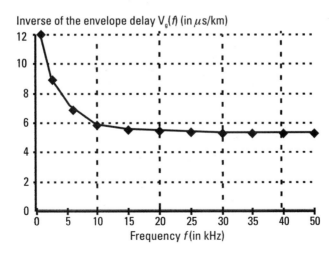

Figure 4.3 Envelope delay versus frequency for a 26-AWG cable. (From: [2]. © 1991 IEEE. Reprinted with permission.)

The curve plotted in Figure 4.3 illustrates the evolution of the inverse of the envelope delay $V_g(f)$ in the case of a 24-AWG (that is, 0.6-mm) copper pair versus frequency f in kHz. One notices that the value of this inverse is low for high frequencies. In other terms, the higher the frequency f, the higher the envelope delay. For $f > 30$ kHz, envelope delay is almost constant and equal to 185,000 km/sec.

4.3   Shannon's Capacity of a Twisted Copper Pair

Section 2.2.1 introduced the concept and the expression (2.1) of the maximum capacity in bits per second of a digital transmission channel under stationary AWGN.[5] A white noise is characterized by a uniform *power spectral density* (psd) on the whole bandwidth of the transmission channel. Such a noise is stationary if the *probability density function* (pdf) characterizing its value is time-independent. A white Gaussian noise is a white noise with a power fluctuating according to a Gaussian (or normal) distribution. Let us recall below the basic principle of Shannon's capacity. Figure 4.4 describes the typical block diagram of a digital transmission channel between a data source and a data receiver.

A line code is used in order to facilitate clock recovery at the receiver or in order to introduce redundant code words for OAM purposes at the physical layer. At the output of the line coder, data may be interleaved to prevent long bursts of errors.[6] A modulator adapts the data received from the output of the interleaver to the useful bandwidth of the transmission channel. Essentially, two types of modulation are possible: baseband modulation and bandpass modulation. In the context of electrical transmission over copper pairs, baseband modulation consists of two operations. First, successive bits are grouped into fixed size blocks. Then, a given voltage level is associated to each of these blocks. PAM is a typical example of baseband modulation.[7] Bandpass modulation corresponds to the modulation (in phase, in frequency, or in amplitude) of a carrier. This carrier is used to shift the natural spectrum of the NRZ signal generated by a data source toward higher frequencies better suited to the SNR characteristics of the transmission medium.

Baseband modulation can be used for short-range data transmission (under a 5-km range) over copper pairs. Let X and Y be the random variables

---

5. This theoretical capacity was discovered by Claude Shannon about 50 years ago.

6. In general, it is easier to detect at the receiver isolated erroneous bits than successive erroneous bits. Fluctuating transmission channels as radio channels are frequently subject to bursts of errors.

7. Chapter 2 described two examples of PAM modulations with HDB3 and 2B1Q used in T1/E1 LLs and in ISDN access, respectively.

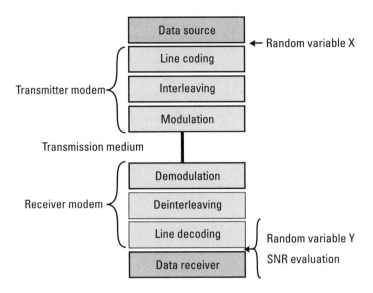

Figure 4.4 Block diagram of a typical digital transmission channel.

associated to the data generated by a source node and received by a destination node, respectively (see Figure 4.4). In practice, X and Y do not follow the same pdf because they are partially decorellated. Numerous noise sources observed between the data source and the data receiver explain this relative independence between X and Y. These noise sources are either inherent to the transmission medium or to the end modems. The on-line signal psd is upper-bounded by W Hz. The pdf of the noise introduced by the transmission channel can be assimilated to a centered Gaussian random variable with constant psd in the $[-W, +W]$ interval and a null psd out of this interval. Claude Shannon has shown that the maximum capacity in bits per second of any data channel with stationary bandwidth-limited AWGN is given by:

$$C_{max} = W \cdot \log_2\left(1 + S/N\right) \tag{4.16}$$

where S and N stand for the average power of the received signal and for the average power of the noise[8] at the output of the line decoder. As an example, let us consider the case of Internet access by means of a V.34bis dial-up modem (see Section 2.2.1). We have seen that such a modem enables full-duplex data rates up to 33.6 Kbps over a 4-kHz bandwidth. In the optimum situation, the only

---

8. The noise N considered in the Shannon's formula cumulates the noise proper to the transmission medium and the noise due to the end modems.

noise to be considered is the quantization error between 38 and 39 dB. This implies a Shannon's capacity equal to 34 Kbps. The value of 33.6 Kbps very close to the Shannon's limit announced by the vendors is in practice achievable in very few cases because of the presence of additional noises due to the nonhomogeneity of the copper lines. As described in the following sections, a subscriber line is in practice made of a succession of point-to-point copper pairs with different characteristic impedances. Even in the case of an ideal twisted pair, Shannon's capacity must not be assimilated to the useful capacity. From the application layer to the physical layer, a certain amount of protocol overheads are necessary. These overheads may consume a nonnegligible fraction of the medium capacity. Claude Shannon has outlined that the random variables X and Y mentioned in Figure 4.4 become totally decorellated if one tries to send at bit rates greater than the maximum capacity C [4].

4.4 Electrical Characteristics of a Subscriber Line

Section 4.2 describes the electrical characteristics of a copper pair isolated in its environment, like in a LAN, for instance. Let us now consider the case of a twisted pair used in an access network. Chapter 3 showed that twisted pairs are progressively grouped into binder cables via successive distribution points from the CPEs to the MDF. In the local loop environment, a certain amount of additional disruptive effects degrade Shannon's capacity. In addition to the attenuation distortion and to phase distortion mentioned in Section 4.2, these disruptive effects are inherent to the following:

- Bridge taps;
- Impulse noises;
- Thermal noise;
- Radio interferences;
- Impedance incompatibility;
- Crosstalk.

4.4.1 Bridge Taps

The lifetime of a subscriber line may be several decades. The oldest pairs were installed about 50 years ago in the local loop. During this long period, the location of the successive users of a same copper pair may have changed. Let us consider, for instance, the case of a subscriber who moves from one apartment to another. The individual pair of this subscriber remains then unused. To save money, the ILEC of this user usually decides to reuse this pair for another subscriber installed in the neighborhood. For that purpose, a passive derivation called a *bridge tap* connected to the unused pair may be set up by the carrier to serve this new subscriber. Many carriers have managed their copper infrastructure in this way for decades without any incidence on telephony services. Figure 4.5 illustrates an example of subscriber line equipped with multiple bridge taps. In Figure 4.5, we can see that successively at least four subscribers have used the same twisted pair to be connected to the same EO. The horizontal section of the cable represented on the figure corresponds to the oldest part of the subscriber line, the first user of the line being installed in building *A*. Once this first user has moved out, a second user living in building *B* has been connected to the same line thanks to bridge tap 1. Again, two other users living in buildings *C* and *D* have successively been connected to the same line. In Figure 4.5, the customer located in house *D* is the current user of the line. One notices that bridge taps 1 and 2 and the original termination *A* have been let by the carrier intact and open-ended. In contrast, bridge tap 3 serving the current user is terminated on the input impedance of the telephone set of this user.

A carrier may also plan to connect in advance bridge taps along its new cables. It is the case, for instance, when a new set of buildings or houses in a same neighborhood are under construction. Carriers take care that at most, a single subscriber is connected on each pair arriving on a line interface card at the EO. In the context of xDSL access for which the useful bandwidth is much over

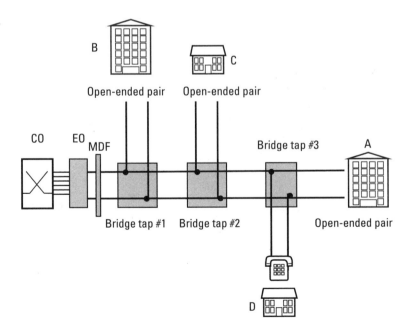

Figure 4.5 Example of a subscriber line with multiple bridge taps.

4 kHz, disruptive effects may occur due to the presence of open-ended bridge taps. Figure 4.6 illustrates a scenario in which an end user receives data from its Web server by means of its xDSL modem. One notices that three open-ended bridge taps are connected to the subscriber line of this user.

When it propagates from the EO to the xDSL modem, the on-line signal also propagates in the bridge taps as is illustrated in the figure (for simplicity, only the signal propagation in bridge tap 2 is shown). Section 4.2 outlines the motivation for ending a copper pair by two pieces of end equipment with an input impedance equal to the characteristic impedance of the line. In the case of bridge tap 2, half of the power of the downstream signal gets into the derivation before being totally reflected at its end. This echo is superimposed with the reference signal when it comes back to the point of derivation. In other words, the average power of the resulting signal after the derivation point remains roughly unchanged. Nevertheless, the echo induces a time dispersion on this resulting signal comparable to ISI. For specific values of the length of a bridge tap d, phase shift between the echo and the reference signal at the point of derivation may be equal to π. Knowing that $\exp(j\pi) = -1$, the average energy of the resulting signal after the derivation is then divided by 2—that is, 3 dB. If such a phenomenon occurs on multiple bridge taps associated to the same subscriber line, a strong reduction in SNR may be observed at the receiver. It can be shown that a phase distortion of π between the echo and the reference signal occurs if the length d of the bridge tap is equal to an odd number of times the fourth of the wavelength of the signal. Let us recall that the phase delay is proportional to frequency f:

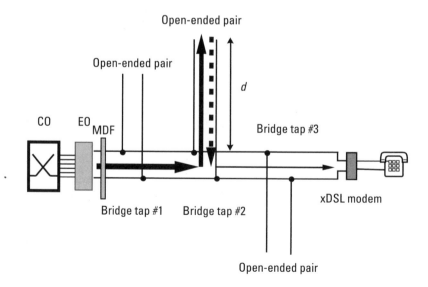

Figure 4.6 Pseudo cutoff frequency due to a bridge tap.

$$V_\phi(f) = \frac{\omega}{\beta(\omega)} = \lambda \cdot f \qquad (4.17)$$

For a given bridge tap, let f_0 be the first discrete value of f for which a 3-dB attenuation is observed:

$$f_0 = \frac{V(f_0)}{\lambda_0} = \frac{V(f_0)}{4 \cdot d} \qquad (4.18)$$

In practice, a finite number of frequencies f_k of the form $f_k = (2 \cdot k + 1) \cdot f_0$ with $k = 1, 2$ and so on is thus potentially associated to each bridge tap of a subscriber line. These discrete frequencies induce an effective 3-dB attenuation only if they belong to the spectrum of the downstream signal. Chapter 6 discusses how a multicarrier modulation technique makes it possible to avoid the disruptive effect of bridge taps.

In conclusion, echoes introduced by bridge taps are in general at the origin of bursts of errors because of ISI. These errors may be corrected either at the physical layer by means of self-correcting codes (Reed-Solomon codes) or in the higher layers—for instance, thanks to error recovery protocols such as TCP. In the worst case, a bridge tap implies a 3-dB attenuation, degrading considerably the Shannon's capacity of the line.

4.4.2 Impulse Noises

An impulse noise is characterized by an abrupt and transient variation of the voltage level observed on a copper pair in the local loop. In 50 years, the nature of impulse noise sources has considerably evolved. At the time of analog telephony, impulse noises were essentially due to mechanical connections established within crossbar switches between an input port and an output port. At connection setup or at connection release, a voltage discontinuity could be observed on the considered subscriber line. In many cases, this voltage discontinuity could also be observed on other pairs transported in the same binder cable. Another source of impulse noise was related to rotary pulse dialing [5]. A telephone set in the on-hook position is remotely power-supplied by a $-$ 48-V battery located at the EO. When a calling party dials the telephone number of a called party, the continuous dc current circulating along the local loop is interrupted according to the dialed digits. These voltage or current pulses circulating from a CPE to the EO on a given pair may induce voltage discontinuities on other pairs adjacent to the considered pair.

Since the digitization of PSTN, signaling, and especially dialing are not any longer transported under the form of voltage pulses but under the form of *dual-tone multifrequency* (DTMF). DTMF dialing associates with each digit a combination of two audio frequencies belonging to the [300 Hz, 3,400 Hz] interval. Remote power supply by means of –48-V batteries located at the EO is still used to make operations of the telephone sets independent from the electric power supply. For years, carriers did not really take care of impulse noises because of their marginal effect on the quality of audio services. With the emergence of the first dial-up modems, one has learned that impulse noises can be at the origin of unpredictable-length bursts of errors in the transported data. Today, with xDSL modems operating at a higher bit rate than dial-up modems, bursts of errors due to impulse noises are even more noticeable.

Since the 1970s, electromechanical switches used in analog PSTN have been replaced by digital electronic exchanges. Several studies have aimed to model the statistical behavior of impulse noise [6–8]. In current digital PSTN networks, impulse noises may be classified into two categories: internal and external. Internal impulse noises are not any longer due to the transmission of dialing signals but to the propagation of a ringing tone from the EO to the calling party indicating the activation of the bell at the called party. External impulse noises find their origin outside the local loop environment. They can be due, for instance, to microcurrents generated by high-voltage systems such as a streetcar line, a subway line, or a railway line that propagate into the ground close to a binder cable.

Section 4.4 mentions six main perturbations observed in the local loop. Except for impulse noises, one has a good knowledge of these perturbations and of the strategies that can reduce their impact. Impulse noises, on the other hand, remained largely unknown until the end of the 1990s. The ways in which impulse noises are unpredictable are described as follows.

- Impulse noises are unpredictable in the time domain; the same environment does not induce systematically the same effects.

- For a given source, impulse noises are unpredictable in duration; the duration of the observed voltage discontinuities may vary from one occurrence to another.

- For a given noise source, impulse noises are unpredictable in frequency; the same cause may induce a noise located at different places in the spectrum.

Since the end of the 1980s, continuous research activities have been carried out to achieve a better understanding of impulse noises. These activities were applied first to ISDN access [9, 10] and then to ADSL access [11–13]. Several of these studies have tried to conceive a model able to reproduce the statistical behavior of impulse noises. To my knowledge, [3] describes the most advanced results in this matter. Impulse noise may be approximated by several random processes, listed as follows:

- The impulse noise interarrival process may be characterized by a Poisson process.

- Impulse noise duration may be characterized by an approximation of a log-normal pdf.

- Impulse noise voltage levels follow a Rayleigh distribution.

The interested reader may consult [14, 15] for a detailed explanation of the Poisson process, log-normal distribution, and Rayleigh distribution. In [3], intensive experiments have shown that in their great majority, impulse noises had a duration under 500 μs. Chapter 6 describes how this better knowledge of impulse noises characteristics has been very rapidly exploited by xDSL modems designers.

4.4.3 Thermal Noise

Thermal noise corresponds to the erratic motion of a certain amount of electrons along a copper pair or within the end modems. The intensity of this motion—that is, the number of erratic electrons and their speed—is strongly dependent on temperature. When the SNR is evaluated between the receiver modem and the data receiver, thermal noise is characterized by weak energy noise compared to the energy of the received signal. Thermal noise is also characterized by very low-speed fluctuations. The statistical behavior of thermal noise can be modeled by an AWGN. Section 4.3 presents the mathematical definition of AWGN. In practice, the on-line signal is always bandwidth-limited whereas AWGN is supposed to have mathematically a constant power spectral density over the entire spectrum [0, +∞]. In physics, there is not any observable signal with a positive psd over an infinite spectrum. This is the reason why one assimilates the transmission medium to an ideal bandpass filter of width B (in hertz) centered on frequency f_c. The psd of a white noise filtered by such a bandpass filter is illustrated in Figure 4.7. A white noise filtered by an ideal bandpass filter is sometimes called bandpass white noise. The resulting power of the bandpass white noise is $N_0 B$. In practice, knowing the operational bandwidth of a modem, the impact of thermal noise remains limited thanks to filtering.

One assumes that the value of B overlaps the bandwidth Δ of the data signal, B being much larger than Δ. Filtered AWGN must be taken into account in the SNR evaluation to determine the Shannon's capacity of a copper pair.

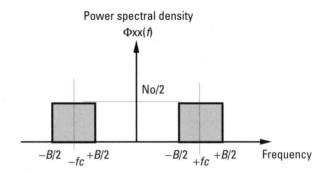

Figure 4.7 Power spectral density of AWGN.

4.4.4   Radio Interferences

In the previous sections, one has assumed that subscriber lines were based on twisted pairs placed in underground binder cables. The oldest subscriber lines are in fact based on bare conductors suspended at pole tops. As the conductors are not protected by plastic gains, it has been shown experimentally that such subscriber lines may interfere with *amplitude modulation* (AM) radio transmitters or with short-wave radio sources. Two types of radio interference must be considered in this case: ingress or egress. Ingress interference corresponds to the electrical noise generated by a radio environment on a subscriber line. Egress interference corresponds to the perturbation induced by the subscriber line on its environment. In the case of xDSL access, the higher the bit rate, the larger the on-line spectrum. At high frequencies, a copper wire tends to behave like an antenna. It irradiates an electromagnetic field in the neighborhood, and it detects electromagnetic fields in the environment. As Chapter 7 discusses, electromagnetic compatibility between the copper infrastructure and its environment is one of the major difficulties in achieving very high bit rates in the local loop.

4.4.5   Impedance Impairments

Section 4.1, which detailed bridge taps, underlined the fact that subscriber lines are in general made of a mix of several sections of copper pairs. These sections are characterized by different diameters, twisted periods, and characteristic impedances. Figure 4.8 illustrates two examples of subscriber lines effectively implemented in a real network [16]. Configuration 1 outlines a subscriber line made of four successive sections and of one bridge tap. The summation of the length of the four horizontal sections is equal to 745m. This means that the end user connected to this line is very close to its EO. All the sections and the bridge

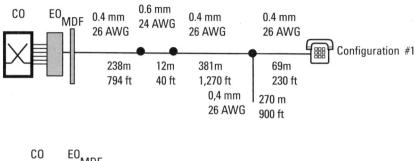

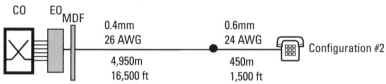

Figure 4.8 Two examples of subscriber lines. (*From:* [16]. © 1991 IEEE. Reprinted with permission.)

tap of line 1 are 26-AWG except for the second section, which is 24-AWG. Configuration 2 looks much more homogeneous. It is made only of two sections of 26-AWG and 24-AWG, respectively. One notices that line 2 is much longer than line 1, with a total length of 5,400m.

Figure 4.9 depicts the attenuation a of each of these two local loop configurations versus frequency *f*. Figure 4.9 confirms that a great disparity may exist between the electrical characteristics of different subscriber lines.

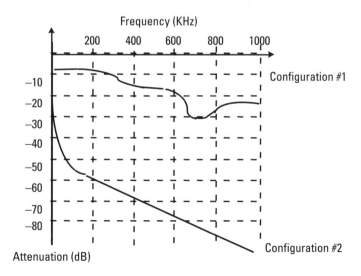

Figure 4.9 Attenuation versus frequency. (*From:* [16]. © 1991 IEEE. Reprinted with permission.)

Attenuation is very irregular in the case of configuration 1 whereas it is almost linear on the whole spectrum in the case of configuration 2. In the context of xDSL access with on-line frequencies up to the megahertz range, configuration 1 needs a much more sophisticated equalizer than configuration 2 where a simple linear equalizer can be used between 150 kHz and 1 MHz.

Cable manufacturers provide twisted pair sections between 150- and 500-m-long. When a carrier has to install a subscriber line of several kilometers (as in the case of configuration 2 in our previous example), this assumes that multiple sections of cables must be set end-to-end. Each solder carried out to join two successive cable sections introduces an impedance impairment. These impairments induce echoes that degrade the SNR at the receiver and then the achievable data rate on the line. In addition, carriers have noticed that solders are very sensitive to corrosion in the presence of humidity. This corrosion worsens the Shannon's capacity reduction of a subscriber line.

4.4.6 Crosstalk

Crosstalk refers to electromagnetic interference between copper pairs transported in a same binder. We have described in the previous sections how the negative impact of crosstalk on the SNR is noticeable only for frequencies above 100 kHz. Several studies were carried out at the end of the 1980s to analyze this impact for frequencies up to 30 MHz [17, 18]. Three major types of crosstalks must be distinguished:

- *Near-end-crosstalk* (NEXT);
- *Far-end-crosstalk* (FEXT);
- *Self-near-end-crosstalk* (SNEXT).

This section describes only the principle of NEXT and of FEXT. SNEXT, which refers to the particular case where a copper pair interferes with itself, is described in Chapter 5.

4.4.6.1 NEXT

Figure 4.10 depicts the principle of NEXT between two copper pairs i and j transported in a same binder. Let N be the amount of pairs transported in this binder. All the users connected on these pairs are supposed to use xDSL modems. Let us consider pair j as a reference pair for which we are going to evaluate the Shannon's capacity at modem R_2. Let us assume that an end user connected to pair j by means of modem R_2 receives high bit rate data from the EO (for instance, a digital video sequence). One assumes that modem E_1 connected to pair i generates high bit rate data (for instance, a file transfer) to the

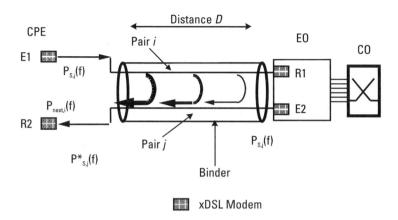

Figure 4.10 Principle of NEXT.

EO. The high frequencies associated with this upstream signal are partially detected by pair i. The resulting energy of this detection is added to the energy of the downstream signal propagating on pair j. NEXT interference corresponds to the noise induced by pair i on pair j. This phenomenon occurs all along the binder cable. As it is illustrated in Figure 4.10,[9] the disruptive energy due to pair i on pair j decreases progressively from the CPEs to the EO.

Let $P_{s,j}(f)$ be the psd of the signal sent by modem E_2 to modem R_2 and $P_{s,i}(f)$ be the psd of the signal sent by modem E_1 to modem R_1. In the absence of any interferer among all pairs $i, j \neq i$, that is, when only reference pair j is active, the psd of the received signal by modem R_2 is given by:

$$P*_{s,j}(f) = \|H_0(f,d)\|^2 \cdot P_{s,j}(f) \tag{4.19}$$

where $H_0(f)$ stands for the transfer function of a twisted pair. With this transfer function given by (4.3), one has then:

$$H_0(f,d) = e^{-\gamma \cdot D} = e^{-(\alpha + j \cdot \beta) \cdot D} \tag{4.20}$$

Reporting (4.20) into (4.19), it yields for the psd of the received signal:

---

9. In Figure 4.10, one has illustrated the amplitude of the perturbation of pair i on pair j by the thickness of the folded arrows.

$$P_{s,j}^{*}(f) = \left\| H_0(f,d) \right\|^2 \cdot P_{s,j}(f) = e^{-2\alpha D} \cdot P_{s,j}(f) \qquad (4.21)$$

One assumes that a common distance separates modems E_k from modems R_k, $k \in \{1, 2, \ldots N\}$. One has then $P_{s,j}(f) = P_{s,i}(f) = P_s(f)$, $\forall\, (i, j) \in \{1, 2, \ldots N\} \times \{1, 2, \ldots N\}$. In [2], it is shown that cumulative NEXT B_j induced by multiple interferers on the reference pair j detected by modem R_2 is given by the product of a transfer function $H_x(f)$ with $P_s(f)$. From experimental measurements, the authors have determined an approximate formulation of B_j:

$$B_j(f) = \left\| H_x(f) \right\|^2 \cdot P_s(f) = \psi \cdot f^{3/2} P_{s,j}(f) \qquad (4.22)$$

In (4.22), the value of coefficient ψ depends on several parameters characterizing a subscriber line: the average diameter,[10] the average twisted period, and the amount of NEXT interferers. It is possible to determine the analytical expression of the Shannon's capacity under cumulative NEXT and thermal noise. In [17], the authors assume that cumulative NEXT added with thermal noise is also a stationary Gaussian noise. Let C_{next} be the Shannon's capacity of reference pair j. In reporting the expression of B_j in (4.16), it comes:

$$C_{next} = \int_{f \in \Omega} \log_2 \left\{ 1 + \frac{\left\| H_0(f) \right\|^2 \cdot P_s(f)}{\left\| H_x(f) \right\|^2 \cdot P_s(f) + \frac{N_0}{2}} \right\} \cdot df \qquad (4.23)$$

where Ω stands for the set of frequencies for which $P_s(f) > 0$. In reporting the expressions of $H_0(f)$ and of $H_x(f)$ in (4.23), it becomes:

$$C_{next} = \int_{f \in \Omega} \log_2 \left\{ 1 + \frac{e^{-2\alpha D} \cdot P_s(f)}{\psi \cdot f^{3/2} \cdot P_s(f) + \frac{N_0}{2}} \right\} \cdot df \qquad (4.24)$$

There is not any closed form of (4.24). The logarithm function included in (4.24) is positive and increases with f. It is possible to find an upper bound of C_{next} in replacing the set Ω by an infinite interval $[0, +\infty]$. Let us call C_{next}^{*} this upper bound:

$$C_{next}^{*} = \int_0^{+\infty} \log_2 \left\{ 1 + \frac{e^{-2\alpha(f) \cdot D} \cdot P_s(f)}{\psi \cdot f^{3/2} \cdot P_s(f) + \frac{N_0}{2}} \right\} \cdot df \qquad (4.25)$$

---

10. The average parameter is used in case of inhomogeneous cable sections.

Depending on the value of f, different expressions of $\alpha(f)$ must be considered in reference to (4.5)–(4.9). Section 4.5 discusses the impact of this result on the maximum achievable data rate on a subscriber line.

4.4.6.2 FEXT

Figure 4.11 illustrates the principle of FEXT between two pairs i and j belonging to a same binder. One considers the same assumptions associated with Figure 4.11 as those associated with Figure 4.10.

Let us consider two end users connected to pair j by means of modem R_2 and to pair i by means of modem R_1, respectively. These two users both receive high bit rate data from the same EO. FEXT interference corresponds to downstream frequencies generated by modem E_1 that are detected by pair j. Like NEXT, FEXT occurs all along the binder cable. As illustrated in Figure 4.11, the disruptive energy due to pair i on pair j decreases progressively from the EO to the CPEs. For the same network configuration, NEXT is then more penalizing than FEXT on the SNR. Indeed, the value of noise N evaluated at modem R_2 is lower with FEXT than with NEXT, the value of S remaining unchanged. In [2], it is shown that the cumulative FEXT noise B'_j induced by multiple interferers on the reference pair j detected by modem R_2 is given by the product of a transfer function $H_y(f)$ with $P_s(f)$.

$$B'_j(f) = \left\| H_y(f) \right\|^2 \cdot Ps(f) \qquad (4.26)$$

In the absence of any interferer, the psd of the received signal at modem R_2 is given by (4.21). In [19], it is shown that SNR (or S/N) at the receiver in presence of the only FEXT and neglecting AWGN is such that:

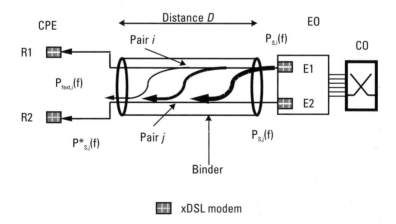

Figure 4.11 Principle of FEXT.

$$S/N = \frac{1}{\chi \cdot f^2 \cdot D} \qquad (4.27)$$

where the coefficient χ depends on several parameters of the subscriber line: the average diameter, the average twisted period, and the amount of FEXT interferers. The expression of S/N given in (4.27) is characterized by a decreasing of 20 dB per decade with f and by a decreasing of 10 dB per decade with D. Such a behavior of S/N has been confirmed on real subscriber lines when $f > 20$ kHz. One has then for Shannon's capacity of pair j in presence of cumulative FEXT:

$$C_{fext} = \int_{f \in \Omega} \log_2 \left\{ 1 + \frac{1}{\chi \cdot f^2 \cdot D} \right\} \cdot df \qquad (4.28)$$

One notices that this expression is independent from the transfer function of individual pairs $H_0(f)$.

4.5   Impact of NEXT on Shannon's Capacity

There is not a closed-form expression of the upper bound C^*_{next} of the Shannon's capacity given in (4.25) if one considers the various expressions of $\alpha(f)$ mentioned in (4.5)–(4.9). Knowing that under high frequencies, the psd of thermal noise can be neglected compared to $\psi \cdot f^{1.5} \cdot Ps(f)$, one has for $f < 10$ kHz and for $f > 150$ kHz:

$$
\begin{aligned}
C^*_{next} &= \int_0^{+\infty} \log_2 \left\{ 1 + \frac{e^{-2k\sqrt{fD}} \cdot P_s(f)}{\psi \cdot f^{3/2} \cdot P_s(f) + \frac{N_0}{2}} \right\} \cdot df \\[2mm]
&\approx \int_0^{+\infty} \log_2 \left\{ 1 + \frac{e^{-2k\sqrt{f}}}{\psi \cdot f^{3/2}} \right\} \cdot df
\end{aligned} \qquad (4.29)
$$

Power series expansion applied to the exponential in the above expression enables one to plot an approximate curve of Shannon's capacity versus the length D of the subscriber line [17]. Figure 4.12 represents a numerical version of this curve when ψ corresponds to a 24-AWG binder cable with 50 pairs and 49 interferers.

We see that data rates up to 30 Mbps in both directions are in theory achievable on the shortest subscriber lines with $D < 1$ km. For a distance of 2 km—that is, for the average length of a subscriber line in urban area—bit rates

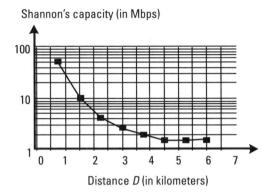

Shannon's capacity (in Mbps)

Distance *D* (in kilometers)

Figure 4.12 Shannon's capacity versus distance *D* for a 50-pair cable. (*From:* [17]. © 1990 IEEE. Reprinted with permission.)

of up to 6 Mbps are achievable. As a comparison, let us recall that dial-up V.90 modems are limited to 33.6 Kbps in the upstream and to 56 Kbps in the downstream.

4.6 Conclusion and Perspectives

In this chapter, we have investigated the electrical characteristics of a twisted copper pair. Two types of environments have been successively considered: a LAN environment with a single isolated pair and the local loop environment where pairs are grouped in binder cables. Various factors such as bridge taps, impulse noise, thermal noise, radio interference, impedance impairments, and crosstalk have an impact of the Shannon's capacity of a subscriber line. Until the 1990s, the influence of most of these factors on the achievable data rates were rather well-known except impulse noises. Impulse noises were intensively investigated in the mid-1990s. Today, one has a better idea of their statistical behavior. In comparison to impulse noises, which are rare events, NEXT and FEXT crosstalks are almost unavoidable in the local loop. We have determined that FEXT is less penalizing than NEXT on the SNR. Concluding that the NEXT impact on SNR is not equivalent seen from the CPEs or from the EO, one may conclude that higher bit rates are preferable for downstream traffic than for upstream traffic. We shall see in Chapter 6 that this conclusion is at the origin of the adjective "asymmetrical" in the acronym ADSL. By chance, such an asymmetry corresponds to the nature of residential Internet access. Thanks to Bellcore (today Telcordia Technologies, Inc.) investigations, one has learned from the Shannon's capacity that copper pairs were largely underused. This discovery in conjunction with the emergence of the Internet has been at the origin of the development of new generation xDSL modems in the early 1990s. In the following chapters, we discuss the various approaches that have been proposed to increase data rates on subscriber lines. These approaches are essentially based on new line codes and new modulation techniques.

References

[1]   Cook, J., and P. Sheppard, "ADSL and VADSL Splitter Design and Telephony Performance," *IEEE Journal on Selected Areas in Communications*, Vol. 13, No. 9, December 1995, pp. 1634–1642.

[2]   Werner, J. J., "The HDSL Environment," *IEEE Journal on Selected Areas in Communications*, Vol. 9, No. 6, August 1991.

[3]   Melinand, R., and M. Kavehrad, *Impulse Noise Modeling and Simulation in the Loop Plant*, Technical Report of Broadband Communications Research Laboratories, Canada, 1996.

[4]   Cohen, G., J.-L. Dornstetter, and P. Godlewski, *Codes correcteurs d'erreurs: une introduction au codage algébrique*, Collection CENT-ENST, Masson, 1992.

[5]   Anttalainen, T., *Introduction to Telecommunications Network Engineering*, Norwood, MA: Artech House, 1999.

[6]   Ungar, S. G., "Testing Basic Access Loops for Service-Affecting Impulse Noise," *Proc.of IEEE Globecom'89*, 1989, pp. 1585–1591.

[7]   Werner, J. J., "Impulse Noise in the Loop Plant," *Proc. of IEEE Globecom'90*, 1990, pp. 1734–1737.

[8]   Fano, R. M., "A Theory of Impulse Noise in Telephone Networks," *IEEE Trans. on Communications*, Vol. 25, 1997, p. 577.

[9]   Nynex Corporation, *Impulse Noise Measurements in New York City and Their Possible Ramifications on a Selection of DSL Line Code Standard*, Contribution T1D1.3/86-144, 1986.

[10] Szechenyi, K., "On the NEXT and Impulse Noise Properties of Subscriber Loops," *Proc. of the IEEE Globecom'89*, 1989, pp. 1569–1573.

[11] Valenti, C. F., and K. Kerpez, "Analysis of Wideband Noise Measurements and Implications for Signal Processing in ADSL Systems," *Proc. of ICC'94*, 1994.

[12] Cook, J. W., "Wideband Impulsive Noise Survey in the Access Network," *British Telecom Technological Journal*, Vol. 11, 1993, pp. 155–162.

[13] Henkel, W., T. Kessler, and H. Y. Chung, "Coded 64-CAP ADSL in an Impulse Noise Environment: Modeling of Impulse Noise and First Simulation Results," *IEEE Journal on Selected Areas in Communications*, Vol. 13, No. 9, 1995, pp. 1611–1621.

[14] Proakis, J. G., *Digital Communications*, 3rd ed., New York: McGraw-Hill, 1995.

[15] Kleinrock, L., *Queueing Systems: Volumes 1 and 2*, New York: John Wiley & Sons, 1975–1976.

[16]   Waring, D. L., J. W. Leichleider, and T. R. Hsing, "DSL Technology Facilitates a Graceful Transition from Copper to Fiber," *IEEE Communications Magazine,* March 1991, pp. 96–103.

[17]   Halet, I., and S. Shamai, "On the Capacity of a Twisted-Wire Pair: Gaussian Model," *IEEE Trans. on Communications,* Vol. 38, No. 3, 1990, pp. 379–383.

[18]   Lawrence, V. B., et al., "Broadband Access to the Home on Copper," *Bell Labs Technical Journal,* Vol. 1, No. 1, Summer 1996, pp. 100–114.

[19]   Chen, W. Y., "Broadcast Digital Subscriber Lines," *IEEE Journal on Selected Areas in Communications,* Vol. 13, No. 9, December 1995.

Selected Bibliography

Bisaglia, P., R. Castle, and S. H. Baynham, "Channel Modeling and System Performance for HomePNA 2.0," *IEEE Journal on Selected Areas in Communications,* Vol. 20, No. 5, June 2002.

Mc Donald, R. A., and C. F. Valenti, *Assumptions for Bellcore HDSL Studies,* Bellcore Contribution T1E1.4/89–066, 1989.

5

HDSL and Its Variants

5.1 Introduction

Chapter 3 discusses the basic principle of xDSL modems. Unlike voiceband modems, xDSL modems operate on a bandwidth much wider than 4 kHz. They must be installed by pair, at the customer premises, and at the EO. The HDSL technique is considered as the first version of xDSL modems. It aims to replace classical LLs that have been described in Section 2.4 for E1/T1 services provisioning. During about 15 years, from 1975 to 1990, LLs have been the only solution for high-speed interconnection between remote LANs. This interconnection may occur either via PSTN or via a packet-switching network such as frame relay or ATM. We have outlined in Chapter 2 the high cost of implementation of T1/E1 LLs. This cost is due to the fact that although it relies on twisted copper pairs, a LL cannot reuse existing pairs already installed in the local loop. According to the HDB3 or AMI line code used in LL modems, electromagnetic compatibility between pairs imposes LL copper pairs to be isolated from regular subscriber lines. We have emphasized that because of their high on-line frequencies, LLs require repeaters at least every 1,800m. Because of this relative complexity of implementation in the local loop, LL provisioning by a carrier needs in general several weeks, or even several months. The HDSL technique was developed by Bellcore[1] research laboratories in the United States at the end of the 1980s [1]. It aims to enable T1/E1 full-duplex point-to-point links at a much cheaper cost than classical LLs. Because they operate on a bandwidth equal to a fourth of the bandwidth required by traditional LLs, HDSL LLs may reuse existing subscriber lines. Thanks to this characteristic, HDSL access

---

1. Today, Bellcore is Telcordia Technologies, Inc.

systems can be provided by the carriers with very short delays (a few days). As it is described further, HDSL systems may operate either on one, two, or three copper pairs. It is highly recommended that these copper pairs are compatible with CSA specifications (see Section 2.6).

HDSL was standardized by the ITU-T in October 1988 under the reference G.991.1. A new version of HDSL operating on a single pair and known as SHDSL has been standardized in 2000 under the reference G.991.2. The European standardization institute ETSI also standardized HDSL under the reference ETSI TS-101-135.

Whereas LLs are used traditionally to interconnect remote LANs, they are also used today in radio-mobile networks to link base stations with dedicated circuit switches named *mobile switching systems* (MSSs). LLs are also used to link the server farms of service providers to the nearest IP router of a NSP. Because HDSL access systems are much cheaper than traditional LLs, they should favor the emergence of small-size servers managed by residential users in countries that have adopted flat-rate billing for local calls. Such a context exists in North America but it is not yet the case in Europe where local calls billing is still based on a call's duration. At the end of 2000, about 2.4 million xDSL lines were in service in the United States and 430,000 in Canada.

This chapter is organized into three sections. Section 5.2 describes the main characteristics of HDSL modems. Section 5.3 presents the various HDSL frame formats when transmission occurs on one, two, or three copper pairs. Section 5.4 is dedicated to four variants of HDSL named IDSL, SDSL, HDSL2, and SHDSL.

5.2 HDSL Modems

Figure 5.1 illustrates the typical configuration of three types of access systems: E1/T1 LLs, T1-HDSL, and E1-HDSL. Each of these three configurations are used to link a gateway or a router located at the premises of an enterprise with a public data network—for instance, a frame relay switch or an IP access router. In the case of E1/T1 LLs, we mentioned that repeaters are in general necessary at least every 1,800m. One notices that both T1-HDSL and E1/T1 LLs require two pairs, whereas E1-HDSL needs three pairs because of its higher bit rate. Each pair operates at 1.54 Mbps in half-duplex in the case of T1 LLs instead of 784 Kbps in full-duplex in the case of E1/T1-HDSL. Full-duplex communication on a single pair consists of using the same frequency band for both upstream and downstream traffic. Like in narrowband IDSN, full-duplex transmission in E1/T1-HDSL is carried out thanks to echo cancellation; the principle of this technique is described in the next section. One notices that the global rates of E1-HDSL and of T1-HDSL equal to 2,352 Kbps and to 1,568 Kbps,

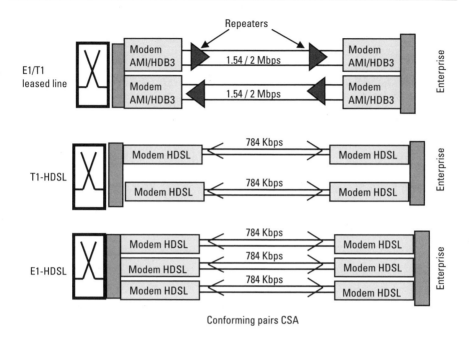

Figure 5.1   T1/E1 LLs, T1-HDSL, and E1-HDSL configurations.

respectively, are slightly greater than the corresponding E1/T1 LL rates of 2,048 Kbps and 1,540 Kbps. The reason for this disparity is explained further.

5.2.1   Comparison Between ISDN and HDSL Modems

Both narrowband ISDN access and HDSL access systems enable symmetrical bit rates in the local loop. For this reason, a certain analogy exists between these two techniques. Thus, both systems operate in baseband, that is with a line code that also serves as a modulation technique. Whereas E1 and T1 LLs use HDB3 and AMI line codes, respectively, HDSL modems rely on the same line code as ISDN modems that is the 2B1Q. Several line codes have been investigated by Bellcore for HDSL access systems [2], including 2B1Q, 3B2T, 4B3T, AMI, and Manchester. The choice of a line code for HDSL has been based on various criteria such as its sensitivity to impulse noise, ISI, and NEXT interference. On the basis of the average copper pair's SNR in the local loop, the standardization bodies have adopted 2B1Q essentially for its simplicity of implementation. Thus, the 2B1Q line code enables it to reach a 784-Kbps bit rate on a single pair over a 3,600-m range without requiring any repeater. Such a bit rate corresponds to 12 DS-0 digital voice channels at 64 Kbps with a certain signaling overhead.

We have described in Figure 2.12 the principle of 2B1Q line code. This code associates to each suite of two bits (such a suite is called a symbol) a voltage level chosen among four possible values: $+V_1$, $V_2 = +3 V_1$, $-V_1$ and $-V_2$. The on-line signal is characterized by a PAM modulation with voltage pulses of duration T, where T is twice the bit duration T_b. In the remainder of this section, we want to determine the power spectral density (i.e., the frequency spectra of the signal associated to the output of a 2B1Q encoder). A 2B1Q encoder may be seen as a filter with impulse response $r(t)$ that associates every T seconds a rectangular voltage pulse to a succession of Dirac delta functions corresponding to the four possible amplitudes of the modulated signal. Let $e(t)$ be a random suite of Dirac delta functions[2] of amplitude V_k. One has:

$$e(t) = \sum_{k=-\infty}^{+\infty} V_k \cdot \delta(t - kT) \tag{5.1}$$

The signal $s(t)$ obtained at the output of the 2B1Q encoder is given by the convolution product between $e(t)$ and $r(t)$:

$$s(t) = e(t) \otimes r(t) = \int_{-\infty}^{+\infty} e(t - \tau) \cdot r(\tau) \cdot d\tau \tag{5.2}$$

It has been shown that the *power spectral density* (psd) $P_{xx}(f)$ of a baseband signal $x(t)$ transmitted on a copper pair depends on the real part of the characteristic impedance Z_c of the line and on the symbol duration T. For a given cable, $P_{xx}(f)$ fluctuates with frequency f according to the *Fourier transform* (FT) of the autocorrelation function $C_{ss}(t)$ of the signal at the output of the encoder:

$$P_{ss}(f) = \frac{2 \cdot FT[C_{ss}(t)]}{R_e(Z_c) \cdot T} \tag{5.3}$$

In order to evaluate the expression of the power spectral density of the 2B1Q code, one has to determine the autocorrelation function $C_{ss}(t)$ of $s(t)$:

$$C_{ss}(\tau) = E[s(t) \cdot s(t - \tau)] \tag{5.4}$$

where $E[x]$ is the statistical expectation of the random variable x. In replacing in (5.4) $s(t)$ by its expression given in (5.2), it yields:

---

2. The product of a function $F(t)$ by the Dirac delta function at point $(t - t_0)$ corresponds to value of $F(t_0)$.

$$C_{ss}(\tau) = E\left[\int_{-\infty}^{+\infty} e(t-\alpha)\cdot r(\alpha)\cdot d\alpha \cdot \int_{-\infty}^{+\infty} e(t-\tau-\beta)\cdot r(\beta)\cdot d\beta\right] \quad (5.5)$$

Thanks to the ergodicity property, we can write:

$$C_{ss}(\tau) = \int_{-\infty}^{+\infty}\int_{-\infty}^{+\infty} E\left[e(t-\alpha)\cdot e(t-\tau-\beta)\cdot r(\alpha)\cdot r(\beta)\right]\cdot d\alpha\, d\beta \quad (5.6)$$

This expression may also be written after a changing of variable $u = t + \beta$:

$$C_{ss}(\tau) = \int_{-\infty}^{+\infty}\int_{-\infty}^{+\infty} C_{ee}(\tau+\alpha-\beta)\cdot r(\alpha)\cdot r(\beta)\cdot d\alpha\, d\beta \quad (5.7)$$

The Fourier transform $S(f)$ of $C_{ss}(t)$ is given by:

$$S(f) = \int_{-\infty}^{+\infty} C_{ss}(\tau)\cdot e^{-2j\pi f t}\cdot dt \quad (5.8)$$

We have then in reporting (5.7) into (5.8):

$$S(f) = \int_{-\infty}^{+\infty}\int_{-\infty}^{+\infty}\int_{-\infty}^{+\infty} C_{ee}(\tau+\alpha-\beta)\cdot r(\alpha)\cdot r(\beta)\cdot e^{-2j\pi f t}\cdot d\alpha\, d\beta\, dt \quad (5.9)$$

After a changing of variable $t = \tau + \alpha - \beta$, the exponential term in (5.9) may be decomposed into three exponential terms relative to the three variables t, α, and β, respectively:

$$S(f) = \int_{-\infty}^{+\infty} C_{ee}(t)\cdot e^{-2j\pi f t}\cdot dt \cdot \int_{-\infty}^{+\infty} r(\alpha)\cdot e^{+2j\pi f\alpha}\cdot d\alpha \int_{-\infty}^{+\infty} r(\beta)\cdot e^{-2j\pi f\beta}\cdot d\beta \quad (5.10)$$

From (5.1), the autocorrelation of the input signal $e(t)$ may be written:

$$C_{ee}(t) = E\left[V_k^2\right]\cdot\delta(t) \quad (5.11)$$

where $E[V_k^2]$ represents the second-order momentum of the random variable V_k. The expression of this momentum can be expressed as a function of the voltage level V_1:

$$E\left[V_k^2\right] = \frac{1}{4}\cdot\left\{V_1^2 + V_2^2 + (-V_1)^2 + (-V_2)^2\right\}$$

$$= \frac{1}{4}\cdot\left\{V_1^2 + (3V_1)^2 + V_1^2 + (-3V_1)^2\right\} \qquad (5.12)$$

$$= 5V_1^2$$

We have then for $S(f)$:

$$S(f) = \int_{-\infty}^{+\infty} 5V_1^2 \cdot \delta(t)\cdot e^{-2\,j\pi ft}\cdot dt \cdot \int_{-\infty}^{+\infty} r(\alpha)\cdot e^{+2\,j\pi f\alpha}$$

$$\cdot d\alpha \cdot \int_{-\infty}^{+\infty} r(\beta)\cdot e^{-2\,j\pi f\beta}\cdot d\beta \qquad (5.13)$$

The first term in (5.13) corresponds to the Fourier transform of the Dirac impulsion. Knowing that $FT[\delta(t)] = 1$, it yields:

$$S(f) = 5V_1^2 \cdot R(f)\cdot R*(f) \qquad (5.14)$$

where $R(f)$ and $R^*(f)$ represent the Fourier transform of the impulse response of the 2B1Q encoder and its complex conjugate. This last expression may also be written:

$$S(f) = 5V_1^2\left|R(f)\right|^2 \qquad (5.15)$$

Let us make the assumption that the impulse response $r(t)$ of the 2B1Q encoder is a rectangular function with unit amplitude, centered on $t = 0$ and with a width T. It yields then:

$$S(f) = 5V_1^2\left|\frac{\sin(\pi fT)}{\pi f}\right|^2 \qquad (5.16)$$

Finally, one obtains the expression of the power spectral density $P_{ss}(f)$ of the on-line HDSL signal in reporting (5.16) into (5.3):

$$P_{ss}(f) = \frac{2 \cdot FT\left[C_{ss}(t)\right]}{R_e(Z_c) \cdot T} = \frac{2 \cdot S(f)}{R_e(Z_c) \cdot T} =$$

$$10V_1^2 \cdot \frac{\sin^2(\pi f T)}{R_e(Z_c) \cdot T \cdot (\pi f)^2}$$

(5.17)

The expression of the power spectral density of a narrowband ISDN signal is similar to (5.17). Only the expressions of the characteristic impedance Z_c and of the symbol duration T differ between these two types of baseband modems. One notices that $P_{ss}(f)$ is null for $f = k/T = k/2\ T_b$ with $k = 1, 2$, and so on. Figure 5.2 compares the bandwidths required by basic ISDN access and T1-HDSL, respectively, when two copper pairs are used. Basic ISDN access requires a bandwidth of 80 kHz corresponding to a gain by a factor 2 on the 160 kHz of the NRZ signal.[3] This gain is due to the 2B1Q line code. One also notices that the cutting frequency of a T1-HDSL receive modem must be equal to 392 kHz. In comparison, an AMI T1 LL requires a bandwidth of 1.54 MHz corresponding to the 1.54-Mbps bit rate per copper pair. There is then a gain in bandwidth by a factor 4 between T1-HDSL and T1-AMI-LL. Both the 2B1Q line code and the echo cancellation technique allow for the reduction of the on-line bandwidth by a factor of 2. The principle of echo cancellation is described in the following section.

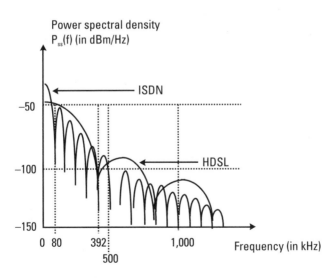

Figure 5.2 ISDN and two pairs T1-HDSL psds.

---

3. We have seen in Chapter 2 that basic access to ISDN corresponds to a 160-Kbps bit rate.

Knowing that HDSL is designed to operate on CSA-compatible sub-scriber lines under 3,600-m range, no specific preconditioning of the copper cables is necessary. One can then expect HDSL access to be more cost-effective than traditional LLs. One has observed a reduction of about 30% on the price of T1 services offered via T1-HDSL access systems in comparison to traditional T1 LLs in North America. By reusing existing subscriber lines, HDSL enables operators to provision E1/T1 with much shorter delays than traditional LLs.

5.2.2   Block Diagram of an HDSL Modem

We mentioned in Section 2.4.4 that narrowband ISDN modems developed in the mid-1980s were using echo cancellation to enable full-duplex transmission in the same frequency band. HDSL modems reuse the same principle of echo cancellation but at higher speeds. Figure 5.3 depicts the block diagram of an HDSL modem.[4] The same block diagram characterizes narrowband ISDN modems. The main elements of this block diagram are the 2B1Q line encoder, the transmission filter, the hybrid, the receive filter, the A/D converter, the echo canceller, and the DFE. The role of each of these elements is described next.

The data generated by a source are first encoded into 2B1Q before being amplified and then transmitted onto the medium. Each arrow represented on the diagram corresponds to a copper pair transporting a voltage between two successive electronic modules of the modem. The electrical signal observed on the line is equal to the algebraic summation of the upstream and of the down-stream signals that propagate in opposite directions. Downstream signal is trans-ferred to a receive filter via a hybrid.[5] The concept of hybrid circuit has been introduced historically with analog telephony to prevent an end user from hear-ing the echo of his or her own voice from the handset. In the context of HDSL, a hybrid prevents a fraction of the energy of the transmitted signal from being

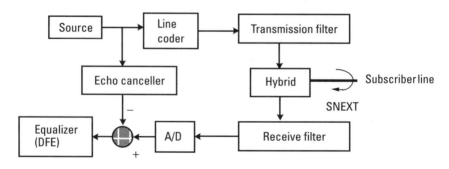

Figure 5.3   Block diagram of an HDSL modem.

---

4. We recall that T1-HDSL and E1-HDSL need two or three modems of that type at each end of the access system.

received as a disruptive noise in the receive filter. Like NEXT crosstalk presented in Chapter 4, this disruptive noise called *SNEXT* degrades the SNR at the receiver and then reduces the achievable throughput on the line. The key element of an HDSL modem is the echo canceller in charge of retrieving in real time SNEXT from the received signal after its A/D conversion. Figure 5.4 recalls the principle of a hybrid in the context of PSTN.

In examining Figure 5.4, one recalls that the EO separates upstream and downstream analog signals by means of another hybrid. This separation occurs before to proceed to the digitization of the upstream channel and after to proceed to the D/A conversion of the downstream signal. In all PSTNs, upstream and downstream digital flows are transported on distinct links and are regenerated by dedicated electronic devices. We have plotted in Figure 5.5 the path followed by a voice signal from the calling party to the called party via PSTN. The thick solid line describes the regular path of a voice call. A hybrid is an electronic circuit made of multiple operational amplifiers. As illustrated in Figure 5.5, a complex impedance Z_h in theory equal to the characteristic impedance Z_c of the line is connected to each hybrid.

Ideally, a hybrid is able to prevent totally SNEXT crosstalk in the receive filter if Z_h is exactly equal to Z_c. This condition cannot be satisfied because, as has been outlined in Chapter 4, it is impossible for economy of scale to adjust the value of Z_h to the specific value of Z_c for each subscriber line. In practice, Z_h is simply an approximation of Z_c. The fact that the four hybrids represented in Figure 5.5 are not ideal induces two disruptive noises for the calling party. A first noise named E_1 on Figure 5.5 is due to the hybrid of the calling party. This

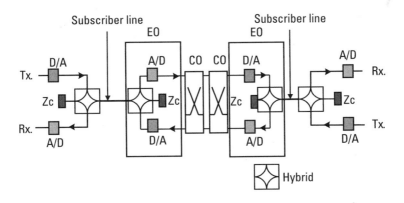

Figure 5.4 Hybrid used in PSTN access.

---

5. In Europe, a hybrid is sometimes called a two-wire/four-wire transformer. The two wires correspond to the subscriber line. The four wires correspond to the two wires used between the transmission filter and the hybrid and to the two wires used between the hybrid and the receive filter.

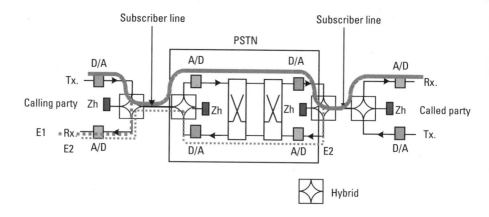

Figure 5.5   Different types of echoes due to the imperfection of the hybrids.

noise corresponding to SNEXT is represented by a thick dotted line. A second noise named E_2 also detected by the receiver of the calling party is due to the hybrid of the EO on which is connected the called party. The path followed by E_2 is represented by a thin dotted line. In telephony, echoes E_1 and E_2 are in most cases imperceptible because of the too-short delays that separate the reference signal generated by the calling party from echoes E_1 and E_2. Nevertheless, echo E_2 becomes perceptible by the calling party in the case of very long-range communications. Such a phenomenon occurs, for instance, when a European calling party is connected to a called party located in Asia or in America. If the phone call transits through a geostationary satellite (see Section 3.6.2.1), the 240-ms round-trip propagation delay between the Earth stations and the satellite makes echo E_2 clearly perceptible by the end user. Long-distance telephone carriers are thus obliged to install expensive equipment in their network to suppress echo E_2.

If echo E_1 is not penalizing for circuit-switched telephony, its impact on HDSL access efficiency cannot be neglected. Indeed, the phase shift between echo E_1 and the transmitted data is not negligible in comparison to the bit duration. In other terms, SNEXT may induce in HDSL several erroneous bits after decoding the received data. The phase shift is even more noticeable between echo E_2 and the transmitted data. In practice, canceling echo E_2 by means of a correlation demodulator is quite simple because of the clear time shift between E_2 and the transmitted data. On the other hand, canceling echo E_1 (SNEXT) cannot be carried out by means of a correlation demodulator because of the quasi-simultaneity between this echo and the transmitted data. Echo cancellation is in fact the only way to reduce the impact of SNEXT.

Narrowband ISDN modems developed in the 1970s required an initialization phase. Before being operational, such modems had to be initialized by a technician of the carrier. The visit of the technician at the customer premises to proceed to this initialization explains partially the high cost of the first narrowband ISDN access. The initialization phase consists of evaluating the energy of SNEXT versus frequency. For that purpose, data sequences are transmitted in the upstream direction on short periods of time Δt. The average frequency of the on-line signal associated with each of these sequences corresponds to a frequency f_k. Knowing that not any data is received in the downstream during the initialization phase, the echo canceler records in a database a table of values associated to a set of discrete frequencies f_k and the energy I_k of the measured SNEXT. In the case of ISDN modems, the range of test frequencies covers [0, 80 kHz]. The modem becomes operational as soon as the database has been completed. Time is slotted into periods of Δt seconds. When it receives data, the A/D converter of the modem stores during Δt the sequence received during the last period. This delay enables the echo canceler to consult its database in order to know the average energy I_k of SNEXT for the last Δt seconds. During the next period of Δt seconds, the data stored in the A/D converter are transmitted to the DFE. As illustrated in Figure 5.3, the echo canceler generates artificially a noise of energy I_k that is subtracted from the received data between the A/D converter and the DFE. The performance of an echo canceler increases when Δt tends to zero. This assumes the availability of high-speed electronics and of a large memory space for the database. Echo cancelers used in HDSL modems operate in the same manner as ISDN echo cancelers. Because of the advances in the speed of electronics, signal processing, and *random access memory* (RAM) technology, HDSL modems are now able to manage higher bit rates than ISDN modems. They do not require a manual initialization, the database of the echo canceler being updated automatically as soon as a period of inactivity is observed on the line. Because of the more or less complex and random distortions observed on the received signal, "intelligent" equalizers called *DFEs* must be used in the HDSL modems. Chapter 6 describes DFEs.

5.3 HDSL Framing

Section 2.4.2 describes the format of E1/T1 frames used in digital LLs. E1/T1 frames are transported on each of the two or three pairs of an HDSL access system by means of HDSL-specific frames. As mentioned in the introduction of this chapter, two versions of HDSL have been standardized by ITU-T: T1-HDSL (G.991.1) for the American continent and E1-HDSL (G.991.2) for Europe because of the distinction between E1 and T1 frames in the PDH hierarchy.

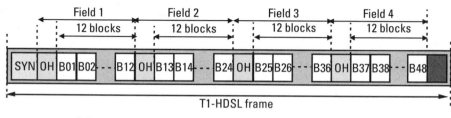

SYN : 14 synchronization bits
OH : 12 header bits (signaling)
■ Stuffing

Figure 5.6   Format of a T1-HDSL frame (two copper pairs).

5.3.1 T1-HDSL

Figure 5.1 shows that T1-HDSL operates on two pairs. Figure 5.6 shows the format of a T1-HDSL frame. A T1-HDSL frame is transmitted in both directions every 6 ms. Each frame starts with a 14-bit synchronization preamble representing seven 2B1Q symbols. The payload of a frame is divided into four fields of same capacity. Each of these four fields begins with a 12 bits overhead (QH) followed by 12 blocks of 97 bits each. The set of the blocks spread out over the four fields are indexed from B_{01} to B_{48}.

For each direction of transmission (upstream or downstream), a T1-HDSL frame is sent by means of an alternate transmission of successive bytes on each of the two pairs of a T1-HDSL access system. Figure 5.7 details the format of the fields B_{01} to B_{48}. Each block of 97 bits is made of 12 information bytes preceded by a single bit (F) used as a flag delimiter.

One notices that DS0 channels[6] 1 to 12 are transmitted on pair 1 whereas DS0 channels 13 to 24 are transmitted on pair 2. A T1-HDSL frame being made of 48 blocks, we can remark that for each of the 24 DS0 channels of a T1-frame, 48 bytes are transmitted every 6 ms. Such a throughput corresponds to 64 Kbps. The capacity of each pair is equal to 48 × 12 bytes every 6 ms, which corresponds to 768 Kbps for 12 DS0 channels per pair and 24 DS0 channels for the two pairs. The set of the bits F, SYN, OH and the stuffing bits used on each pair represent a total of 96 bits. The gross bit rate of a T1-HDSL access system is then equal to 784 Kbps per pair.

5.3.2 E1-HDSL

We have seen in Figure 5.1 that E1-HDSL operates on three pairs. In addition to the G.991.2 ITU-T standard, E1-HDSL has also been standardized by ETSI

---

6. A DS0 channel at 64 Kbps corresponds to 48 × 8 bits every 6 ms.

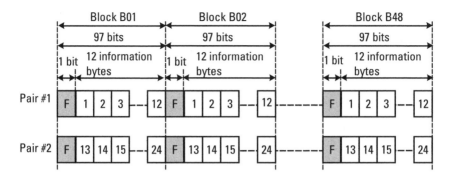

Figure 5.7 Format of the blocks within a two-pair T1-HDSL frame.

under the reference TS-101-135. The frame format adopted for E1-HDSL is strongly inspired by the format initially proposed by Bellcore for T1-HDSL. Globally, an E1-HDSL frame is organized as a T1-HDSL frame. It begins with a 14-bit synchronization preamble followed by four fields of 12 blocks each. An E1-HDSL frame is transmitted in both directions every 6 ms. The ETSI standardization body specifies that the amount of pairs required to offer an E1-HDSL service depends in fact on the range of the access system. Thus, E1-HDSL may operate on one, two, or three pairs. The rule of thumb is that one pair is used over a very short distance under 1 km. Two pairs are used between 1 and 3 km, and three pairs are required between 3 and 4 km. Depending on the adopted configuration, the format of the blocks within an E1-HDSL frame varies. Figure 5.8 illustrates this format in the case of three pairs E1-HDSL.

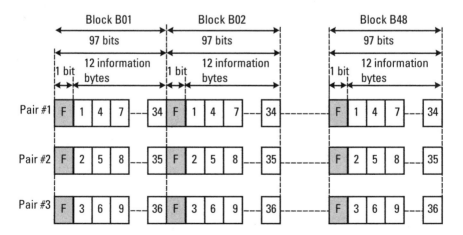

Figure 5.8 Format of the blocks within a three-pair E1-HDSL frame.

On each of the three pairs, an E1-HDSL frame is made of 48 blocks of 97 bits. Each block starts with a single flag bit (F) followed by 12 bytes of information. Like in T1-HDSL, useful data rate on each pair is equal to 768 Kbps, that is 12 DS0 channels. Knowing that the set of the SYN, F, OH, and stuffing bits represents 96 bits per pair every 6 ms, the gross bit rate on each pair is then equal to 784 Kbps.[7] Figure 5.9 describes the format of the blocks in the case of a two-pair E1-HDSL frame.

On each pair, each block of 145 bits starts with a single flag bit (F) followed by 18 information bytes. The useful data rate on each pair is then equal to 1,152 Kbps—that is, 48×18 bytes every 6 ms representing 18 DS0 channels. Like for T1-HDSL, the set of the SYN, F, OH, and stuffing bits represents 96 bits per pair. We have then a gross bit rate per pair equal to 1,168 Kbps—that is, 48×144 bits + 96 overhead bits transmitted every 6 ms. Figure 5.10 describes the format of the blocks in the case of a single-pair E1-HDSL.

On each pair, each block of 289 bits starts with a single flag bit (F) followed by 36 bytes of information. The useful data rate is then equal to 48×36

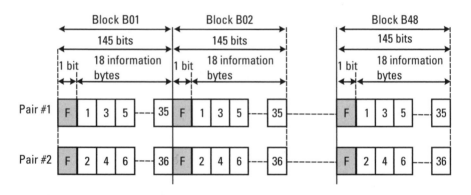

Figure 5.9 Format of the blocks within a two-pair E1-HDSL frame.

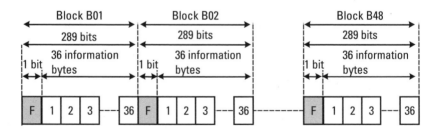

Figure 5.10 Format of the blocks within a single-pair E1-HDSL.

---

7. One has in each 6-ms frame on each pair 12×48 bytes + 96 overhead bits—that is, 4,704 bits representing a data rate of 784 Kbps.

bytes transmitted every 6 ms, that is, 2,304 Kbps. The set of the SYN, F, OH and stuffing bits representing 96 bits, the gross bit rate is in this last configuration equal to 2,320 Kbps. The three-pair version of E1-HDSL is used in the United Kingdom, whereas the two-pair version has been adopted by the other European countries.

5.4   HDSL Variants

Once the various versions of T1/E1-HDSL access systems had been standardized, new variants known as IDSL, SDSL, HDSL2, and SHDSL were proposed. This section describes the main characteristics of each of these four variants.

5.4.1   IDSL

Basic access or primary access to narrowband ISDN operating with gross bit rates equal to 160 Kbps and to 2,048 Kbps, respectively, were standardized about 20 years ago by ITU-T and ETSI. Both these standardization bodies have supported the specification of an adaptation of HDSL in order to facilitate primary ISDN access. Thus, the two-pair version of E1-HDSL has been adopted between CPEs and ISDN switches. As outlined above, this version authorizes 1,168 Kbps in full-duplex on each pair representing a global bit rate of 2,336 Kbps.

5.4.2   SDSL

Regular HDSL remains an expensive and too rigid solution for very small enterprises that cannot afford the installation of a second or third subscriber line. The range of HDSL systems is restrictive because of the fixed data rate of E1/T1 services. The SDSL technique reuses the 2B1Q line code. Whereas single-pair HDSL is limited to about 3 km in terms of range for the provision of T1/E1 services, SDSL aims to provision data rates between ISDN and T1/E1 on a single pair but on larger distance thanks to DLC systems.[8] DLC systems have been widely deployed in North America. Their popularity is rather disparate in Europe. In the context of the American market, Bellcore laboratories have shown the feasibility of symmetrical data rates between 128 Kbps (the equivalent of the two B channels of basic ISDN access) and 1.54 Mbps or 2 Mbps (the equivalent of a T1/E1 LLs) on a single copper pair. As illustrated in Figure 5.11, two SDSL modems are used at the CPE (the HTU-R) and at the remote terminal (the HTU-C), respectively.

---

8. DLC systems are described in Section 2.6.

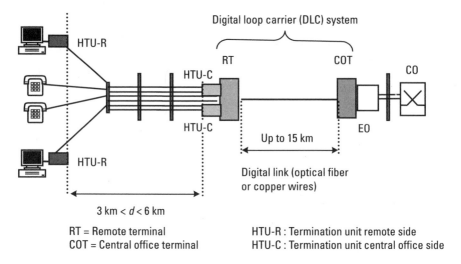

Figure 5.11   SDSL access system in the context of CSA-compatible subscriber lines.

Because it is not standardized, several proprietary versions of SDSL operating at different bit rates have been developed by the vendors for a range between 3 and 6 km.

5.4.3   HDSL2

The previous section determined that SDSL access systems are designed to operate in conjunction with DLC systems. However, DLC systems are not always available in carriers' networks, and SDSL modems are proprietary. Because of these two limitations, ETSI and ANSI standardization bodies have both specified a single-pair version of HDSL named HDSL2 able to provide the same symmetrical bit rate on a single pair but with the same range as regular HDSL. The range of a single-pair HDSL2 access system is at most 3,500m, the subscriber line being CSA-compatible (see Section 2.6). To take into account SNR degradation inherent to single-pair HDSL in comparison to the two- or three-pair versions, autocorrecting codes (Reed-Solomon codes) and trellis-coded modulation[9] are considered for HDSL2. Reed-Solomon encoding and trellis-coded modulation present two limitations, described as follows, in the context of interactive services over copper wires:

- Both these techniques introduce additional bits in the data stream for data protection. Longitudinal and transversal parity bits are inserted in the data stream with Reed-Solomon encoding. Trellis coding assumes a

---

9. The principle of trellis-coded modulation is described in Chapter 6.

redundancy in the alphabet specifying the codewords used in the system. These additional bits imply then a higher bandwidth capacity on the subscriber lines.

- Reed-Solomon encoding relies on a bit or byte interleaving/deinterleaving in the end modems. Trellis-coded modulation assumes more or less complex computations in the receive modem to carry out the maximum likelihood algorithm. In the case of real-time traffic, processing delays inherent to these two techniques must be kept under a certain threshold.

Chapter 2 discussed the fact that 2B1Q is both a line-code and a baseband modulation technique. For HDSL2, more sophisticated modulation techniques may be adopted, including multicarrier modulation with DMT and *carrierless amplitude and phase modulation* (CAP). The DMT and CAP modulations are detailed in the chapter dedicated to ADSL. CAP modulation allows for a significant reduction in the on-line bandwidth. Thus, the 784 Kbps on each pair of a T1-HDSL access system require a 392 kHz bandwidth with 2B1Q code whereas only 175 kHz are necessary with CAP modulation.[10]

Figure 5.12 compares the bandwidths required by three different T1 access systems: 2B1Q-HDSL, CAP-HDSL, and AMI LL. One notices that CAP modulation does not use the lower part of the spectrum under 10 kHz. This unused bandwidth enables the simultaneous provision of T1 services with an

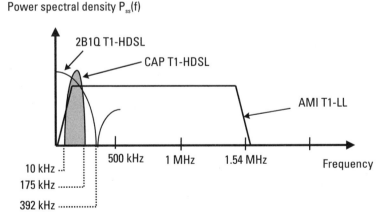

Figure 5.12   2B1Q-T1-HDSL, CAP-T1-HDSL, and T1-LL frequency spectrums.

---

10. The great capacity of CAP modulation to reduce the on-line bandwidth is detailed in Chapter 6.

analog voice channel in the [0,4 kHz] band. One of the interesting properties of
CAP modulation is its high spectral efficiency. Thus, CAP T1-HDSL needs
42% and about 11% of the bandwidths required by 2B1Q-T1-HDSL and
AMI-T1 LL, respectively. Some vendors and carriers exploit the benefits of
HDSL2 in the context of two-pair transmission in order to double the performance of the system. Thus, two-pair HDSL2 can achieve data rates around 4
Mbps on a range a 4 km. A group of DSL modems manufacturers and vendors
known as *overlapped PAM transmission with interlocking spectra* (OPTIS) is promoting a new line code named *trellis-coded PAM* (TC-PAM) for HDSL2. This
code is in fact at the basis of another HDSL variant known as SHDSL.

5.4.4   SHDSL

The SHDSL is the most recent variant of HDSL. An SHDSL access system
operates on one or two subscriber lines with a 0.4-mm diameter (or 26 gauge in
the United States). SHDSL was standardized by ITU-T in February 2001 in the
G.991.2 recommendation. ETSI also standardized SHDSL under the reference
ETSI TS-101.524 in June 2000. Like the various HDSL variants, the SHDSL
market is focused on business applications characterized by a mix of symmetrical
(e.g., for multiple voice line delivery) and asymmetrical traffic (e.g., Internet
access).

In the case of a single-pair configuration, SHDSL supports symmetric useful data rates from 192 Kbps to 2,304 Kbps. In the case of the dual-pair configuration, SHDSL enables data rates between 384 Kbps and 4,624 Kbps. Among
the HDSL variants, SHDSL and SDSL are the only ones to be rate-adaptive.
The rate granularity offered by SHDSL systems is equal to 8 Kbps. The expression "fractional T1/E1" service is sometimes used in this context. For the highest bit rate, the range of single pair SHDSL is limited to 2 km (about 6.6 kft) or
4 km in the case of the dual-pair version. Carriers may extend the range of an
SHDSL system up to 7,800m by means of repeaters. Like the other variants of
HDSL presented above, SHDSL enables full-duplex transmission in the same
frequency band thanks to echo cancellation. SHDSL relies on a PAM. The
SHDSL PAM modulation uses 3 bits per symbol. A voltage level chosen among
16 ($2^4 = 16$) is associated with each symbol. This means that SHDSL coding
considers twice the number of voltage levels required by the set of bit combinations per symbol. Such a redundancy is used in order to compensate for SNR
degradation introduced by the greater proximity between the on-line voltage
levels. Two techniques known as trellis coding and Viterbi algorithm applied at
the transmitter and at the receiver, respectively, exploit this redundancy of the
code for a higher reliability of the received data after decoding. We briefly
describe below the basic principle of trellis precoding and of the Viterbi algorithm, a deeper presentation being given in Chapter 6.

5.4.4.1 SHDSL Line Code: TC-PAM

SHDSL is based on a TC-PAM. Trellis coding[11] consists of introducing a redundancy in the line code in associating to each input block of n bits a certain amount of output symbols of m bits with $m > n$. In the context of SHDSL access systems, each of these symbols corresponds to one of the 16 possible voltage levels. In other terms, an SHDSL encoder associates with each input block of 3 bits two possible output symbols. The choice of a symbol for a given input block depends on the value of the n bits of the current kth input block but also on the recent past of the encoder. Indeed, the choice of the kth output symbol at the transmitter depends on the value of the x previous generated symbols—that is, on the values of symbols of order $k - 1$, $k - 2$, …, $k - x$. The logic behind the choice between the two possible values for the kth symbol relies on a trellis diagram. As it is described by means of several examples in Chapter 6, the Viterbi algorithm used at the receiver enables the alleviation of ambiguity at decoding in case of noisy channel thanks to this trellis diagram. The Viterbi algorithm consists of determining at the receiver the most probable sequence among the possible values for the $x + 1 \times n$ bits received bits. The gain inherent to the Viterbi algorithm in the context of SHDSL is equal to about 5 dB. Unlike regular HDSL modems, SHDSL modems cannot use DFEs that are incompatible with TC-PAM. A Tomlinson-Harashima precoder (see Chapter 6) is used in order to compensate for the attenuation distortions induced by the line. Such a coder operates by anticipation at the transmitter side in introducing a channel-dependent predistortion in the transmitted signal.

5.4.4.2 Frequency Spectrum of SHDSL Access Systems

The fact that input blocks are made of 3 bits instead of 2 like in HDSL2, SHDSL enables the division by 3 of the required bandwidth in comparison to traditional AMI LLs. Figure 5.13 compares spectral occupancies of TC-PAM SHDSL and of 2B1Q SDSL for T1 service provision [3] over two pairs. In this context, the bit rate on each pair is equal to 768 Kbps in full-duplex. The 2B1Q line code allows for the division by two of the required bandwidth for SDSL (i.e., 384 kHz instead of 768 kHz). In comparison, TC-PAM allows for the division by three of the frequency spectrum (i.e., 256 kHz).

5.4.4.3 SHDSL Frame

Figure 5.14 illustrates the format of SHDSL frames. We have mentioned above that SHDSL is rate-adaptive. The format of SHDSL frames depends on the required data rate, between 192 Kbps and 2,304 Kbps with a granularity

---

11. In more sophisticated transmission systems like some of the wireless access systems described in Chapter 11, trellis coding is carried out before an additional line encoding. This is why the term trellis precoding is also frequently used in the literature.

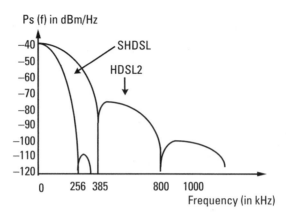

Figure 5.13 Power spectral density of SHDSL and of SDSL. (*From:* [3]. © 2000 Adtran, Inc. Reprinted with permission.)

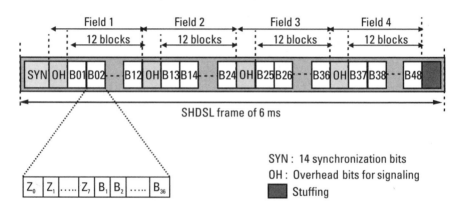

Figure 5.14 Format of SHDSL frames.

between these two bounds of 8 Kbps. Whatever the data rate, the duration of an SHDSL frame is always equal to 6 ms.

Globally, an SHDSL frame is organized as a T1-HDSL frame (see Figure 5.6). It starts with a 14-bit synchronization preamble followed by 48 blocks B_{01} to B_{48}. Every 12 blocks, a 2-bit overhead field (OH) is inserted. The detailed configuration of a block is given in Figure 5.14. It is made of at least 3 and of at most 36 B-channels at 64 Kbps. If n stands for the amount of B-channels in the system, one has then $3 < n < 36$. In each block, eight auxiliary channels referred as Z-channels and index Z_0 to Z_7 represent a bit rate of 8 Kbps each.[12] The other bits in the frame referred as OH bits represent a total of 32 bits per frame. After

---

12. For each Z-channel, 48 bits are transmitted every 6 ms, representing an 8-Kbps bit rate.

the last block B$_{48}$, two spare time bits are used to manage clock dispersion between the end modems.[13] The minimal payload of an SHDSL frame is then three B-channels—that is, 192 Kbps—the maximum payload being 36 B-channels—that is, 2,304 Kbps.

5.5 Conclusion and Perspectives

Among the four variants of HDSL presented in this chapter (IDSL, SDSL, HDSL2, SHDSL), SHDSL is the most recent and the most advanced. IDSL and HDSL2 operate at fixed data rates whereas SDSL and SHDSL are rate-adaptive. Apart from IDSL, all these techniques are applicable to single-pair or two-pair access configurations. The market of IDSL systems should remain disparate, depending on the deployment of ISDN services in the different countries. The main drawbacks of SDSL are the absence of standardization and the strong dependence on DLCs, which are deployed only in certain countries. HDSL2 can be viewed as a transient version of HDSL on which is based SHDSL. One of the advantages of SHDSL is its electric compatibility with ADSL systems, which are described in Chapter 6. ADSL systems operate in two separate bands: [20 kHz, 140 kHz] for upstream traffic and [150 kHz, 1,100 kHz] for downstream traffic. NEXT crosstalk between an SHDSL line sharing the same bundle with an ADSL system remains limited. Indeed, the overlap between an upstream SHDSL band and a downstream ADSL band is limited to the [150 kHz, 256 kHz]. SHDSL can be used either for data transfers or for the transport of multiple digital voice channels. Up to 36 voice channels are achievable on a single- or a dual-pair SHDSL configuration, depending on the range of the system and on the quality of the subscriber lines.

References

[1]   Lechleider, J. W., "High Bit Rate Digital Subscriber Lines: A Review of HDSL Progress," *IEEE Journal on Selected Areas in Communications,* Vol. 9, No. 6, 1991, pp. 769–784.

[2]   Lechleider, J. W., "Line Codes for Digital Subscriber Lines," *IEEE Communications Magazine,* September 1989, pp. 25–32,

[3]   Adtran Inc., *Get Ready for G.shdsl,* Adtran Corporation, http://www.adtran.com, accessed 2001.

---

13. Clock dispersion due to propagation and to the gradient of temperature between the modems is known as *plesiochronism.*

Selected Bibliography

Koen Van de Vel, "Avoiding xDSL Provisioning Pitfalls: How to Mass-Market and Roll Out xDSL Services," *Proc. of the 40th European Telecommunications Congress,* Barcelona, Spain, August 22–25, 2001.

Paradyne Corporation, *The DSL Source Book: Plain Answers About Digital Subscriber Line Opportunities,* 1997.

Townsend, R. L., and J. J. Werner, "Using Technology to Bring ATM to the Desktop," *AT&T Technical Journal,* July/August 1995, pp. 25–37.

6

ADSL Physical Layer

6.1 Introduction

The HDSL technique described in Chapter 5 enables symmetrical bit rates over twisted copper pairs. We have outlined that depending on the considered variant of HDSL, one, two, or even three copper pairs could be necessary for HDSL access. The ADSL technique was developed in the early 1990s. It is an extension of HDSL designed initially for the provision of asymmetrical bit rates over a single subscriber line. Like HDSL, ADSL requires the use of two modems located at the CPE and the EO. The main reason for this asymmetry was justified by the objective of ADSL to provide VoD services to residential users. In fact, it appeared in the first half of the 1990s that such services were not technically and economically mature because of four major constraints:

- The first constraint was due to the necessity to develop costly centralized video servers. Such servers were supposed to be installed in core networks, connected either to an IP router or to an ATM switch.

- The second constraint was due to the unsuitability of IP networks to manage simultaneously multiple real-time video streams from the central video server to multiple access networks.

- The third constraint was related to the poor quality of MPEG-1 digital video available at that time.

- Last, the commercial success of VoD was strongly penalized by the high cost of set-top boxes in charge of providing interactivity to the end users.

For all these reasons, ADSL objectives were reoriented at the end of the 1990s toward the simultaneous provision of high-speed Internet access and analog telephony [1] to end users. Today, several advances in Internet protocols, IP router architecture, video coding, and video servers allow us to reconsider the initial target of ADSL services.

Figure 6.1 illustrates the typical configuration of an ADSL access system. The two ADSL modems located at the EO and at the CPE are called *ADSL transmission unit* (ATU) at the *central office side* (ATU-C) and *ATU at the remote side* (ATU-R), respectively. Two filters called POTS splitters are used at each end of the subscriber line to separate or to combine POTS signals with ADSL signals. We showed in Chapter 2 that analog voice signals could be truncated to a limited [0, 4 kHz] bandwidth before being digitized at the EO. Upstream and downstream data signals corresponding to high-speed access to the Internet or to VoD signals are assumed to be transmitted at upper frequencies. A POTS splitter cannot be assimilated to a simple bandpass filter. It is in fact simultaneously a [0, 4 kHz] lowpass filter for POTS services and a bandpass filter in the [20 kHz, 1,100 kHz] for bidirectional data services.

In the very first versions of ADSL modems, POTS splitters were directly implemented within the ATU-C and ATU-R. In Chapter 3, we determined that two different administrative entities could offer to the same end user a POTS service and an interactive data service, respectively. We saw that such a separation of the responsibilities in the local loop refers typically to partial unbundling (e.g., where the ILEC provides POTS while at the same time a CLEC provides Internet services). A third party can also be in charge of the wired infrastructure in the local loop. In such a context, the provider of POTS asks to be able either to be directly the provider of POTS splitters, or at least to have the possibility to check the conformance of these equipment. Indeed,

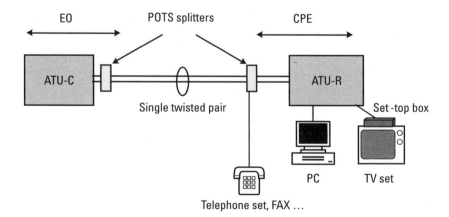

Figure 6.1 Configuration of an ADSL access.

POTS splitters condition directly the quality of the audio signal perceived by the receiver. Let us remark that a visit of a representative of the ILEC to check in situ this conformance is at a high cost for the customer. This representative cannot proceed to the test of a POTS splitter if this one is directly implemented within the ATU-R. Let us also note that POTS splitters inserted inside ADSL modems were based on active devices (i.e., transistors and operational amplifiers). Such devices need to be powered locally to be operational. If electrical power is disabled at the CPE, the ATU-R and by consequence its POTS are out of order. In other terms, a failure in the electrical power supply of the customer means a loss of availability of the phone line.[1] For these various reasons, POTS splitters have been physically separated from ADSL modems as illustrated in Figure 6.1.

Two approaches are possible in order to provide bidirectional communications on a subscriber line: FDM and EC. These two techniques correspond to the two successive versions of ADSL modems. The former dates from the early 1990s; the latter has been in use since the mid-1990s.

6.1.1 Frequency Multiplexing

Frequency multiplexing illustrated by Figure 6.2 consists of transporting upstream data flows from the ATU-R to the ATU-C in the [25, 140 kHz] and the downstream data flow from the ATU-C to the ATU-R in the [150 kHz, 1,104 kHz]. We have represented on Figure 6.2 a carrier situated between the analog band and the digital band (upstream arrow). This frequency is used by some

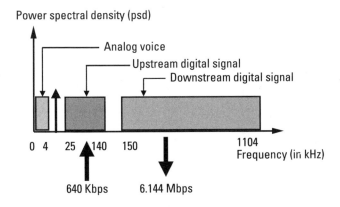

Figure 6.2 Bandwidth utilization in FDM-ADSL.

---

1. Let us recall that a telephone set is powered remotely by means of a continuous 48-V voltage generated at the EO. In the case of narrowband ISDN access, a single telephone-set installed at the CPE may be powered remotely by the CO.

carriers to propagate from the CO to the telephone set of the customer periodic pulses corresponding to the real-time charge of the phone call.

Depending on the adopted modulation technique and on the line characteristics, target upstream data rates between 16 Kbps and 640 Kbps and downstream data rates between 1.5 Mbps and 6.1 Mbps are considered for FDM-ADSL.

6.1.2 EC

The principle of EC, which permits the transmission of both an upstream flow and a downstream flow in the same frequency band, is described in Chapter 5. We have underlined that an echo canceler requires a modem initialization in order to test the weight of SNEXT in the band where upstream and downstream flows overlap. We outlined the fact that the efficiency of an echo canceler is directly related to the speed of its electronics. The faster the electronics, the better the echo cancellation. According to the state of technology, echo cancellation may be applied at most on a [25 kHz, 130 kHz] band.[2] Figure 6.3 illustrates bandwidth utilization in the case of EC-ADSL. One notices that downstream traffic occupies the [25 kHz, 1104 kHz] band with an overlap on the [25 kHz, 130 kHz] band used by upstream traffic.

FDM is simpler and less expensive to implement than EC but it uses less efficiently the available bandwidth on the medium. It prevents NEXT between adjacent pairs belonging to a same binder cable. One notices from Figure 6.3

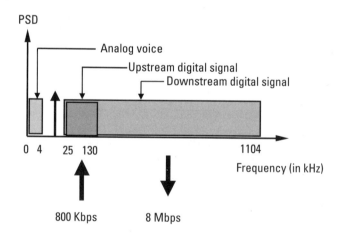

Figure 6.3 Bandwidth utilization in EC-ADSL.

---

2. The upper bound of 130 kHz for EC is indicative. This value may slightly fluctuate depending on the vendors.

that EC authorizes higher throughputs for downstream traffic than FDM. Target bit rates of 800 Kbps and 8 Mbps for upstream and downstream traffic respectively are considered for EC-ADSL. Such bit rates strongly depend on the line characteristics and refer in fact to the most favorable situation. The main limitation of EC is that subscriber lines are in this case subject to NEXT in the [20 kHz, 13 kHz] band between adjacent pairs belonging to a same binder cable. Today, the great majority of ADSL modem vendors have adopted EC.

6.1.3 ADSL Forum and Standardization

In 1994, several electronic devices manufacturers, modems vendors, carriers, and service providers formed the ADSL Forum [2]. The main objective of the forum is to find a consensus on recommendations guaranteeing interoperability between the modems of different vendors and between the networks of different carriers. Such an interoperability concerns the following:

- System architectures;

- Protocols;

- Interfaces for the main ADSL services such as high-speed Internet access, VPNs (see Chapter 1), voice over DSL, and VoD.

Currently, the ADSL Forum counts about 300 members. The ADSL technique has been standardized by ANSI [3, 4] and by ITU-T [5] under the references ANSI T1.413 and G.992.1, respectively, and by ETSI [6]. We have mentioned above the strong dependency between the target bit rates of FDM-ADSL and EC-ADSL with subscriber line characteristics. As an indication, Table 6.1 gives the order of a few achievable downstream data rates for typical ranges between the CPE and the CO [2].

The G.992.1 standard defines three classes of ADSL modems: 2M-1, 2M-2, and 2M-3 by order of increasing bit rate. Table 6.2 summarizes the

Table 6.1
Examples of Achievable Throughputs for Different Ranges

Throughput	Range
1.54 Mbps (T1)	5,400m–17,716 ft
2.048 Mbps (E1)	4,800m–15,748 ft
6.312 Mbps (DS2)	3,600m–11,811 ft
8.448 Mbps (E2)	2,700m–8,858 ft

Table 6.2
The Three Classes of ADSL Modem Specified in the G.992.1 Standard

Class of Modem	2M-1	2M-2	2M-3
Downstream rate	6.144 Mbps	4.096 Mps	2.048 Mps
Upstream rate	64 Kbps	64 Kbps	16 Kbps
Diameter: 0.5 mm	5,500m–18,044 ft	Not specified	3,700m–12139 ft
Diameter: 0.4 mm	4,600m–15,092 ft	Not specified	2,700m–8,858 ft

performance of each class for different subscriber lines characteristics such as the range and the copper wire diameter [7].

6.1.4   ADSL Market

To illustrate the progress of the ADSL market, statistics were published in 2001 by the Point Topic consulting cabinet (United States). Figure 6.4 illustrates the ADSL market penetration at the end of 2000. The number of ADSL subscribers expressed in thousands is given for different countries.[3] Due to a governmental

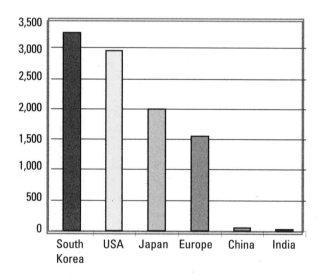

Figure 6.4   Number of ADSL subscribers (in thousands) at the end of 2000.[4]

---

3.   Western European countries have been considered globally in these statistics.

4.   The number of subscribers mentioned in this figure covers in fact HDSL and ADSL users, the latter being the majority.

initiative in favor of a large deployment of ADSL services, South Korea was the world leader with 3.25 million subscribers. At this same date, there were 2.953 million subscribers in the United States, 2 million in Japan, 1.55 million in Europe, 50,000 in China, and 15,000 in India. Globally, there were 9.825 million ADSL subscribers in the world at the end of 2000. About 25 million subscribers on the world market were expected for the end of 2001.

Concerning the European market, it is interesting to notice that ADSL is deployed more rapidly in Germany than in the other European countries because of the availability of a great number of narrowband ISDN access lines. Indeed, due to a governmental will in the 1980s to facilitate the access of SMEs to data networks, many ISDN access lines have been installed by Deutsche Telekom in the former Western Germany. These lines are today an excellent environment for ADSL. Whereas standard ADSL over POTS uses the [25 kHz, 1,104 kHz] band, ADSL over ISDN access lines uses the [138 kHz, 1,104 kHz] band. Because of the limited range and the better quality of ISDN access lines mentioned in Chapter 2, ADSL over ISDN local loops enables the same upperbound bit rates as ADSL over POTS. At the end of 2000, 500,000 subscribers were equipped by Deutsche Telekom with ADSL access. This figure increased to 2.6 million at the end of 2001. In comparison, 177,000 users were equipped by France Telecom at the end of 2000 and about 700,000 at the end of 2001. In Japan, the number of xDSL subscribers has increased from 186,000 to 1 million from the end of 2000 to the end of 2001.

6.2 Reference Model

The ANSI T1.413 standard dating from 1995 [3] defines the different reference points specifying the various interfaces of an ADSL access system. The TM6 committee of ETSI standardization body has been inspired by this ANSI document to propose an adaptation for the European market in 1996 [6]. We describe in the two following sections ADSL reference points and ADSL protocol architecture.

6.2.1 Reference Points

Standardization documents [3, 6] precise line test procedures that can be carried out at the various reference points. Figure 6.5 illustrates the localization of these reference points (or interfaces) for which are specified the physical junctions and the exchange modes between the various pieces of equipment used in an ADSL access system.

The V_c interface is located between the ATU-C and the nearest packet switch. It specifies in fact the connection of the ATU-C with different kinds of

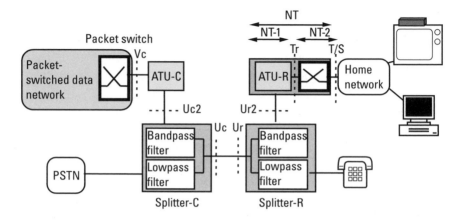

Figure 6.5 Reference points of an ADSL access system.

equipment, including an ATM switch, an IP router, an X.25 switch, and a frame relay switch. In the case of simultaneous access to several of these pieces of equipment, the interface enables the multiplexing of different messages on the same physical link. The lowpass filter output is in charge of the junction between the subscriber line and the EO or the CO[5] (this side of the interface is referenced by the letter "C") and the telephone set of the end user (this side of the interface is referenced by the letter "R"). The U_{c2} and U_{r2} interfaces separate the POTS splitters referenced "splitter-C" and "splitter-R" from the ATU-C and the ATU-R, respectively. The U_c and U_r interfaces separate the subscriber line from the splitter-C and the splitter-R, respectively. In the case of ADSL-lite modems (the ADSL-lite technique is presented in Chapter 7), the bandpass filter of the splitter-C and of the splitter-R may be integrated within the ATU-C and ATU-R, respectively. In such a context, the U_{c2} and U_{r2} interfaces are merged with the U_c and U_r interfaces, respectively.

On the subscriber side, the ATU-R may be coupled to a private LEX. In that case, we recognize a configuration comparable to the one adopted for ISDN access (see Chapter 2). The ATU-R and the private local switch correspond to the *network termination of type 1* (NT-1) and to the *network termination of type 2* (NT-2), respectively. The T/S interface separates the NT-2 from the home network. The analog telephone set represented on the right-hand side of Figure 6.5 is remotely powered by the CO. This is not the case for the ATU-R, which is powered locally by the existing electrical wiring. The T_r interface separates the ATU-R from the NT-2. In the 1990s, two options were considered for the home network:

---

5. We saw in Chapter 2 that traffic concentration occurs in the same building as the CO in urban areas with high population density.

- A low-cost configuration based on a 10-baseT Ethernet hub allows an ADSL link to be shared between several PCs. Local communications are of course possible via this Ethernet hub, which serves as an NT-2. Dedicated wiring has nevertheless to be installed between each PC and a 10-baseT interface of the NT-2.

- A more expensive configuration consisting of an ATM-25 ATM switch may be used to share the ADSL access between different machines (i.e., computers or printers). In practice, the great majority of the operators and of the vendors propose actually the 10-baseT option.

Currently, several alternatives are considered for the NT-2 and the home network. These alternatives assume either the use of existing in-house wiring, or the installation of newly structured in-house wiring, or a wireless home network. In the case of the in-house wiring, two suboptions are possible: the use of the installed UTP copper wires or the use of the electrical in-house wiring. The former of these two suboptions is supported by the Home Phoneline Networking Alliance (HomePNA) consortium. The latter is supported by the HomePlug Powerline Alliance. Five suboptions are also considered for wireless NT-2: the IEEE 802.11 wireless LAN (also known as wireless Ethernet); the HomeRF approach, which is also a wireless LAN approach but with a more limited range than IEEE 802.11; Bluetooth, which can be compared to HomeRF; HiperLAN, which is a wireless LAN comparable to IEEE 802.11 in terms of range; and ultrawideband [8]. Today, the T/S and T_r interfaces are specified only for the most mature of all these options—that is, IEEE 802.11 and Bluetooth. Chapter 8 details the concept of home networking.

6.2.2 Protocol Architecture

Traditionally, a modem only refers to the physical layer. Different standardization study groups have decided to divide into two sublayers the physical layer of ADSL modems: the *physical medium–dependent* (PMD) and the *transmission convergence* (TC) sublayers.

6.2.2.1 PMD Sublayer

The PMD sublayer specification depends on the type of copper pair used for the subscriber line. According to the electrical characteristics of this line (i.e., diameter, twisted period, and impedance impairments), a certain type of line code is adopted. ADSL transmission is baseband-oriented. That means that a direct dependency exists between the adopted codewords and the shape of the electrical signal on the line. This shape is defined in terms of amplitude and phase time fluctuations of the electrical signal. This is the reason why in ADSL coding and modulation both correspond to the same operation.[6] As underlined in Section 6.11, ADSL access requires a line initialization carried out by the ATU-R and the ATU-C. This phase of test is described in the PMD specifications. The PMD sublayer is in charge of clock recovery at the bit level at the receiver. As outlined in Chapter 4, several electromagnetic perturbations may occur along the line and introduce bursts of erroneous bits after demodulation at the receiver. The longer the bursts of errors, the more difficult the erroneous bit detection and correction. This is the reason why, like in radio communications, one specifies in the PMD sublayer the use of an interleaving/deinterleaving technique at the transmitter and at the receiver, respectively. By interleaving the bytes or bits before modulation at the transmitter, one spreads out erroneous bits after deinterleaving at the receiver. The PMD sublayer is also in charge of the supervision of the POTS splitters (splitter-C and splitter-R).

6.2.2.2   TC Sublayer

The TC-sublayer specifies the on-line bit rate. Three types of data formats are possible between the ATU-C and the ATU-R: E1/T1 frames, ATM cells, and IP packets. In current implementations, modem vendors have all adopted the ATM cell format at the TC sublayer. ATM cell delineation is carried out at the TC sublayer. On top of ATM, asynchronous IP packets may be transported via AAL5 or E1/T1 circuit emulations may be provided via AAL1. Multiple logical connections may be multiplexed on the same ADSL link. The detailed protocol stack corresponding to the ATM-based ADSL implementation is detailed in Chapter 9. In order to prevent retransmissions in case of errors, FEC based on Reed-Solomon encoding may be adopted. Reed-Solomon encoding, which assumes the use of redundant bits and its proper interleaving technique, is described in Section 6.7. Two institutions, the *Ethernet in the First Mile Alliance* (EFMA) [9] and the IEEE, currently promote the emergence of a new standard interface for ADSL modems. Instead of being ATM-oriented, this new interface is Ethernet-based [10]. The main idea is to replace the ATM-based transmission between the ATU-C and the ATU-R and ATM traffic concentration at the EO by Ethernet-based transmission and Ethernet traffic concentration, respectively. This new approach known as *Ethernet in the first mile over Copper*[7] (EFMC) is detailed in Chapter 12.

6.3   NEXT Predominance Under High Frequencies

Chapters 2 and 4 defined the different types of crosstalks appearing in the local loop: NEXT, FEXT, and SNEXT. Let us consider as an example the case of a

---

6. Section 6.4 is dedicated to the description of the main modulation techniques adopted for ADSL. In the literature, the same operations may be referred to as line coding.

7. The expression *Ethernet DSL* (EDSL) is also used.

binder cable made of 50 UTP-3 copper pairs[8] used between CPEs and an EO. All the pairs of this binder cable are supposed to have the same length d equal to D. We assume that the channel frequency response $H_0(f, d)$ is given by[9]:

$$H_0(f,d) = e^{-(\alpha + j\cdot\beta)\cdot d} \qquad (6.1)$$

where α and β stand for the attenuation in decibels per kilometer and the phase distortion in rad/s.km. Let us assume an idealistic attenuation proportional to the square root of f (see Chapter 4) and that all disturbers have the same power spectral density PSD(f). It has been shown in [11] that under such hypothesis, NEXT and FEXT signals can be modeled by (6.2) and (6.3), respectively:

$$PSD_{NEXT}(f) = PSD(f)\cdot\left(\frac{N}{49}\right)^{0.6}\cdot A\cdot f^{3/2} \qquad (6.2)$$

where A is constant proper to the line characteristics (diameter, twisted period, and length) and N corresponds to the number of NEXT interferers,

$$PSD_{FEXT}(f) = PSD(f)\cdot\|H_0(f,d)\|^2\cdot\left(\frac{N}{49}\right)^{0.6}\cdot B\cdot D\cdot f^2 \qquad (6.3)$$

where B is another constant proper to the line characteristics (diameter and twisted period). Figure 6.6 shows the average NEXT and FEXT coupling

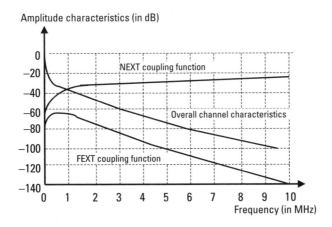

Figure 6.6 NEXT/FEXT coupling functions and overall channel characteristic. (*From:* [11]. © 2002 IEEE. Reprinted with permission.)

---

8. Let us recall that UTP-3 corresponds to a 0.4-mm diameter.

9. The parameters f and d stand for frequency and distance.

functions and the amplitude characteristic of the of the overall channel transfer function when $D = 1$ km.

The same types of curves could be obtained for UTP-5 cables, the point of intersection between the NEXT coupling function and the overall channel characteristic being in this case shifted toward higher frequencies. For a given range D of the subscriber line, the higher f, the lower the power of the reference signal and the higher the power of NEXT detected at the receiver. In other terms, the higher f, the lower the SNR and the lower the Shannon's capacity. The results plotted in Figure 6.6 will be used in Section 6.4.2 to explain the principle of CAP used in ADSL modems. On the whole spectrum up to 10 MHz in our example—that is, at frequencies much larger than the 1.104 MHz specified for ADSL—one notices that FEXT is less penalizing than NEXT. The higher f, the more negligible the impact of FEXT in comparison to the impact of NEXT on SNR.

6.4 Modulation Techniques

Section 4.4 lists the multiple factors that may degrade the capacity of a copper pair in the local loop, including impedance impairments, crosstalk, impulse noise, and thermal noise. Above 1 MHz, real measurements reveal that for lines longer than 4 km, α may reach 90 dB. Several coding and modulation techniques have been developed for ADSL. Two of them are today implemented in the modems of the market: CAP and DMT[10]. The majority of CAP-ADSL modems adopt FDM, whereas the majority of DMT-ADSL modems use EC. Nevertheless, both CAP and DMT may operate with FDM or EC. CAP modulation is inspired from the QAM already used in voiceband modems (e.g., V.34 and V.34bis).

6.4.1 QAM

Before describing the principle of QAM, let us recall the constraints imposed to the equalizer used at the receiver to minimize error probability due to ISI in the context of baseband transmission. These constraints are summarized in the Nyquist criteria that must be satisfied by the filter equivalent to the transmission channel coupled to the equalizer.

6.4.1.1 Nyquist Criteria

Let us consider a data source at the input of a transmission line. This source generates a suite of bits $d(t)$ with a period of T seconds. Let c_n be the nth generated bit such as $c_n = d(t = nT)$. Baseband transmission consists in associating, either

---

10. These two techniques have already been cited in Chapter 5 dedicated to HDSL.

to each bit c_n or to a set of m successive bits $(c_n, c_{n+1}, \ldots c_{n+m-1})$ a stable state of the on-line signal according to a given coding table. This stable state may correspond to a fixed voltage, to a fixed current intensity, or to a fixed amplitude A and phase Φ of an alternative voltage or current of the form $Ae^{+j(\omega t + \Phi)}$. As an example, NRZ baseband encoding (or modulation) associates a voltage $-V$ to c_n $= 1$ and a voltage $+V$ when $c_n = 0$. If we assimilate the suite of information bits to a suite of Dirac impulses $\delta(t)$ with a ± 1 amplitude spaced by T seconds,[11] one has:

$$c(t) = \sum_{k=0}^{+\infty} c_k \cdot \delta(t - k \cdot T) \tag{6.4}$$

This suite of Dirac impulses is inserted at the input of a transmission line. One assumes that subscriber lines' characteristics are stationary.[12] A transmission system may be modeled by three filters in cascade as is represented in Figure 6.7.

In Figure 6.7, $R(f)$, $H(f)$, and $Y(f)$ correspond to the frequency response of the transmitter, of the line of transmission, and of the receiver modem. Let $r(t)$, $h(t)$, and $y(t)$ be the impulse response associated to $R(f)$, $H(f)$, and $Y(f)$[13], respectively. In practice, one may assimilate the set of these three filters to a low-pass filter with frequency response $Q(f)$ such that:

$$Q(f) = R(f) \cdot H(f) \cdot Y(f) \tag{6.5}$$

with $Q(f) = 1$ when $f\ [-B, +B]$ and $Q(f) = 0$ elsewhere. The impulse response of such a filter is given by its inverse Fourier transform:

$$q(t) = \int_{-\infty}^{+\infty} e^{+2i\pi ft} \cdot Q(f) \cdot df = \int_{-B}^{+B} e^{+2i\pi ft} \cdot df = \frac{\sin(\pi \cdot 2B \cdot t)}{\pi \cdot t} \tag{6.6}$$

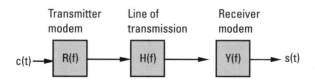

Figure 6.7 Model of a baseband line of transmission.

---

11. T is the bit duration.

12. The frequency response (including attenuation and phase distortion) of a stationary line of transmission does not fluctuate with time.

13. $R(f)$, $H(f)$, and $Y(f)$ are the Fourier transforms of $r(t)$, $h(t)$, and $y(t)$, respectively.

Figure 6.8 illustrates the shape of $q(t)$ and of $Q(f)$, $Q(f)$ being the *Fourier transform* (FT) of $q(t)$. The shape of $q(t)$ is known as a "sinc" function.

The impulse response $q(t)$ of this lowpass filter is given by $s(t) = q(t) \otimes \delta(t)$ = $q(t)$. As an example, let us consider the suite of generated bits $c_1 = 1$, $c_2 = -1$, $c_3 = -1$, and $c_4 = 1$ corresponding to the information sequence "1001." One assumes that the sum of propagation delay along the transmission line with processing delay in the transmitter and receiver modems is equal to Δ seconds. The response $s(t)$ to this input $c(t)$ is given by the convolution of $q(t)$ and $c(t)$:

$$s(t) = q(t) \otimes c(t) = \int_{-\infty}^{+\infty} c(\tau) \cdot q(t - \tau) \cdot d\tau = \frac{\sin(\pi \cdot 2B \cdot t)}{\pi \cdot t} \tag{6.7}$$

$$\otimes \sum_{n=1}^{4} c_n \cdot \delta(t - n \cdot T)$$

Knowing that:

$$q(t) \otimes \delta(t - n \cdot T) = q(t - n \cdot T) \tag{6.8}$$

one has for $s(t)$:

$$s(t) = \sum_{n=1}^{4} c_n \cdot \frac{\sin(\pi \cdot 2B \cdot (t - nT - \Delta))}{\pi \cdot (t - nT - \Delta)} = \sum_{n=1}^{4} c_n \cdot \sin c(t - nT - \Delta) \tag{6.9}$$

Figure 6.9 illustrates the shape of $s(t)$. For the sake of simplicity, one has represented in Figure 6.9 only the primary lobes of the four "sinc" functions included in the above expression. For each "sinc" function, one estimates that the two secondary lobes are negligible in terms of amplitude in comparison to the primary central lobe.

The receiver modem proceeds after equalization to signal sampling in order to determine the value of the received bits. One has represented in Figure 6.9 the

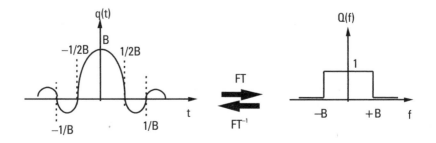

Figure 6.8 Time and frequency characteristics of a lowpass filter.

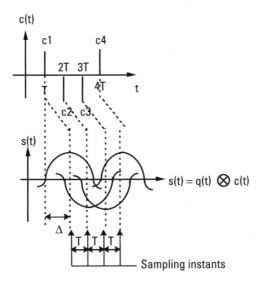

Figure 6.9 Example of sampling at the receiver.

ideal sampling instants of $s(t)$. At such instants, the observed amplitude of $s(t)$ is the summation of the four "sinc" functions associated with the four transmitted bits. For instance, we see that at sampling instant $t = 4T + \Delta$, the amplitude of $s(t)$ is low due to the phase difference between the "sinc" functions associated with the third and fourth bits, respectively. In this situation, the voltage level of $s(t)$ being low compared to background noise $b(t)$, an erroneous bit may be introduced after sampling. In order to prevent such errors, the value of $q(t)$ must be such that at the sampling instant of the kth bit, the summation of the amplitudes of the "sinc" functions associated with the other bits (the bits before or after this kth bit) is null. This constraint can be summarized by the following:

$$\begin{cases} q(0) = 1 \\ q(kT) = 0, \forall k, k \neq 0 \end{cases} \tag{6.10}$$

Equation (6.10) corresponds to the first Nyquist criteria that guarantees the absence of ISI. In the frequency domain, this criteria becomes:

$$\sum_k Q\left(f - \frac{k}{T}\right) = T \tag{6.11}$$

Ideally, (6.11) is satisfied for a lowpass filter such that $Q(f) = T$ for f $[-1/2T, +1/2T]$ and $Q(f) = 0$ elsewhere. In practice, it is not possible to design such ideal lowpass filters for which attenuation is infinite for discrete frequencies. It is nevertheless possible to implement nonideal lowpass filters that respect the Nyquist criteria. Such filters, known as *raised cosine filters*, are such that:

$$Q(f) + Q\left(\frac{1}{T} - f\right) = T \qquad (6.12)$$

This means that for real lowpass filters, the first Nyquist criteria is satisfied if $Q(f)$ presents an odd symmetry around the point with coordinates $(1/2\,T, T/2)$, as illustrated in Figure 6.10.

In order to evaluate the performance of a lowpass filter, one defines the roll-off factor β of a filter at a cutoff frequency f_c such that:

$$\beta = \frac{2 \cdot f_c - 1/T}{1/T}, \text{with } 0 \le \beta \le 1 \qquad (6.13)$$

We see from Figure 6.10 that the lower β, the more efficient the filter. The first Nyquist criteria imposes to $q(t)$ specific values at instants $t = k \cdot T$. A thinner approximation of the ideal lowpass filter is obtained by imposing specific values to $q(t)$ at instants $t = (2 \cdot k + 1) \cdot T/2$ that corresponds to the second Nyquist criteria. A modem manufacturer knows the frequency response $R(f)$ of its

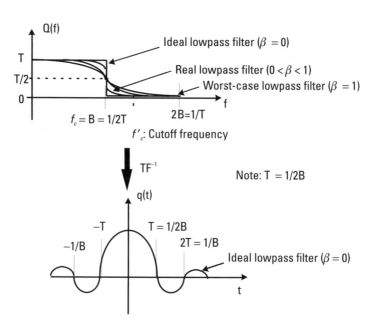

Figure 6.10 Principle of a raised cosine lowpass filter.

transmitter modem. Chapter 4 outlined approximate expressions $H(f)$ of the frequency response of a subscriber line. Imposing $Q(f)$ to satisfy Nyquist criteria, we are able to determine from (6.5) the frequency response $Y(f)$ of the equalizer:

$$Y(f) = \frac{Q(f)}{R(f) \cdot H(f)} \qquad (6.14)$$

6.4.1.2 Single-Carrier Modulation: QAM

The NRZ baseband technique mentioned in the previous section is very sensitive to noise and may then operate only on very short range. In the local loop environment, one uses either single-carrier or multicarrier modulations. The QAM corresponds to a single-carrier modulation. It associates to each suite of m bits called a symbol a particular point M within a constellation of points represented in a two-dimensional space. QAM consists of an amplitude and phase modulation of a single carrier at frequency ω_0 generated by a quartz implemented within each transmitter modem. For every m bits duration, one evaluates for the corresponding symbol an amplitude and a phase with which the carrier ω_0 has to be modulated. The resulting on-line signal $m(t)$ is given by:

$$m(t) = A(t) \cdot \cos\{\omega_0 \cdot t + \Phi + \theta(t)\} = A(t) \cdot \cos\{\theta(t)\}$$
$$\cdot \cos\{\omega_0 \cdot t + \Phi\} - A(t) \cdot \sin\{\theta(t)\} \cdot \sin\{\omega_0 \cdot t + \Phi\} \qquad (6.15)$$

where $\omega_0 = 2\pi \cdot f_0$. If one sets $x(t) = A(t)\cos\{(\theta t)\}$ and $y(t) = A(t)\sin\{\theta(t)\}$, then $x(t)$ and $y(t)$ represent the coordinates of M in the constellation. One notices that $x(t)$ and $y(t)$ are orthogonal, that is,

$$\int_0^T \sin\{\omega_0 \cdot t\}\cos\{\omega_0 \cdot t\}dt = 0 \qquad (6.16)$$

where $T = 2/\omega_0$. We have then for $m(t)$:

$$m(t) = x(t) \cdot \cos\{\omega_0 \cdot t\} + y(t) \cdot \cos\left\{\omega_0 \cdot t + \frac{\pi}{2}\right\} \qquad (6.17)$$

The module $x^2(t) + y^2(t)$ of the vector linking the origin of the two-dimensional space with point M determines the amplitude of the on-line signal. The phase of the on-line signal is given by $\text{Arctg}(y(t)/x(t))$. Figure 6.11 describes some of the constellations used in xDSL modems called 4-QAM, 8-QAM, 16-

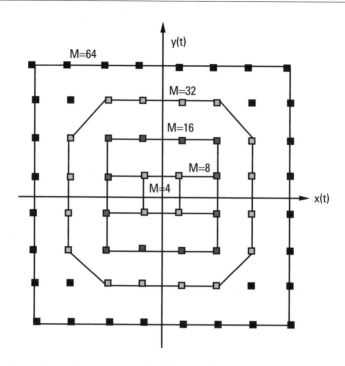

Figure 6.11 Example of different types of QAM constellations.

QAM, and 64-QAM. The points of these constellations are distinguished by different colors, the points of the 2^k-QAM constellation being included in the 2^{k+1}-QAM constellation.

The 4-QAM constellation is limited to the four nearest points around the origin; this modulation corresponds to a simple phase modulation. All the other constellations represented in Figure 6.11 consider both amplitude and phase modulation. The larger the constellation, the higher the number of considered amplitudes and phases, and the more efficient the frequency utilization expressed in bits per second per hertz. As an example, Figure 6.12 details the value of the symbols of a 16-QAM used in V.32 modems.

Figure 6.13 illustrates the principle of a QAM modulator that associates with each suite of m bits received at instant t (this instant corresponds to the date of arrival of the first of these m bits at the input of the modulator) the values of $x(t)$ and $y(t)$ by means of an encoder. On the basis of these two values, one proceeds to the analog modulation of the in-phase clock ω_0 and of its quadrature $\omega_0 + \pi/2$ by means of $x(t)$ and $y(t)$, respectively. The two components $x(t)\cos(\omega_0 \cdot t)$ and $y(t)\sin(\omega_0 \cdot t)$ are superimposed by means of a logic adder. As mentioned in the previous section, the transmission line may be assimilated to a *lowpass filter* (LPF).

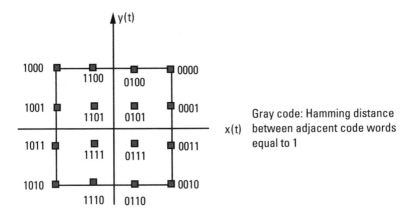

Figure 6.12 The 16-QAM constellation.

Gray code: Hamming distance between adjacent code words equal to 1

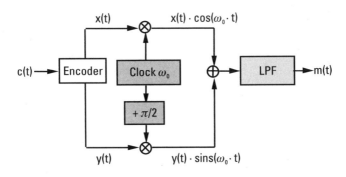

Figure 6.13 Block diagram of a QAM modulator.

Figure 6.14 depicts the principle of a QAM demodulator. Let $m'(t)$ be the received signal that does not correspond exactly to the transmitted signal $m(t)$ because of the various impairments of the line.

The typical shape of the power density spectrum of a QAM signal is illustrated in Figure 6.15.[14] One has represented on the horizontal axis the value of $(f - f_0)$, where $f_0 = \omega_0/2\pi$.

One has at the input of the QAM demodulator:

$$m'(t) = x'(t) \cdot \cos\{\omega_0 \cdot t + \Phi\} + y'(t) \cdot \cos\left\{\omega_0 \cdot t + \Phi + \frac{\pi}{2}\right\} \quad (6.18)$$

---

14. The values indicated on the horizontal and vertical axis of this figure do not refer to an xDSL modem.

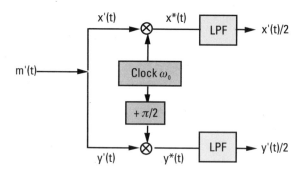

Figure 6.14 Block diagram of a QAM demodulator.

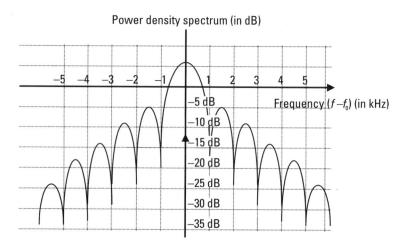

Figure 6.15 Typical shape of the power density spectrum of a QAM signal.

The multiplier used in the upper branch of the demodulator gives the expression $x * (t) = m'(t) \cdot \cos(\omega_0 \cdot t + \Phi')$ where $+\Phi'$ represents the phase distortion due to propagation. One has then after the multiplier $x'(t) / 2 \cdot \cos(\Phi' - \Phi)$ followed by other harmonics at upper frequencies. The phase Φ' is introduced to specify the phase difference between the phase of the local clock proper to the receiver from the phase at the transmitter.[15] One uses an LPF in this upper branch to suppress the high-order harmonics of $x * (t)$. In other terms, the signal $x'(t) / 2 \cdot \cos(\Phi' - \Phi)$ is obtained at the output of this LPF. If the receiver clock is in phase with the received signal, which is not the case in reality, one has $\Phi' = \Phi$. In that case, one obtains at the output of the lowpass filter of the upper

15. We assumed in fact a null phase at the transmitter.

branch of the demodulator $x'(t)/2$. The same remarks may be done for the lower branch of the demodulator. Knowing the values of $x'(t)$ and of $y'(t)$, which are approximate values of $x(t)$ and $y(t)$, respectively, it is then necessary to use a decoder in order to determine the value of the m bits associated with the received symbol.

6.4.2 CAP Modulation

CAP modulation is strongly inspired by QAM. Thus, the power density spectrums of a CAP signal and of a QAM signal look quite similar (see Figure 6.15). Meanwhile, CAP modulation does not need like QAM the transmission of a carrier ω_0. We have mentioned in the previous section that this carrier is only used in QAM to proceed to demodulation but does not carry information. We have underlined in Chapter 4 the phase distortion phenomenon inherent to propagation. In the case of QAM, the modulation carrier $e^{+j\omega_0 t} = \cos(\omega_0 t) + j \cdot \sin(\omega_0 \cdot t)$ is in fact transmitted with the modulated signal in order to be reused by the demodulator. The phase shift $\Phi' - \Phi$ mentioned in the previous section induces a slight random rotation of the constellation as it is detected by the receiver around the axis orthogonal to the $x(t)$ and $y(t)$ axis. This random rotation may degrade considerably the SNR at the receiver, and then the achievable bit rate on the line. The basic motivation of CAP modulation is to reduce the complexity and cost of the equalizer to be used at the receiver. Two techniques are adopted for that purpose. First, pseudo-QAM modulation in CAP does not require the analog carrier at the receiver for demodulation. By this artifact, CAP is much less sensitive to phase distortion of the line. Second, CAP modulation aims to optimize spectral efficiency to limit the impact of nonlinearities in amplitude distortion. Thanks to the advances in the field of electronics and *digital signal processing* (DSP), it is nowadays possible to realize digital filters enabling the emulation of a QAM modulation without requiring analog modulation. Figure 6.16 depicts the principle of a CAP modulator [12]. The bit stream $c(t)$ to be transmitted is inserted into an encoder. Every T seconds (T is the symbol duration), one associates by means of this encoder with each symbol of m bits two coefficients a_k and b_k corresponding to the terms $x(t)$ and of $y(t)$, respectively, mentioned for QAM modulation in the previous section.[16]

Like QAM, CAP modulation uses the concept of constellation with $k = 2^m$ points, each point A_k in the constellation being represented as a complex symbol $A_k = a_k + j \cdot b_k$. For each coefficient a_k and b_k, two impulses $a_k\delta(t - kT)$ and $b_k \cdot \delta(t - kT)$ are fed to the input of two digital shaping filters. The aim of these filters is to minimize the spectral occupancy of the on-line signal. For that purpose, one imposes the time shape of the on-line signal transmitted for each

---

16. These coefficients correspond to the coordinates of a point of a QAM constellation.

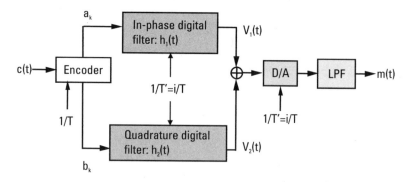

Figure 6.16 Block diagram of a digital CAP transmitter. (*From:* [12]. © 1995 IEEE. Reprinted with permission.)

symbol to be quite similar to a truncated "sinc" function. We have seen in Figure 6.8 that the Fourier transform of a "sinc" function corresponds to a square frequency response. In practice, real implementation of the digital shaping filters correspond to raised cosine filters. The output $V_1(t)$ and $V_2(t)$ of these two filters are then added up and transferred to a D/A converter. The two digital shaping filters and the D/A converter both operate at a sampling rate $1/T' = i/T$ where i is a suitably chosen integer. Let $h_1(t)$ and $h_2(t)$ be the impulse responses of the in-phase and the quadrature filters, respectively. We have at the output of these two filters:

$$\begin{cases} V_1(t) = a_k \cdot \delta(t - k \cdot T) \otimes h_1(t) = a_k \\ \cdot \int_{-\infty}^{+\infty} h_1(t - \alpha)\delta(\alpha - k \cdot T)d\alpha = a_k \cdot h_1(t - k \cdot T) \\ V_2(t) = b_k \cdot \delta(t - k \cdot T) \otimes h_2(t) = b_k \\ \cdot \int_{-\infty}^{+\infty} h_2(t - \alpha)\delta(\alpha - k \cdot T)d\alpha = b_k \cdot h_2(t - k \cdot T) \end{cases} \quad (6.19)$$

These two signals are added up giving for each symbol:

$$m(t) = a_k \cdot h_1(t - kT) + b_k \cdot h_2(t - kT) \quad (6.20)$$

Let us now compare the output of a CAP modulator with the output of a QAM modulator, both modulators being based on a same-size constellation and a same symbol period T. We have seen in (6.15) that the output of a QAM modulator is of the form $a_k \cdot \cos(\omega_0 t) - b_k \cdot \sin(\omega_0 t)$. In practice, each square voltage pulse considered by the in-phase and by the quadrature branches of the QAM modulator is reshaped by means of a raised cosine filter with impulse response $\Phi(t)$ comparable to the function $q(t)$ depicted in Figure 6.10. We have then at the output of the QAM modulator:

$$m(t) = a_k \cdot \Phi(t - kT) \cdot \cos\{\omega_0 \cdot t\} - b_k \cdot \Phi(t - kT) \cdot \sin\{\omega_0 \cdot t\} \qquad (6.21)$$

To have an equivalence between the output of the QAM modulator and the output of the CAP modulator, we must have:

$$\begin{cases} h_1(t - kT) = \Phi(t - kT) \cdot \cos\{\omega_0 \cdot t\} \\ h_2(t - kT) = -\Phi(t - kT) \cdot \sin\{\omega_0 \cdot t\} \end{cases} \qquad (6.22)$$

Knowing that $\omega_0 T = 2\pi$, it comes:

$$\begin{cases} h_1(t - kT) = -\Phi(t - kT) \cdot \cos\{\omega_0 \cdot (t - kT)\} \\ h_2(t - kT) = -\Phi(t - kT) \cdot \sin\{\omega_0 \cdot (t - kT)\} \end{cases} \qquad (6.23)$$

We can deduce from (6.23) the expressions of the impulse responses of the two digital shaping filters of the CAP modulator:

$$\begin{cases} h_1(t) = \Phi(t) \cdot \cos\{\omega_0 \cdot t\} \\ h_2(t) = -\Phi(t) \cdot \sin\{\omega_0 \cdot t\} \end{cases} \qquad (6.24)$$

The previous equations refer to a unique symbol; one has more generally for a suite of symbols at the output of the CAP modulator:

$$V_1(t) + V_2(t) = \sum_{-\infty}^{+\infty} [a_k \cdot h_1(t - kT) - b_k \cdot h_2(t - kT)] \qquad (6.25)$$

The discrete signal $V_1(t) + V_2(t)$ is passed through a D/A converter that associates with each symbol a dedicated voltage level and phase. The obtained signals are sent to an interpolating LPF.

Figure 6.17 depicts the block diagram of digital CAP demodulator [13]. The received signal $m'(t)$ passes through an A/D converter followed by two adaptive digital filters.[17] The A/D converter and the adaptive digital filters operate at a sampling rate $1/T' = i/T$, with the same value of the parameter i as for the CAP transmitter. The outputs of the two adaptive filters are sampled at the symbol rate $1/T$ to feed a decision device followed by a decoder in charge on converting symbols into bits.

---

17. Such filters are called *fractionally spaced linear equalizers* (FSLEs).

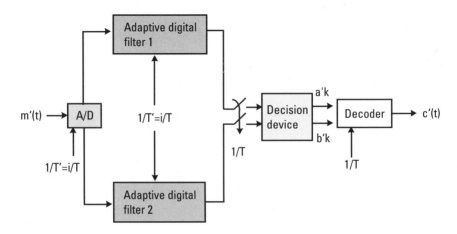

Figure 6.17 Block diagram of a digital CAP receiver. (*From:* [12]. © 1995 IEEE. Reprinted with permission.)

6.4.3 DMT Modulation

Single-carrier modulation techniques operating with large on-line spectrum require complex equalizers at the receivers. Although it operates under high bit rates, the CAP single-carrier modulation tries to circumvent this difficulty by optimizing spectral efficiency. DMT is a multicarrier modulation that adopts a totally different approach to solve the problem of equalization. It is inspired by *orthogonal frequency division multiplexing* (OFDM), which has been widely adopted in radio communications.[18] The basic idea of DMT modulation is to divide the operational ADSL bandwidth into very small subchannels. The bandwidth associated with each subchannel is sufficiently limited to consider the frequency response of the transmission line as linear in each subchannel. In other terms, one assumes that in each subchannel, phase and amplitude distortions are sufficiently limited to be assimilated to linear distortions. Thus the [0, 1,104 kHz] ADSL bandwidth is divided into 256 subchannels indexed by increasing order from low frequencies to high frequencies. Each subchannel is 4.3125-kHz-wide. The first six subchannels are reserved for the existing analog telephony channel and its associated signaling. Several subchannels are in general used by the carriers as signaling channels or as guard-band. Upstream data can occupy 24 subchannels. FDM-ADSL and EC-ADSL were introduced in Sections 1.1 and 1.2. The bandwidth occupancies proper to each of these systems are illustrated in Figures 6.2 and 6.3, respectively. Downstream data use either up to 222 or up to 248 subchannels in the case of FDM-ADSL or in the case of EC-ADSL, respectively.

---

18. The theoretical principles of OFDM date from the 1960s. Meanwhile, implementation of OFDM in real equipment is much more recent.

6.4.3.1 Principle of DMT Modulation

Discrete carriers (or tones) are used in the center of each data subchannel. These carriers are used to transmit data independently in each subchannel by means of a specified QAM modulation. The discovery of the *discrete Fourier transform* (DFT) in the 1970s has been a key achievement for enabling the practical implementation of OFDM in real equipment [14]. In DMT, the upstream and downstream data flows are dispatched over a certain amount of subchannels. DMT-ADSL (either the FDM-ADSL or the EC-ADSL versions) assumes a line initialization. For that purpose, both the ATU-R and the ATU-C exchange test sequences in order to analyze the quality of the subscriber line in terms of SNR. Line initialization consists of evaluating and storing in the ATU-C and ATU-R the value of the SNR for each of the subchannels from 7 to 256. On the basis of the value of SNR, an ADSL modem adapts automatically the size of the QAM constellation that can be used in the considered subchannel in accordance with the achievable bit rate. Figure 6.18 illustrates the basic principle of DMT modulation.

The upper part of Figure 6.18 depicts the average values of SNR observed in each subchannel. The lower part of Figure 6.18 gives an example of capacity allocation per subchannel.[19] In this example, the reader can see that the higher the SNR, the larger the size of the constellation—that is, the higher the bits per

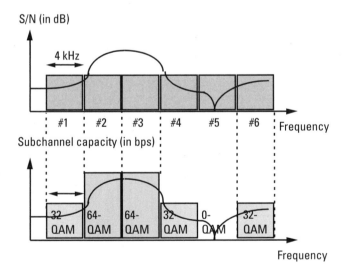

Figure 6.18 Basic principle of DMT modulation.

---

19. The curves and values mentioned in Figure 6.18 are only illustrative and do not correspond to a real implementation.

second/hertz capacity. One notices that SNR is very low for a discrete frequency within subchannel 5. The observed cutoff frequency in this subchannel may be due to some of the line impairments described in Chapter 4 (e.g., presence of multiple bridge taps with an undesirable length or presence of load coils introducing a real cutoff frequency). Subchannels 1, 4, and 6, which are of medium quality, may be used with a 32-QAM corresponding to a capacity of C bps/Hz. Subchannels 2 and 3 are of very good quality and use 64-QAM with a capacity of $6C/5$ bps/Hz. Subchannel 5 is not used at all.

The ADSL standard [5] specifies transmission power in each subchannel.[20] As an example, Figure 6.19 specifies the transmission power and its admissible fluctuations in each of the upstream and downstream subchannels of a FDM-ADSL system. One has plotted on Figure 6.19 the shape of the bandpass filter enabling the isolation of data traffic from analog telephony.

6.4.3.2 Subchannels' Capacity Allocation: Water-Filling Algorithm

The objective of this subsection is to describe the principle of the water-filling algorithm used to determine the optimum capacity to assign to each of the subchannels of a multicarrier transmission system (see Figure 6.18). Let W be the on-line bandwidth used by an ADSL system and $H(f)$ be the frequency response of the line. We assume that noise $b(t)$ due to the various line impairments is Gaussian with power density spectrum $\Phi_{bb}(f)$. Considering DMT-ADSL, let Δf be the width of each subchannel. The value of Δf is sufficiently low to estimate that in each subchannel one has:

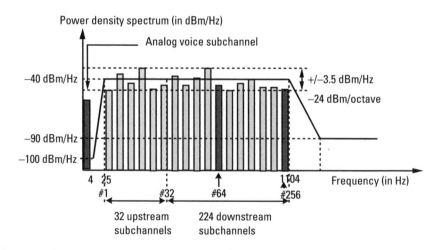

Figure 6.19 Power density spectrum of FDM-ADSL. (One subchannel = 4.3125 kHz; subchannel #64 used as pilot test, subchannel #256 used as Nyquist frequency.)

---

20. These powers are expressed in reference to a power of 1 mW.

$$\frac{\partial \left\{ \dfrac{|H(f)|^2}{\Phi_{bb}(f)} \right\}}{\partial f} \approx 0 \qquad (6.26)$$

Let P_a be the average transmission power on W and $p(f)$ be the power density spectrum of an ADSL modem. We have:

$$\int_{f \in W} p(f) \cdot df \leq P_a \qquad (6.27)$$

We know that the Shannon's capacity C in bps of a transmission channel with limited bandwidth W in presence of AWGN is given by:

$$C = W \cdot \log_2 \left\{ 1 + \frac{S}{N} \right\} = W \cdot \log_2 \left\{ 1 + \frac{P_a}{W \cdot N_0} \right\} \qquad (6.28)$$

Let us recall that AWGN is characterized by a power spectral density $N_0/2$ in two symmetrical bands of B Hz width and centered on two discrete frequencies $+f_c$ and $-f_c$ (see Figure 4.7). In the case of multicarrier modulation, Shannon's capacity proper to a given subchannel k can be expressed as:

$$C_k = \Delta f \cdot \log_2 \left\{ 1 + \frac{\Delta f \cdot P_a(f_k) \cdot |H(f_k)^2|}{\Delta f \cdot \Phi_{bb}(f_k)} \right\} \qquad (6.29)$$

where f_k represents the central carrier of subchannel k and $P_a(f_k)$ is the average transmission power in subchannel k. One assumes that in each subchannel, the power density spectrum of the transmitted signal, the power density spectrum of noise $b(t)$, and the frequency response $H(f)$ of the line are constant over Δf. Let N be the number of subchannels such that:

$$N = \frac{W}{\Delta f} \qquad (6.30)$$

The total capacity C of the line is then given by:

$$C = \sum_{k=1}^{N} C_k = \Delta f \cdot \sum_{k=1}^{N} \log 2 \left\{ 1 + \frac{P_a(f_k) \cdot |H(f_k)|^2}{\Phi_{bb}(f_k)} \right\} \qquad (6.31)$$

Equation (6.31) becomes when $\Delta f \to 0$:

$$C = \int_{f \in W} \left(\log_2 \left\{ 1 + \frac{P_a(f) \cdot |H(f)|^2}{\Phi_{bb}(f)} \right\} \right) \cdot df \qquad (6.32)$$

We want to find the expression of $P_a(f)$ that enables to maximize C. From the theory of variations calculus, one knows that C is maximum for $P_a(f)$ such that the value of the following integral is maximum:

$$C = \int_{f \in W} \left(\log 2 \left\{ 1 + \frac{P_a(f) \cdot |H(f)|^2}{\Phi_{bb}(f)} \right\} + \lambda \cdot p(f) \right) \cdot df \qquad (6.33)$$

where the parameter λ is called the *Lagrange multiplier*. To simplify (6.33), let us set:

$$\alpha(f) = \frac{|H(f)|^2}{\Phi_{bb}(f)} \qquad (6.34)$$

In replacing $P_a(f)$ by $p(f)$ for the sake of simplicity, it becomes for (6.33):

$$\int_{f \in W} \left(\log_2 [1 + p(f) \cdot \alpha(f)] + \lambda \cdot p(f) \right) \cdot df \qquad (6.35)$$

For a weak variation of $p(f)$ noted $p(f) + \varepsilon \cdot p$, (6.35) becomes:

$$\int_{f \in W} \left(\log_2 [1 + p(f) \cdot \alpha(f) + \varepsilon \cdot p \cdot \alpha(f)] + \lambda \cdot \varepsilon \cdot p + \lambda \cdot p(f) \right) df \qquad (6.36)$$

Equation (6.36) can be expressed as:

$$\int_{f \in W} \left(\log_2 \left[\{1 + p(f) \cdot \alpha(f)\} \left\{ 1 + \frac{\varepsilon \cdot p \cdot \alpha(f)}{(1 + p(f) \cdot \alpha(f))} \right\} \right] \right) \cdot df$$
$$+ \int_{f \in W} \{ \lambda \cdot \varepsilon \cdot p + \lambda \cdot p(f) \} df \qquad (6.37)$$

Let us introduce function $\theta(p(f))$ such that:

$$\theta(p(f)) = \int_{f \in W} \left[\log_2 \{ 1 + p(f) \cdot \alpha(f) \} + \lambda \cdot p(f) \right] df \qquad (6.38)$$

Knowing that $\log_2(AB) = \log_2(A) + \log_2(B)$, (6.37) becomes then:

$$\theta(p(f)) + \lambda \cdot \varepsilon \cdot \int_{f \in W} p \cdot df + \int_{f \in W} \log_2 \left(\frac{\varepsilon \cdot \alpha(f) \cdot p}{(1 + p(f) \cdot \alpha(f))} \right) \cdot df \qquad (6.39)$$

One knows that $\lim_{x \to 0} \log(1 + x) = x$; it becomes then for (6.39):

$$\theta(p(f)) + \lambda \cdot \varepsilon \cdot \int_{f \in W} p \cdot df + \varepsilon \cdot \int_{f \in W} \frac{\alpha(f) \cdot p}{[1 + p(f) \cdot \alpha(f)]} \cdot df \qquad (6.40)$$

Equation (6.36) refers to the expression of $\theta(p(f) + \varepsilon \cdot p)$. One has then:

$$\frac{[\theta\{p(f) + \varepsilon \cdot p\} - \theta\{p(f)\}]}{\varepsilon} = \qquad (6.41)$$
$$\lambda \cdot \int_{f \in W} p \cdot df + \int_{f \in W} \frac{\alpha(f) \cdot p}{[1 + p(f) \cdot \alpha(f)]} \cdot df$$

We want (6.41) to tend to 0 for any value of p when ε tends to 0. This imposes that:

$$\lambda + \frac{\alpha(f)}{[1 + p(f) \cdot \alpha(f)]} = 0 \qquad (6.42)$$

In reusing the expression of $\alpha(f)$ given in (6.34), one has then:

$$p(f) = K - \frac{\Phi_{bb}(f)}{|H(f)|^2}, \forall f \in W \qquad (6.43)$$

where $K = -1/\lambda$ = constant. An interpretation of this result is illustrated in Figure 6.20.

A multicarrier modulation system like DMT modulation may be compared to a fluid system. For a given value of S/N such that:

$$\frac{S}{N} = \frac{|H(f)|^2}{\Phi_{bb}(f)} \qquad (6.44)$$

the global data rate that can be transported on a multicarrier transmission system is maximum if the power density spectrum $p(f)$ satisfies (6.43) for any f in W. This problem can be assimilated to the determination of the depth $p(f)$ of liquid in a pool with a profile of its bottom given by $(S/N)^{-1}$ described in Figure 6.20. One assimilates the volume of liquid contained in the pool to the average power of transmission P_a. In other terms, the signal power $p(f)$ should be high when $K - (S/N)^{-1}$ is high [i.e., when $(S/N)^{-1}$ is low, or when S/N is high] [14]. This algorithm is known as the *water-filling algorithm*. Knowing the discrete values of $p(f_k)$ for each subcarrier f_k, one may deduce the number of bits per symbol to associate to the QAM constellation used in subchannel k. In general, the number of bits assigned to the various QAM constellations is such that error probability per symbol is identical between different subchannels.

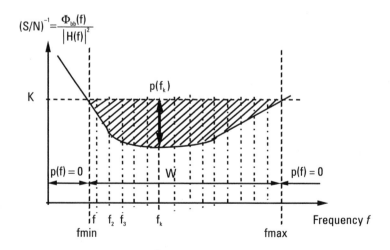

Figure 6.20 The optimum water-filling spectrum for multicarrier modulation. (*From:* [19].
© 1995 McGraw-Hill, Inc.)

The reader may remain interrogative concerning the economical interest of the water-filling algorithm used in DMT modulation. Indeed, physical channel optimization should be carried out by the use of 256 subchannels requiring 256 pairs of xDSL modems. Thanks to advances in digital signal processing and high-speed electronics, this is not fortunately the case. As it is described in the following sections, DMT modulation relies on DFT. DFT is implemented in ADSL modems in using the *fast Fourier transform* (FFT) algorithm. The FFT algorithm enables the implementation in the same chip of the equivalent of the modems necessary for the set of the subchannels of a DMT system. By considering a set of discrete and regularly spaced tones, FFT also facilitates the use of different QAM constellations in the different subchannels centered on these tones. The equivalent modems associated with the different subchannels are synchronized on a common symbol period. The aim of the two following sections is to describe the principle of transmitter (Tx) and receiver (Rx) DMT-ADSL modems, respectively.

6.5 ADSL System Configuration

Figure 6.21 depicts the end-to-end block diagram of an ADSL access system, either DMT-based or CAP-based. As it appears on the figure, the modulator and demodulator are the last and the first elements of the block diagram at the Tx side and at the Rx side, respectively. The three other blocks used on each side of the line of transmission aim to prevent erroneous bit detection at the receiver. They refer to FEC. Two variants of FEC techniques are used in an ADSL

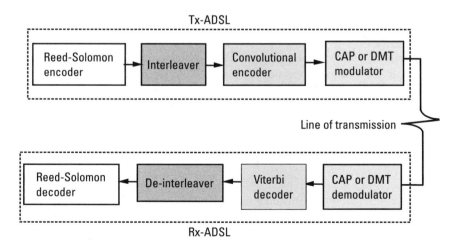

Figure 6.21 Block diagram of an ADSL access system.

modem: Reed-Solomon codes and convolutional codes. Convolutional codes are also known as trellis-coded modulation.

In an ADSL modem, the following functional modules are considered in order to reduce erroneous bit probability at the receiver: Reed-Solomon encoder, interleaver, convolutional encoder, Viterbi decoder, deinterleaver, and Reed-Solomon decoder. For the sake of readability, the description of these different modules has been divided into two sections dedicated to Tx-DMT-ADSL and to Rx-DMT-ADSL, respectively. This separation is in fact a bit artificial because, as will be outlined in the following, several functionalities carried out either at the Tx side or at the Rx side are closely related. We will not come back to CAP-ADSL, which was covered in Section 6.4.2.

6.6   DMT-ADSL Modulator

In digital communications, the association between Reed-Solomon encoding and convolutional encoding is called concatenated encoding. Concatenated encoding is frequently used in radio systems for which the very fluctuating quality of the transmission medium justifies the redundancy of two complementary FEC techniques. One qualifies Reed-Solomon encoding and convolution encoding of external and internal encoders, respectively. We have underlined the partially unpredictable behavior of subscriber lines in Chapter 4. This is the reason why concatenated encoding has also been adopted for ADSL access systems. As an illustration, Figure 6.22 depicts a simplified version of a DMT-ADSL transmission chain using only four subchannels.

We describe in the following each of the six stages of the Tx-DMT-ADSL modem mentioned in Figure 6.22. Let D be the bit rate at the input of the modem. These bits first feed a serial-to-parallel converter corresponding to the first stage of the system. This converter groups bits into blocks (or frames) of M bits. The M bits of a frame are grouped into N sub-blocks $\{B_1, B_2, ..., B_N\}$ corresponding to the N subchannels of the system. The size M_i in bits of sub-block B_i is determined, thanks to the water-filling algorithm. The size of a frame is:

$$M = \sum_{i=1}^{N} M_i \qquad (6.45)$$

Figure 6.23 illustrates an example of sub-blocks allocation corresponding to the four subchannels mentioned in Figure 6.22 with $M_1 = 2$, $M_2 = 4$, $M_3 = 3$, and $M_4 = 1$. Let T_0 be the symbol period corresponding to a 10-bit frame duration. In practice, $T_0 = 500 \ \mu s$. Serial-to-parallel conversion assumes a buffering and a reordering of the bits representing a delay greater than T_0 seconds. By convention, the bits are ordered according to the size of their mother

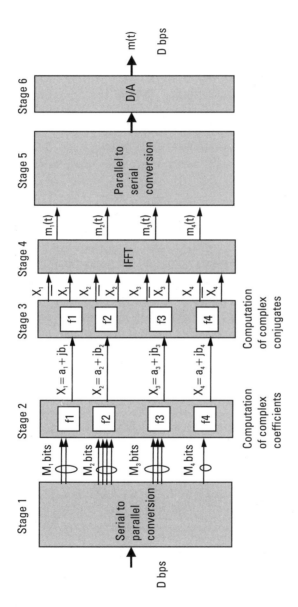

Figure 6.22 Principle of a Tx-DMT-ADSL access system.

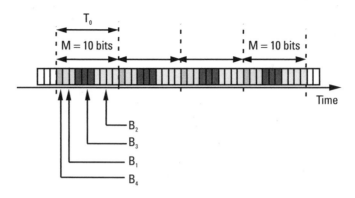

Figure 6.23 Example of frame formatting into sub-blocks.

constellation. Thus, the bits belonging to the smallest constellation are placed at the head of a buffer, followed by the bits belonging to larger size constellations. In the case of constellations with the same size, the associated bits are queued into the buffer according to the value of the associated tone, the constellation with the lowest tone being queued first.

Each sub-block B_N is then passed through the second stage in charge of the computation of the $a_i(kT_0)$ and $b_i(kT_0)$ that correspond to the coordinates of the associated point in the constellation of subchannel i. The tones $f_1, f_2, ..., f_N$ refer to the central carriers proper to each subchannel. Let $X_i(kT_0) = a_i(kT_0) + jb_i(kT_0)$ be the complex coefficient associated to a point in constellation i during a given symbol period. As it will be justified in a few lines, stage 4 must be fed with digital samples with real values. For that purpose, stage 3 determines for each X_i its complex conjugate such that:

$$\overline{X_i(kT_0)} = \overline{a_1(kT_0) + jb_i(kT_0)} = a_i(kT_0) - jb_i(kT_0) \quad (6.46)$$

Stage 4 is in charge of the computation of the samples of the signal in each subchannel. DFT and *inverse DFT* (IDFT) are used at the receiver and at the transmitter, respectively. DFT and IDFT are implemented in hardware circuits by means of FFT and IFFT, respectively. FFT is an algorithm enabling a rapid computation of DFT by means of dedicated electronic circuits. According to their inherent complexity, we give in Sections 6.6.1 and 6.6.2 only the basic principles of DFT and IDFT. For a more detailed description, the interested reader may refer to [15]. Let us simply state that DFT complexity is in $o(N^2)$ whereas the complexity of FFT is in $o(N\log_2 N)$ where N stands for the number of subchannels. In the context of DMT-ADSL, let us recall that N is on the order of 32 for upstream traffic and of 250 for downstream traffic.

6.6.1 DFT

The DFT of a function $g(t)$ enables one to determine numerically from a set of samples (these samples corresponding to real numbers) of this function taken every T_s seconds a set of complex numbers representing the amplitude and the phase of the various harmonics associated with the Fourier series expansion of $g(t)$. Inversely, IDFT enables to rebuild numerically the successive samples of $g(t)$ from the discrete coefficients of its Fourier series expansion. Let us recall a few results of Fourier series expansion.

A periodical signal $g(t)$ with period T_0 may be expressed as a Fourier series expansion (i.e., under the form of a summation of a continuous signal a_0 with a set of two harmonics in quadrature). These orthogonal harmonics with period T_0/n are modulated by real coefficients a_n and b_n, respectively, with $n = 1, 2 \ldots \infty$. The Fourier series expansion of $g(t)$ is given by:

$$g(t) = a0 + \sum_{n=1}^{\infty} \left[a_n \cdot \cos\left(n\frac{2\pi}{T_0} \cdot t \right) + b_n \cdot \sin\left(n\frac{2\pi}{T_0} \cdot t \right) \right] \quad (6.47)$$

A Fourier series expansion may also be obtained from a set of complex terms such that:

$$g(t) = \sum_{n=-\infty}^{\infty} \left[D_n \cdot e^{jn\frac{2\pi}{T_0} \cdot t} \right] \quad (6.48)$$

where coefficients D_n are given by:

$$D_n = \frac{1}{T_0} \cdot \int_{t \in T_0} g(t) \cdot e^{-jn\frac{2\pi}{T_0} \cdot t} \cdot dt \quad (6.49)$$

Complex coefficients D_n are related to real coefficients a_n and b_n by the following equations: $D_0 = a_0$ and for any $n > 0$:

$$\begin{cases} D_0 = a_0 \\ |D_n| = |D_{-n}| = \frac{1}{2} \cdot \sqrt{a_n^2 + b_n^2}, \forall n > 0 \end{cases} \quad (6.50)$$

DFT uses the samples of $g(t)$ at instants kT_s where T_s stands for the sampling period. Assuming that T_s is very small compared to T_0, let N_0 be the number of samples over a period T_0 such that $N_0 = T_0 /T_s$. As a conclusion, DFT may be interpreted as a discrete, and then approximated, version of complex Fourier transform given by:

$$D_n = \frac{1}{N_0} \sum_{k=0}^{N_0-1} \left[g(k \cdot T_s) \cdot e^{-jn\frac{2\pi}{T_0} \cdot kT_s} \right]$$

$$= \frac{1}{N_0} \sum_{k=0}^{N_0-1} \left[g(k \cdot T_s) \cdot e^{-jn\,\omega_0 kT_s} \right]$$

(6.51)

This discrete version of the complex Fourier transform enables one to get numerically the approximated value of the first terms a_n and b_n of the real Fourier series expansion mentioned in (6.47). The reader may notice from (6.51) that the values of coefficients D_n calculated by means of DFT are cyclic such that:

$$D_{n+N_0} = D_n, \forall n > 0$$

(6.52)

The larger the N_0, the better the DFT approximation. In practice, one shows that DFT is a good approximation of a complex Fourier transform if the module of coefficients D_n becomes negligible when $n \geq N_0/2$.

6.6.2   Analogy Between DMT and IDFT

We have outlined in the previous sections the expression $m(t)$ of the modulated signal at the output of a QAM modulator (6.15). Let us consider now the case of a particular constellation of index i among the N subchannels. We have noted T_0 the common symbol duration for the various subchannels. For constellation i whose tone is centered on $i\omega_0$, the modulated signal proper to this subchannel is given by:

$$m_{i,p}(t) = x_{i,p}(t) \cdot \cos(i\omega_0 \cdot t) + y_{i,p}(t) \cdot \sin(i\omega_0 \cdot t)$$

(6.53)

where $\omega_0 = 2\pi/T_0$ and p represents the index of the pth generated symbol. Let us recall that $x_{i,p}$ and $y_{i,p}$ correspond to the coordinates of the point associated with the current symbol in a two-dimensional space. Let us outline that it is possible to approximate the expression of $m_{i,p}(t)$ over a duration T_0 by means of a set of discrete values corresponding to its sampling over this same period by using IDFT. For that purpose, let us consider a sampling period T_s much smaller than T_0. Let N_0 be the number of sampling instants over T_0. Let $m_{i,p,k}(t)$ be the kth sample of $m_{i,p}(t)$ at instant $t = kT_s$. The sampling period T_s is supposed to satisfy the Nyquist criteria (see Section 6.4.1.1):

$$\frac{1}{T_s} = f_s = 2N \cdot \frac{1}{T_0} = 2N \cdot f_0 \qquad (6.54)$$

Knowing that $x_{i,p}(t_k)$ and $y_{i,p}(t_k)$ do not depend on k over a sampling period, it becomes for $m_{i,p,k}(t)$:

$$m_{i,p,k}(t) = x_{i,p} \cdot \cos(i\omega_0 t_k) + y_{i,p} \cdot \sin(i\omega_0 t_k) = x_{i,p} \cdot \cos(i\omega_0 kT_s)$$

$$+ y_{i,p} \cdot \sin(i\omega_0 kT_s) = x_{i,p} \cdot \cos\left(i\omega_0 \frac{k}{2N \cdot f_0}\right) \qquad (6.55)$$

$$+ y_{i,p} \cdot \sin\left(i\omega_0 \cdot \frac{k}{2N \cdot f_0}\right) = x_{i,p} \cdot \cos\left(\frac{i\pi \cdot k}{N}\right) + y_{i,p} \cdot \sin\left(\frac{i\pi \cdot k}{N}\right)$$

k being an integer such as $0 < k \le 2N$. Let us consider now the truncated DFT of $m_{i,p}(t)$ with only the $2N$ first terms of its Fourier series expansion. It comes for the nth term of this Fourier series expansion:

$$D_n\big(m_{i,p}(t)\big) = \frac{1}{2N} \cdot \sum_{k=0}^{2N-1} m_{i,p}(kT_s) \cdot e^{-jn\omega_0 kT_s} = \frac{1}{2N}$$

$$\cdot \sum_{k=0}^{2N-1} \big[x_{i,p} \cos(i\omega_0 \cdot kT_s) + y_{i,p} \cdot \sin(i\omega_0 \cdot kT_s)\big] \cdot e^{-jn\omega_0 kT_s} \qquad (6.56)$$

Knowing that $i\omega_0 \cdot kT_s = i\pi k/N$, it comes:

$$D_n\big(m_{i,p}(t)\big) = \frac{1}{2N} \sum_{k=0}^{2N-1} \left[\begin{array}{l} x_{i,p} \cos\left(\dfrac{i\pi \cdot k}{N}\right) + \\[2mm] y_{i,p} \sin\left(\dfrac{i\pi \cdot k}{N}\right) \end{array} \right] e^{-jn\omega_0 kT_s} \qquad (6.57)$$

In order to determine the expression of $D_n(m_{i,p}(t))$, let us consider the summation of four terms T_1, T_2, T_3, and T_4 issued from (6.57) after having replaced $\cos(t)$ and $\sin(t)$ functions by their equivalent complex expression:

$$\cos(x) = \frac{e^{+jx} + e^{-jx}}{2}$$

$$\sin(x) = \frac{e^{+jx} - e^{-jx}}{2j} \qquad (6.58)$$

One has then:

$$
\begin{aligned}
T_1 &= \sum_{k=0}^{2N-1} \frac{x_{i,p}}{2} \cdot e^{+j\pi k \frac{(i-n)}{N}} \\
T_2 &= \sum_{k=0}^{2N-1} \frac{x_{i,p}}{2} \cdot e^{-j\pi k \frac{(i+n)}{N}} \\
T_3 &= \sum_{k=0}^{2N-1} \frac{y_{i,p}}{2j} \cdot e^{+j\pi k \frac{(i-n)}{N}} \\
T_4 &= \sum_{k=0}^{2N-1} \frac{y_{i,p}}{2j} \cdot e^{-j\pi k \frac{(i+n)}{N}}
\end{aligned}
\tag{6.59}
$$

The expression of $D_n(m_{i,p}(t))$ based on expressions T_1, T_2, T_3, and T_4 is:

$$
D_n\left(m_{i,p}(t)\right) = \frac{1}{2N}\left[T_1 + T_2 + T_3 + T_4\right]
\tag{6.60}
$$

If $i = n$, we see that:

$$
T_1 = \sum_{k=0}^{2N-1} \frac{x_{i,p}}{2} = N \cdot x_{i,p}
\tag{6.61}
$$

$$
T_2 = \sum_{k=0}^{2N-1} \frac{x_{i,p}}{2} \cdot e^{-j\pi k \frac{2i}{N}}
\tag{6.62}
$$

Let us introduce a variable $q = \exp(-j\pi k 2i / N)$. Equation (6.62) is then an integer series expansion with a sum equal to $(1 - q^{2N})/(1 - q)$. Now, $q^{2N} = 1$, and then $T_2 = 0$.

$$
T_3 = \sum_{k=0}^{2N-1} \frac{y_{i,p}}{2j} = -jN \cdot y_{i,p}
\tag{6.63}
$$

With the same considerations as for (6.62), it becomes:

$$
T_4 = \sum_{k=0}^{2N-1} \frac{y_{i,p}}{2j} \cdot e^{-j\pi k \frac{2i}{N}} = 0
\tag{6.64}
$$

It becomes:

$$D_n\left(m_{i,p}(t)\right) = \frac{1}{2N}\left[T_1 + T_3\right] = \frac{1}{2N}\left[N \cdot x_{i,p} - jN \cdot y_{i,p}\right] \quad (6.65)$$

Finally, it becomes when $i = n$:

$$D_n\left(m_{i,p}(t)\right) = \frac{x_{i,p} - j \cdot y_{i,p}}{2} \quad (6.66)$$

and in the case of $i = 2N - n$:

$$D_n\left(m_{i,p}(t)\right) = \frac{x_{i,p} + j \cdot y_{i,p}}{2} \quad (6.67)$$

In all the other cases, that is when $i \neq n$ and $i \neq 2N - n$, $D_n(m_{i,p}(t)) = 0$. We have put into evidence the fact that it is possible to determine an approximate expression of $m_{i,p}(t)$ corresponding to the output signal of a DMT-ADSL modem with the help of DFT. This signal can be assimilated to the summation of the signals obtained at the output of N QAM modulators with their individual tones regularly spaced. One notices that DFT gives in fact the set of values $x_{i,p}$ and $y_{i,p}$ corresponding to each symbol associated with the N subchannels. In order to rebuild the shape of the output signal of a DMT modulator from these coefficients, we have then to use IDFT, which is mentioned in the fourth stage of a Tx-DMT-ADSL modem illustrated in Figure 6.22.

In summary, we can say that IFFT implemented on stage 4 computes discrete real numerical samples of the modulated signal for each subcarrier (or tone) from the complex coefficients determined in the previous stages (2 and 3). Knowing that in practice N is of the order of 250, IFFT computes every T_0 seconds (remember that T_0 is about 500 μs) 250 real numbers $x_{i,p}$ and 250 real numbers $y_{i,p}$ from 250 complex numbers. These 500 real numbers specify the voltage level and the phase of the on-line signal for a symbol duration. Let us recall that DMT modulation assumes a perfect orthogonality between subcarriers to prevent *interchannel interference* (ICI). This orthogonality is expressed for any couple of subchannels of index n and m with $n \neq m$ by:

$$\int_0^{T_0} e^{+jn\omega_0 t} \cdot \overline{e^{+jn\omega_0 t}}\, dt = 0 \quad (6.68)$$

The fact that we use IFFT to determine the shape of the transmitted signal introduces, as it has been mentioned, a light bias inherent to sampling. It is then important that sampling frequency is high enough and satisfies Nyquist criteria.

6.6.3 Cyclic Prefix

In reference to Figure 6.22, stage 5 of a Tx-DMT-ADSL modem is in charge of parallel-to-serial conversion of the samples obtained by IFFT at the output of stage 4. Independence between subchannels at the receiver imposes a symbol resynchronization in the Rx-DMT-ADSL modem. This resynchronization is used to compensate envelope delay disparity described in Section 4.2.3. This resynchronization is carried out by means of the insertion of a cyclic prefix within stage 5. For any set of $2N$ samples $x_{i,p}$ and $y_{i,p}$ generated by stage 4 during the pth symbol period (by sake of coherence, variables i and p keep the same meaning as in above calculations), one inserts a copy of the first η couples of samples as a prefix of these $2N$ samples as illustrated in Figure 6.24.

As an illustration, current FDM-ADSL implementations with 25 upstream subchannels consider $\eta = 4$ (in the case of Figure 6.24). The same type of framing is used in the case of 250 upstream subchannels with $\eta = 32$. At the other end of the line, the receiver extracts these prefixes. The benefits of this prefix mechanism are summarized as follows:

- It enables the introduction of a guard time between successive transmitted symbols. In this way, one reduces IIS.

- We have seen in Section 6.4.1.1 that the received signal $s(t)$ is given by the convolution of the transmitted impulses $c(t)$ with the equivalent impulse response $q(t)$ of the line. In real implementations, the whole ADSL system (transmitter, transmission line, and receiver) induce a nonnegligible response time to short impulses. This response time being dependent with f, it varies from one subchannel to another. This cause although different from envelope delay also contributes to IIS. It is possible to prevent such a form of IIS by choosing a prefix duration at least equal to the response time of the system [16].

- The prefix also introduces a guard space in the frequency domain and contributes to the reduction of ICI. Indeed, without any prefix, frequency spacing Δf between adjacent subchannels is equal to the band-

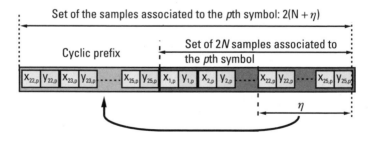

Figure 6.24 Example of synchronization prefix for DMT-ADSL upstream traffic.

width required for each of them. Knowing the sampling frequency T_s the symbol period T_0 and the number N of subchannels, one has:

$$\Delta f = \frac{1}{T_0} = \frac{f_s}{2N} \qquad (6.69)$$

If one uses a prefix, frequency spacing between subchannels becomes Δf^* such that:

$$\Delta f^* = \frac{f_s}{2N + \eta} < \Delta f \qquad (6.70)$$

Let us remark that for a given transmission power, prefix insertion slightly reduces SNR at the receiver. According to current implementations, this attenuation is on the order of 0.35 dB.

Stage 6 represented in Figure 6.22 receives serial bits corresponding to a suite of blocks like the one given in Figure 6.24. This last stage of a Tx-DMT-ADSL modem proceeds to D/A conversion in replacing discrete values by a time continuous signal thanks to a bandpass filter.

6.7 Reed-Solomon Encoding

6.7.1 Information Theory Reminder

Let x and y be two binary words belonging to a same code C. Hamming distance $d(x, y)$ between x and y is given by the number of bits on which x and y differ. For instance, if $x = (1, 1, 1, 0, 1, 0, 0)$ and $y = (0, 1, 0, 1, 1, 0, 0)$, $d(x, y) = 3$. The Hamming weight $\omega(x)$ of a binary word x corresponds to the distance between x and the null word, that is, $\omega(x) = d(x, 0)$. In our example, $\omega(x) = 4$ and $\omega(y) = 3$. The minimum distance $d_{min}(C)$ of a code C corresponds to the shortest possible Hamming distance between any two codewords belonging to this code. If a code C enables the detection of p simultaneous erroneous bits (in the case, for instance, of the RS code described in the following section) in a codeword, one may prevent any error at the receiver by imposing the received erroneous codeword not to belong to C. In other terms, the minimum distance of this code must satisfy $d(C) \geq p + 1$. As an example, let M_1' be the erroneous version of a transmitted codeword M_1 as it is detected at the receiver. Let M_2 be the nearest codeword (according to Hamming distance) of M_1. If the correction capacity of C is of q simultaneous erroneous bits, M_1' is interpreted by the receiver as M_1 if M_1' belongs to the disk of radius q centered on M_1. Thus, $d(M_1, M_2)$ must be greater or equal to $2 \cdot q + 1$ (this distance is always an integer value). Figure 6.25 illustrates this property in the case of a 4-QAM modulation.

A condition for enabling a code C to detect any single error in a received code word is that $d(C) \geq 2.1 + 1$—that is, $d(C) \geq 3$. Knowing the definition of the minimum distance of a code C, this code also detects any double errors in a received code word.

6.7.2   Principle of Reed-Solomon (RS) Encoding

We describe in this section the basic principle of an RS encoder. As illustrated in Figure 6.21, the RS encoder is the first functional block of an ADSL Tx-modem. We have underlined in Chapter 4 the difficulty of preventing the negative impact of impulse noise. We have mentioned that such noise is unpredictable in time, in frequency, and in amplitude. We learned in the mid-1990s that impulse noise induces in most cases a voltage perturbation on one or several copper pairs belonging to a same binder cable during time periods of at most $500 \, \mu s$. RS encoders are traditionally well known for their efficiency in reducing the impact of grouped error bits in radio-mobile communication. Historically, different techniques summarized in Figure 6.26 have been used for error protection [17]. Cyclic codes or *cyclic redundancy checksum* (CRC) are widely adopted for layer 2 protocols like HDLC, LLC, frame relay, or ATM. RS encoders belong to the block codes family whereas the convolutional encoders that we describe in Section 6.9 belong to the trellis code family. Unlike cyclic encoders that operate in a modulo 2 Galois field, RS encoders operate on the basis of an alphabet that consists of associating a symbol to each suite of k information bits.[21] The amount of symbols in the alphabet is then $q = 2^k$. An RS(N, K) encoder associates with an input block made of K information symbols $N - K$ parity check symbols, with $N = q - 1$ and with K being a positive integer strictly smaller than N. The output of an RS coder is made of N/K sub-blocks of K

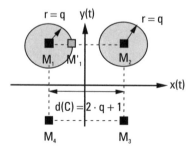

Figure 6.25   Illustration of the minimum distance of a code.

---

21. Such codes are comprised of nonbinary block codes.

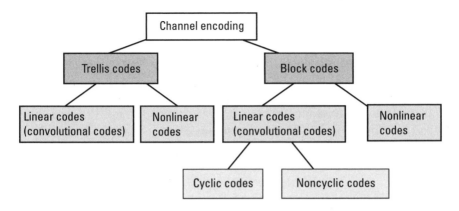

Figure 6.26 An error correction techniques classification. (*From:* [17]. © 1999 IEEE. Reprinted with permission.)

symbols each. Whereas cyclic codes only enable error detection, RS codes also enable error correction. The minimum distance d_{min} of an RS code is equal to $N - K + 1$.

Without going into the details,[22] let us simply say that an RS code is also characterized by its correction capacity L. It is shown that up to L erroneous symbols may be detected and corrected among the set of the N symbols of the N/K output sub-blocks. The value of L is given by $(d_{min} - 1)/2 = (N - K)/2$. As an example, let us consider an RS(12, 4) code with 48-bit input blocks. Let us also assume that an input block is made of four symbols of 12 bits. For each input block, the output of the RS encoder gives three distinct versions (i.e., three sub-blocks) of the input block. The size of a sub-block is equal to the size of the input block (i.e., 48 bits). The minimum distance of the RS(12,4) code being 9, the correction capacity L is then equal to 4. In other terms, a block transmission is successful if at least four symbols are received error-free among the 12 received symbols of the three received sub-blocks. One notices that RS encoding implies a nonnegligible overload on the transmission channel. In the context of ADSL modems, one uses RS(240, 224) encoders. Such coders are able to detect and correct up to eight erroneous symbols per input block.[23] Let us recall that RS encoders are used as outer codes in the ADSL-concatenated codes.

6.8 Interleaver

The second stage of an ADSL transmission chain after an RS(N, K) corresponds to an interleaver (see Figure 6.21). The aim of this interleaver is to facilitate

---

22. The interested reader may refer to [15].

23. Other values of N, K, and L may be used by different vendors.

error detection at the receiver in case of long bursts of errors. Thanks to a certain bit interleaving rule at the transmitter, it is possible to spread out contiguous erroneous bits into isolated erroneous bits after deinterleaving at the receiver. Interleaving operates on a certain amount Ψ (Ψ is, for instance, equal to N/K) of output sub-blocks of $n = K \cdot k$ bits. The simplest interleaving consists in writing in a memory a matrix of Ψ lines and n columns. Up to Ψ successive sub-blocks issued by the RS(N, K) encoder are enqueued into Ψ parallel buffers of n bits. The first sub-block is inserted into the first buffer, the second sub-block into the second buffer, and so on. Once the matrix is full, bits are dequeued column by column as represented in Figure 6.27. Another interleaving rule can consist of dequeueing the matrix byte by byte instead of bit by bit. The thinner the interleaving rule, the better the protection against bursts of errors.

Interleaving/deinterleaving requires a certain delay. According to the state of the technology, such an operation is achieved[24] in about 20 ms in the case of ADSL modems. The ADSL Forum members have estimated that such a delay is incompatible with real-time applications like VoIP.[25] Today, two types of hardware configurations are considered within ADSL modems, the *fast buffer* and

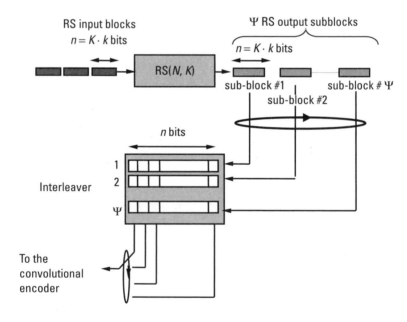

Figure 6.27 Example of interleaver.

---

24. The thinner the granularity of interleaving, the longer the delay inherent to interleaving/deinterleaving.

25. We consider here the case of a user that needs a second phone line in addition to its standard POTS access.

the *interleaved buffer* configurations. The former is dedicated to real-time traffic that cannot support interleaving/deinterleaving delays. The latter is dedicated to nonreal-time traffic like TCP/IP sessions that are less sensitive to end-to-end delays.

6.9 Convolutional Coder

The third stage of an ADSL transmission chain corresponds to a convolutional encoder (see Figure 6.21). Associated to the RS encoder, a convolutional encoder is qualified of inner code of the ADSL concatenated code. Like the RS encoder and the interleaver, the convolutional encoder contributes to error protection. By grouping both DMT-ADSL and CAP-ADSL access systems on a same illustration, the reader may find Figure 6.21 a bit ambiguous. Indeed, CAP modulation is single-carrier–oriented whereas DMT is multicarrier-oriented. In other words, if convolutional encoding is applied to a unique constellation in the case of CAP, separated convolutional encoders are in fact necessary for each of the DMT subchannels; this multiplicity of the convolutional encoders does not appear in Figure 6.21.

6.9.1 Objectives of Convolutional Codes

The main objective of a convolutional encoder is to prevent any ambiguity at decoding. Let us illustrate in a very simplified approach the impact of noise on a received signal. The average impact of additive noise in terms of erroneous bits after decoding may be represented in the two-dimensional space inherent to QAM modulation by several disks with radius r centered on each point M of the constellation. Let us consider for example a 16-QAM constellation. We assume that the amplitude and phase of the received signal during a symbol period corresponds to point M' represented in Figure 6.28. One notices that M' belongs to the intersection of two adjacent noise disks centered on M_1 and M_2. If M stands for the codeword associated to the current symbol at the transmitter, an error at decoding occurs if M' is, for instance, interpreted as M_2, whereas $M = M_1$.

From Figure 6.28, it appears that error probability at decoding is related to the Hamming distance between the received signal and the reference points of the constellation. Let x_i and y_i be the coordinates of each point M_i in the constellation ($i \in [1, 2, 3 \dots 2^k]$). Let us assume a uniform probability of occurrence of the various codewords. For codewords with k bits, the average transmission power of a QAM signal is then given by:

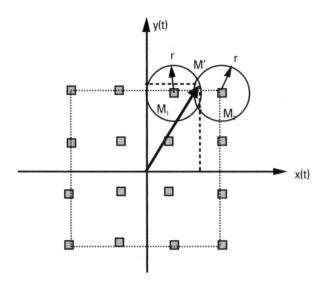

Figure 6.28 Noise illustration in a 16-QAM.

$$P_a = \frac{1}{2^k} \cdot \sum_{i=1}^{2^k} \sqrt{x_i^2 + y_i^2} \qquad (6.71)$$

Two constellations with the same size and the same minimum Euclidian distance may be distinguished by means of their average transmission power. For a given error probability, the constellation enabling the lowest average transmission power is preferable. For a given maximum admissible transmission power (we must take care not to saturate the sensitivity of receivers), SNR decreases with the number of bits per symbol. The objective of a convolutional encoder is to be able to recover with a high probability the right information associated with each received signal during successive symbol periods. For that purpose, a convolution encoder must keep track of the most recent transmitted symbols. On the basis of this memory, a convolutional encoder is able to bypass any ambiguity at decoding. For that purpose, a convolutional encoder proceeds to an extension and to a partition of the original constellation. Thanks to a constellation redundancy, several symbols are possible for a given input information block.[26] The choice of the symbol among different candidates is determined on the basis of an algorithm known by both the transmitter and the receiver. The decoding of the current signal also depends on the values of the Γ previously received information blocks.

---

26. With RS encoders, a single value of the output symbol is assigned to each input block.

6.9.2 Partition and Extension of a QAM Constellation

6.9.2.1 Limits of Hard Decision Decoding

Cyclic codes (see Figure 6.26) are based on the adjunction of p redundant bits to each input information block of k bits. For this type of code, modulation (for instance, QAM) and channel encoding are independent. Traditional error-correcting codes are based on "hard decision decoding" that associates with the received signal (an amplitude and a phase for QAM) the nearest codeword. In the example of Figure 6.28, for a transmitted symbol M_1, the received signal M is interpreted as symbol M_2 if $d(M, M_2) \leq d(M, M_1)$. To maintain the same useful data rate in spite of the overrate induced by the p redundant bits, two options are possible:

- One may increase the rapidity of modulation and then enlarge the on-line spectrum. We know that the larger the on-line spectrum, the greater the sensitivity to amplitude and phase distortion.
- We can also improve spectral efficiency by increasing the size of QAM constellation. This operation degrades SNR.

In practice, ADSL modem manufacturers prefer the second approach, the on-line spectrum being specified in the standard (1,104 kHz).

6.9.2.2 Principle of Soft Decision Decoding

Hard decision decoding may induce irreversible errors in the received data.[27] Soft decision decoding makes it possible to prevent this irreversibility. Let a_n be the successive bits inserted into the transmission system. Let r_n be the successive bits received after demodulation. The r_n suite does not coincide exactly with the suite a_n because of noise. Soft decoding consists of determining the suite a^*_n that is at a minimum Euclidian distance from suite a_n such that:

$$\left| r_n - a^*_n \right|^2 = \underset{a^*_n \in C}{Min} \sum_n \left| r_n - a_n \right|^2 \qquad (6.72)$$

where a^*_n describes the set of all the bits that may be generated randomly by the source. In other words, soft decoding consists of building a code C in coherence with the adopted modulation that maximizes the minimum Euclidian distance between a_n and r_n. Such a maximization technique has been developed by Gottfried Ungerboek [18]. Soft decoding efficiency is paid by a greater computing complexity essentially carried out in the Rx-ADSL modem.

---

27. Irreversibility only concerns the physical layer. Fortunately, error recovery techniques are also considered in upper layer protocols such as LLC and TCP.

For a given constellation, it is possible to associate a codeword to each point of the constellation in order to maximize the Euclidian distance between each other by partitioning judiciously the constellation. In that case, the term of *coded modulation* is frequently adopted because of the strong dependence between modulation and channel encoding. Figure 6.29 illustrates, for instance, how it is possible to partition an 8-PSK constellation [19].

Each point of the constellation corresponds to a 3-bit codeword. Let E be the energy of the modulated signal corresponding to the square of the Euclidian distance between each point from the origin (the eight points of the constellation are equidistant from the origin). It is possible to partition step by step this constellation into subconstellations such that, within each subconstellation, the minimum Euclidian distance increases step by step. Thus, let us observe the subconstellations given in Figure 6.29 from the top to the bottom. At the top of Figure 6.29, one has represented the original 8-PSK constellation for which the minimum Euclidian distance d_0 is given by:

$$d_0 = \sqrt{\left(2 - \sqrt{2}\right)E} \qquad (6.73)$$

A first possible partition consists of separating this constellation into two subsets of four points each with a minimum Euclidian distance d_1 given by:

$$d_1 = \sqrt{2 \cdot E} > \sqrt{\left(2 - \sqrt{2}\right)E} \qquad (6.74)$$

In a second step, each four-point subconstellation may in its turn be partitioned into two subconstellations of two points each with a minimum Euclidian distance d_2 given by:

$$d_1 = 2\sqrt{E} > \sqrt{2 \cdot E} > \sqrt{\left(2 - \sqrt{2}\right)E} \qquad (6.75)$$

We describe in Section 6.10.4 how it is possible to couple such a QAM partitioning technique with convolutional encoding.

6.9.3 Principle of a Convolutional Coder

In reference to the block diagram of a Tx-ADSL modem (see Figure 6.21), the serial bits arriving from the interleaver are passed to the convolutional encoder. By convention these bits are grouped into blocks of k bits. As already mentioned, a convolutional encoder is characterized by a memory of the last Γ transmitted blocks. The value of the codeword assigned to the ith input block depends on this ith input block but also on input blocks $(i - 1)$, $(i - 2)$, ..., $(i - \Gamma)$. A

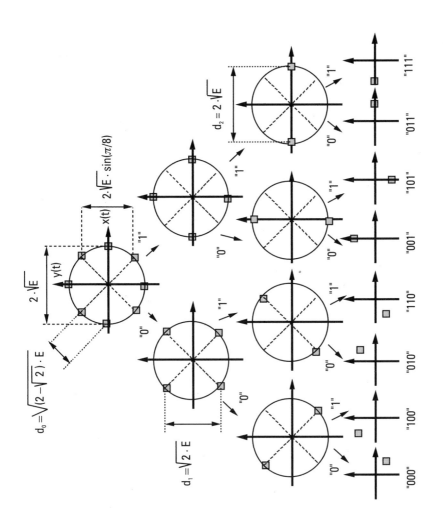

Figure 6.29 Example of constellation partitioning for an 8-PSK. (*From:* [19]. © 1995 McGraw-Hill, Inc.)

convolutional encoder is made of a set of $\Gamma + 1$ shift registers of size k. These shift registers are connected between each other by means of n algebraic additioners corresponding to the n outputs of the convolutional encoder. Figure 6.30 gives an example of convolution encoder with $\Gamma = 2$, $k = 3$, and $n = 4$.

After each bit duration, n output bits are calculated by the convolutional encoder. This means that for every k-bit input block, one gets an output sequence of $k \cdot n$ bits that are passed to the next stage—that is, to the CAP or DMT modulator. Simultaneously, the $(\Gamma + 1)$ blocks are passed to the right from one shift register to the next one.[28] The code rate and the constraint length of a convolutional encoder are defined as the ratio k/n and the value of $(\Gamma + 1)$, respectively.

6.9.4 Trellis Diagram

The trellis diagram of a convolutional encoder describes the different possible states of the encoder and the possible transitions between these states that occur after every k bit durations. To detail the principle of the trellis diagram concept, one considers for the sake of simplicity a convolutional encoder with $\Gamma = 2$, $k = 1$, and $n = 2$. The code rate and the constraint length of this encoder are then $1/2$ and 3, respectively.

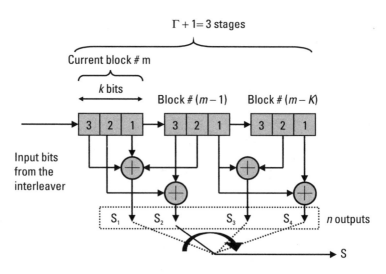

Figure 6.30 An example of convolutional encoder.[29]

---

28. The k bits placed in the most right shift register are lost at their output of this device.

29. This convolutional encoder serves as an illustration and is not the one implemented in ADSL modems.

Let us recall that in DMT-ADSL modems, the size of the constellation associated to each subchannel is reevaluated at each reinitialization of the line. In other terms, the hardware implementation of convolutional encoders must be adaptive. From Figure 6.31, it is possible to determine the trellis diagram inherent to the considered convolutional encoder. This diagram is made of four possible states corresponding to the possible values of (b_{t-1}, b_{t-2}). Transitions from one state to another are activated at each bit b_t arrival. Figure 6.32 depicts the trellis diagram of the encoder of Figure 6.31. In Figure 6.32, the variable e stands for the value of the bit b_t. For instance, if $(b_{t-1}, b_{t-2}) = (1, 0)$ and if $e = 0$, then $(b_{t-1}, b_{t-2}) = (0, 1)$ and $(S_1, S_2) = (1, 0)$ at the next time bit.

Let us now describe in Figure 6.33 the time evolution of the states of the encoder according to the possible sequence of input blocks. The horizontal axis

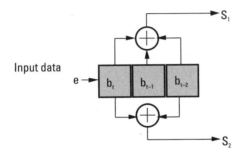

Figure 6.31 Example of a convolutional encoder with code rate 1/2.

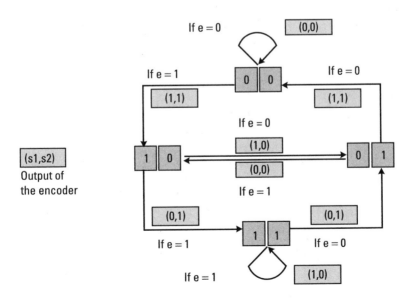

Figure 6.32 Example of trellis diagram.

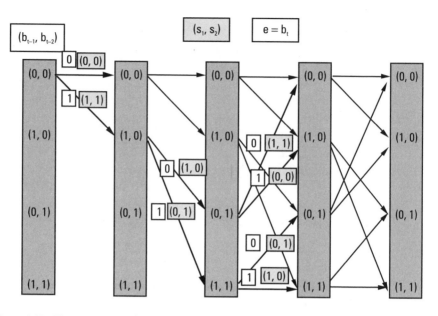

Figure 6.33 Time evolution of the convolutional code of Figure 6.31.

corresponds to discrete time evolution, the increment being equal to an input block duration (i.e., in our simple example to a bit duration). The vertical axis describes the possible state of the first Γ shift registers. In our example, $\Gamma = 2$, the state of the two shift registers with a single bit position being characterized by the couple (b_{t-1}, b_{t-2}). Directed edges between couples (b_{t-1}, b_{t-2}) illustrate the possible transitions between two successive states of the encoder. One has associated with some of these directed edges the value of $e = b_t$ and of (S_1, S_2). After three bit durations, one notices that the automata becomes repetitive.

As an example, let us consider an input sequence of 5 bits equal to "11010." We have plotted this sequence of Figure 6.34. The reader may notice that we have added to the input sequence $\Gamma \cdot k$ bits equal to "0" (i.e., 2 bits in our example) in order to guarantee that the last bit of the input sequence (a "0" in our example) has left the last stage of the encoder. In other terms, one considers in fact the input sequence "1101000." The successive values of (S_1, S_2) associated to this sequence are: (1, 1), (0, 1), (0,1), (0, 0), (1, 0), (1, 1), (0, 0).

6.10 DMT-ADSL Receiver

In reference to Figure 6.21 illustrating the block diagram of an ADSL access system, we have described in the previous sections the four main modules of a

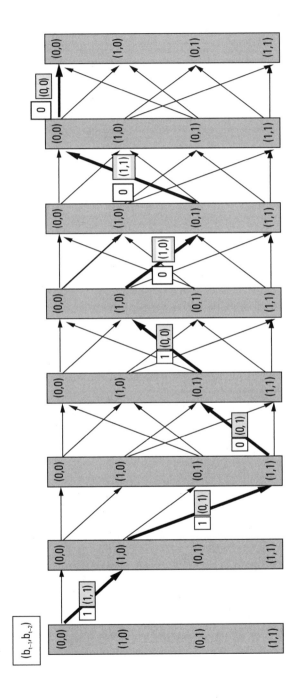

Figure 6.34 An example of transitions of the state automata of a convolutional encoder.

Tx-ADSL modem. For that purpose, we first analyzed in Section 6.6 the principle of a DMT-ADSL modulator. We then studied in Sections 6.7–6.9 the principles of the RS encoder, the interleaver, and the convolutional encoder. The objective of a digital receiver is to extract the information sequence from a discrete time signal. This discrete time signal is obtained from the CAP or DMT demodulator. This section is dedicated to the description of Rx-DMT-ADSL modem. First we present in Section 6.10.1 the main characteristics of a DMT-ADSL demodulator before describing in Sections 6.10.2—6.10.6 the various functionalities carried out in a Rx-ADSL modem.

6.10.1 DMT-ADSL Demodulator

Figure 6.35 describes the principle of a Rx-DMT-ADSL access system made of six stages. The first stage eliminates frequencies higher than $W = 1,104$ kHz by means of a lowpass filter. The second stage proceeds to A/D conversion in sampling the received signal at a frequency f_s equal to at least twice W. From this sampling, one gets a suite of real coefficients $x'_{i,p}$ and $y'_{i,p}$ at instants kT_s. The values of these samples are approximations of the values of coefficients $x_{i,p}$ and $y_{i,p}$ mentioned in Sections 6.6.1 and 6.6.2 (these sections refer to the Tx-DMT-ADSL access system).

Once it has been detected, the cyclic preamble is eliminated by the stage 2, which can then rebuild the block structure mentioned in Section 6.6.2 (see Figure 6.24). Stage 3 converts from serial to parallel the suite of $x'_{i,p}$ and $y'_{i,p}$ coefficients. This operation allows the rebuilding during a same symbol duration T_0 the values of the set of samples associated to each of the N subchannels. DMT demodulation is carried out by means of stage 4. This stage uses FFT whose principle has been described in Section 6.6. With FFT, it is possible to

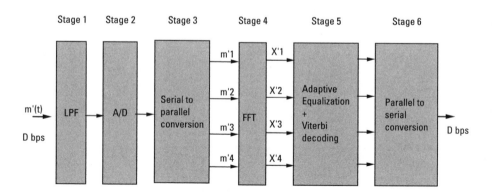

Figure 6.35 Principle of a Rx-DMT-ADSL access system.

determine from the samples of the received signal $m_{i,p}(t)$ during the pth symbol period in subchannel i the coefficients $a_{i,n,p}$ and $b_{i,n,p}$ of the complex Fourier series expansion of these signals. As shown in Figure 6.35, coefficients $a_{i,n,p}$ and $b'_{i,n,p}$ ($X'_{i,n,p} = a'_{i,n,p} + jb'_{i,n,p}$) are passed to stage 5. Let us recall that IFFT and FFT operate on truncated Fourier series expansion up to a limited order. Stage 5 is in charge of adaptive equalization and Viterbi decoding. These two operations are detailed in the following sections.

6.10.2 Adaptive Equalization

Channel distortion results in ISI, which itself induces erroneous bits at decoding. The objective of an equalizer is to reduce the impact of ISI on the received signal. Both CAP and DMT modulations aim to facilitate equalization at the receiver by minimizing on-line bandwidth utilization. Three types of equalizers can be used in digital communication systems. From the least to the best performing, these equalizers are called adaptive transversal equalizers, DFEs, and *maximum-likelihood sequence estimation* (MLSE) equalizers [20].

Adaptive transversal equalizers are based on linear transversal filters followed by a decision device. The logic behind such linear filters is based on a discrete-time model of the transmission channel with ISI. Indeed, a Tx-ADSL modem sends discrete-time symbols every T_s seconds. The received signal at the Rx-ADSL modem passes through a filter matched to the response of the noisy transmission channel to the input signal. Ideally, the set of the filters associated with the Tx-modem, with the noisy transmission channel, and with the Rx-modem should satisfy the Nyquist criteria in order to prevent ISI. This property assumes a regular adaptation of the coefficients of the Rx-modem filter to the fluctuations of the transmission channel. The output of the matched filter at the receiver is sampled every T_s seconds. It is shown that the matched filter can be represented by an equivalent discrete-time transversal filter with tap gain coefficients [19]. In order to adapt the receiver to the fluctuating characteristics of the line, different types of adaptive algorithms may be implemented in an equalizer like the zero-forcing or the mean-square error algorithms. For instance, the zero-forcing technique aims to minimize the worst-case ISI at the output of the equalizer. For that purpose, the transfer function of the equalizer is characterized by a z transform equal to the inverse of the z transform of the transfer function of the transmission channel.

A DFE is made of two filters called *feedforward* and *feedback* and of a decision device. These two filters operate on a discrete time basis corresponding to the symbol period T_s. A detailed description of these equalizers is out of the scope of this book and can be found in [19]. For the sake of simplicity, let us consider a simple NRZ modulation. Let x_k be the samples[30] obtained at the

---

30. Typically, these samples correspond to positive or negative voltage levels $\pm V$.

output of stage 4 of the Rx-modem (see Figure 6.35). A DFE relies on a closed-loop system enabling the evaluation of the value of the received sample at instant n on the basis of the value determined for the previous sample at instant $(n - 1)$ [21, 22]. Figure 6.36 depicts the block diagram of a DFE.

Let us describe the functionalities of the various elements of a DFE:

- The feedforward filter receives successively the signal samples x_k. The role of this filter is to compensate for the evaluation of each sample x_k any memory effect due to the various line impairments. Some noise sources cannot be assimilated to white noise (they present a time dependency evaluated by a non-null autocorrelation at different instants). The feedforward filter tries to modify the characteristics of such noise in order to suppress these autocorrelations.[31] Let I_k be an estimated value of the symbol associated to x_k after noise whitening.

- The feedback filter tries to reduce ISI generated by the previous samples (the ones with an index lower than k). For that purpose, the feedback filter memorizes a sequence $\{I^*_k\}$ of the estimated symbols obtained from the K last received signals. The feedback filter subtract these estimations from the current symbol I_k. The value of I^*_k is obtained by means of a decision device.

Although the feedforward filter and the feedback filters are both linear, a DFE is globally nonlinear because of its closed loop. Two other types of DFE may also be used: predictive DFE and hybrid DFE. The former does not try to whiten the noise but aims at reducing its predictable impact. With hybrid DFE, one tries to associate the benefits of DFE and predictive DFE simultaneously. The main drawback of DFEs is that in case of errors at decoding, this error has an impact on the following symbols. Such a behavior may then favor bursts of errors. Tomlinson-Harashima equalization filters are another class of equalizers

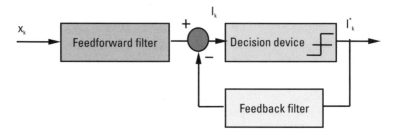

Figure 6.36 Block diagram of a DFE.

---

31. The expression of "noise whitening" is sometimes used in the literature.

aimed at preventing bursts of errors [23]. The main difference between Tomlinson-Harashima filters and DFEs is that they use a feedback filter, not in the Rx modem but in the Tx modem. A few versions of CAP-ADSL on the market have implemented Tomlinson-Harashima equalizers.

6.10.3 Convolutional Decoding

Convolutional decoding is based on the Viterbi algorithm, this algorithm relying itself on the trellis diagrams described above. Similar to Tomslinson-Harashima equalizers, the Viterbi algorithm aims to reduce the impact of an erroneous decoding on the kth symbol on the decoding of symbols $(k+1)$, $(k+2)$, and so on. Let us come back to Figure 6.33 related to the trellis diagram of the convolution encoder of Figure 6.31. One notices that once the encoder is initiated (after three bit durations), two edges arrive on each vertex (b_{t-1}, b_{t-2}). The Viterbi algorithm consists of associating to each edge of the trellis diagram a weight called *weight of Hamming*. This weight corresponds to the Hamming distance between (S'_1, S'_2) evaluated at the receiver and (S^*_1, S^*_2) associated to each edge of the trellis diagram. As an example, let us consider the input sequence $\{b_i\}$ = "1001." This sequence generates at the transmitter the following values $\{(S_1, S_2)\}$ = (1, 1), (1, 0), (1, 1), (1, 1). Let us assume that the receiver decodes the received signal as the following sequence: $\{(S'_1, S'_2)\}$ = (1, 1), (0, 0), (1, 1), (1, 1). There is then an error on bit S'_1 after two bit durations. On Figure 6.37, one

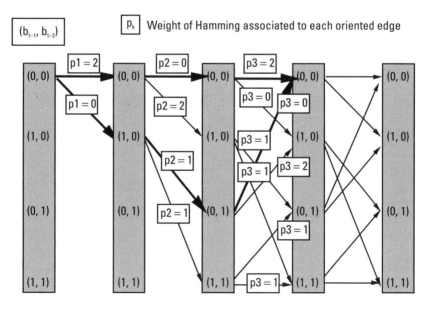

Figure 6.37 Examples of weights of Hamming.

has written the values of the weights of Hamming p_k associated with each edge of the trellis diagram. Let us comment, for instance, on the value of p_2 associated to the edge between $(0, 0)$ and $(1, 0)$ after the second bit duration. One has for this edge $(S'_1, S'_2) = (0, 0)$ and $(S^*_1, S^*_2) = (1, 1)$ then $p_2 = 2$.

In order to prevent an error at decoding,[32] the Viterbi algorithm considers the cumulated weights of Hamming on any path arriving on any state (b_{t-1}, b_{t-2}) from the original state $(0, 0)$. As soon as a path arriving on a state (b_{t-1}, b_{t-2}) is characterized by a cumulated weight of Hamming greater than the cumulated weight of a concurrent path, this path is abandoned in the remainder of the algorithm. Let us consider the same input sequence $\{(S_1, S_2)\} = (1, 1), (1, 0), (1, 1), (1, 1)$ with the same error at the receiver with $\{(S'_1, S'_2)\} = (1, 1), (0, 0), (1, 1), (1, 1)$. Thus, after three bit durations, the path $\{(b_{t-1}, b_{t-2})\} = (0, 0), (0, 0), (0,0), (0,0)$ is characterized by a cumulated weight of Hamming $p_1 + p_2 + p_3 = 2 + 0 + 2 = 4$ which is greater than the cumulated weight of Hamming of the path $\{(b_{t-1}, b_{t-2})\} = (0, 0), (1, 0), (0,1), (0,0)$ for which $p_1 + p_2 + p_3 = 0 + 1 + 0 = 1$ (these two trajectories are represented with bold arrows on Figure 6.37). On the basis of this logic, we are able to determine at each step the maximum likelihood sequence for any state (b_{t-1}, b_{t-2}). As an example, the paths with maximum likelihood after four bit durations are represented in Figure 6.38. For instance, the most probable path is the one arriving on state $(b_{t-1}, b_{t-2}) = (1, 0)$ with $p_1 + p_2 + p_3 + p_4 = 1$.

By reading the supposed values of b_t step by step up to the origin, we may conclude the most probable input sequence represented in Figure 6.38, or $\{b_t\}$ = "1001," which is effectively the right response in spite of the erroneous received bit.

6.10.4 Constellation Partitioning and Convolutional Coding

We have underlined in Section 6.9.2.2 the benefit of QAM constellation partitioning. Coded modulations also called *trellis-coded modulations* (TCMs) associating a QAM modulation with a convolutional encoder are used in both DMT-ADSL modems and CAP-ADSL modems. In the case of DMT-ADSL, one first determines the best QAM for each subchannel according to the value of SNR. One then determines the best line code with the highest minimum distance. Convolutional decoding relies on the analysis and comparison of the distance between the coordinates of the received symbol and the estimated coordinates of the transmitted symbol. Such an analysis is possible only if a redundancy is introduced in the line code. One increases the size of the constellation by keeping the same information rate. Figure 6.39 describes the principle of a TCM encoder associating an 8-PSK modulator with a convolutional

---

32. *T* stands for the bit duration.

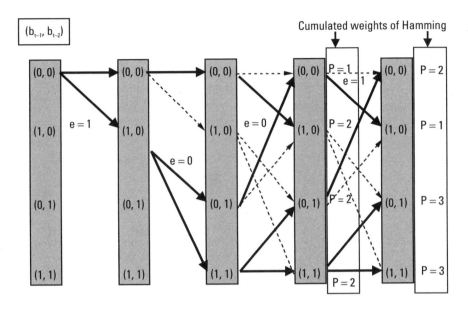

Figure 6.38 Example paths with maximum likelihood at $t = 4T$.

encoder with code rate 1/2. The code rate of the TCM is 3/2 with two input bits e_1 and e_2 and three output bits S_1, S_2, and S_3.

Let us assume that input e_1 is the input bit of the convolutional encoder whereas input bit e_2 is not precoded and feed directly the 8-PSK modulator. The two precoded output bits S_1 and S_2 are used in order to choose among the four subsets of the Hungerboek partition of 8-PSK modulation described in Figure 6.29. The output bit S_3 can choose in each of these four subsets a particular point among two possibilities. The trellis diagram of the TCM encoder depicted in Figure 6.39 is given in Figure 6.40.

6.10.5 Deinterleaving and RS Decoding

The principle of deinterleaving in a Rx-DMT-ADSL modem is described in Section 6.8. Concerning RS decoding, we invite the reader to refer to [24].

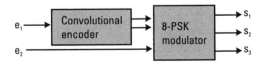

Figure 6.39 An example of TCM encoder.

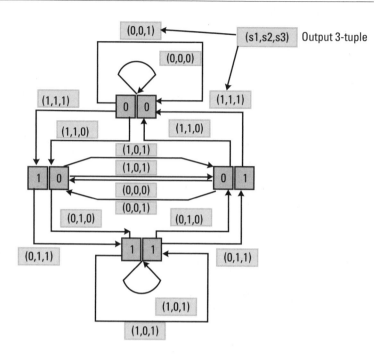

Figure 6.40 Trellis diagram of a TCM encoder.

6.10.6 Parallel-Serial Conversion

The symbols determined by the stage in charge of equalization and Viterbi decoding are passed to the sixth and last stage of a Rx-DMT-ADSL modem (see Figure 6.35). Stage 6 is in charge of parallel-to-serial conversion with the same bit rate of D bps as at the input of the Tx-DMT-ADSL modem.

6.11 Line Initialization

We have mentioned on multiple occasions in the previous sections the necessity to proceed to a test of the transmission line. This operation is necessary to determine the parameters of the equalizer in the Rx-modem. It is also mandatory in DMT-ADSL to set the size of the constellations in each subchannel. In the very first versions of DMT-ADSL modems, this test of the line was carried out only once, in general manually. Today, thanks to the advances in high-speed electronics and DSP, this procedure is automatically reactivated as soon as an inactivity of the line is detected. This flexibility of ADSL modem characteristics is at the origin of the term *rate-adaptive DSL* (RADSL) sometimes used in the literature.

6.11.1 Line Activation

When an ADSL access system is activated, this is either the ATU-R or the ATU-C that takes the initiative to activate the automatic test of the transmission channel. For that purpose, the modem that takes the initiative to activate the ADSL access system sends a request for an activation signal to its correspondent by means of a continuous tone between 200 and 300 kHz. The exact value of this tone is specified in the ITU-T standard T1.413.

6.11.2 Power Control

Once they are activated, the ATU-C and ATU-R must determine their average power of transmission and the level of their decision device. Let us remember that line attenuation strongly depends on frequency. In the case of DMT-ADSL, the power of transmission and the level of the decision device is adjusted for each subchannel. Power budgets of an ADSL access system also depend of the length of the subscriber line. In practice, the modem that takes the initiative to activate an ADSL access system begins to transmit with a low power. It then increases step by step its transmission power until the other modem sends back via a signaling channel a message of acknowledgment. This acknowledgment message specifies to the transmitter that an acceptable power budget has been reached at the receiver. The current transmission power P_0 is then adopted for the duration of the connection. The receiver determines the gain G_0 to apply to the received signal to optimize the efficiency of the A/D converter (see Figure 6.35). An *automatic gain control* (AGC) technique is applied at the receiver to adjust the power level of the signal to be passed to the A/D converter. The AGC relies on a linear adjustment of the gain G over successive time periods. Thus, the gain G_k to be applied during period k is given by the summation of the gain G_{k-1} applied during the previous period with a ponderated value G^*_{k-1}, with G^*_{k-1} being equal to ($G_k - G_0$):

$$G_k = G_{k-1} + \Omega \cdot \left(G_{k-1} - G_0 \right) \qquad (6.76)$$

where Ω is a real characteristic constant of the considered line. The value of Ω is determined empirically.

6.11.3 Synchronization

At the receiver, accurate timing recovery is critical to obtain performance at decoding close to that of the optimal receiver. DMT and OFDM systems are very sensitive to clock distortion. This distortion is noticeable in DMT-ADSL because of the lack of synchronization carriers in the system. Two main types of synchronization are considered: per-block and per-sample. Initially, ADSL designers considered the use of two dedicated subcarriers among the 256 of a DMT-ADSL system. These two subcarriers are used for upstream and downstream channels' synchronization, respectively. Because they are not used for the transport of users' information, these two synchronization subcarriers are generated independently from the IFFT hardware. This approach has not been adopted by the standardization bodies because of its inherent waste in bandwidth capacity. Another synchronization approach that is per-block–oriented uses the prefix mentioned in Figure 6.24. A "block" corresponds to the set of the samples $x_{i,p}$ and $y_{i,p}$ computed by the IFFT for the N subchannels during a symbol period T_s with the associated prefix. As an illustration, the adopted technique making it possible to keep block synchronization at the receiver for upstream traffic is illustrated in Figure 6.41. Assuming 25 upstream subchannels in a DMT-ADSL access system, up to 25 samples $(x_{i,p}, y_{i,p})$ of the pth symbol are considered in a block.

During the line initialization phase, the ATU-R and ATU-C exchange test sequences. By means of a sliding window with size η (η being the size of a prefix), the receiver determines over each symbol period the cumulated gap between the value of a sample at instant t and the value of a sample instant $t - 2N$. These samples refer to the same received signal on a given subchannel during a symbol period. By considering received data during a delay much larger than a block duration, it appears that there are some windows for which the cumulated gap is close to zero. One deduces from the detection of such windows the time positioning of the beginning of blocks.[33] One knows a priori the number of samples $(x_{i,p}, y_{i,p})$ within a block for upstream traffic. One deduces

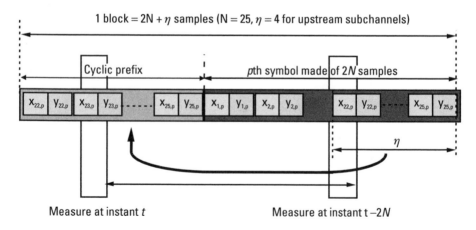

Figure 6.41 Block synchronization for upstream traffic.

33. This technique recalls the cell delineation technique used in ATM networks for recovering cell periodicity.

from this information the samples' synchronization. Two types of synchronizer structures are currently implemented in Rx-DMT-ADSL modems: *data-aided* (DA) and *nondata-aided* (NDA). The former operates on the basis of a training sequence in computing the timing estimates. The latter does not use the transmitted information sequence but averages the likelihood function over random information variables [25].

6.11.4 Subchannel Characterization

During the initialization phase, the Rx-modem determines the impulse response of the line and the power density spectrum of noise in the band [25 kHz, 1,104 kHz]. The Rx-modem then evaluates SNR along this band. The set of these measurements is carried out twice: after line activation, power control, and synchronization and once adaptive equalizers have tested the line before to proceed to their configuration.

6.12 Conclusion and Perspectives

Because of industrial and commercial interests, many debated in the first half of the 1990s the pros and the cons of DMT and CAP. The following concludes this chapter with a few objective elements of comparison between these two possible approaches for the design of ADSL modems:

- As mentioned several times in this chapter, DMT relies on narrowband subchannels that authorize a simpler equalization than in the case of CAP modulation.

- In a first approach, single-carrier modulation adopted by CAP seems to be more high-performing than multicarrier modulation. Indeed, in reference to Shannon's capacity, the division of the medium capacity into many narrowband independent subchannels requires guard bands (obtained by means of cyclic prefix) that consume a fraction of this bandwidth.

- We have mentioned that many disruptive factors must be considered in an ADSL access system (e.g., impulse noise, crosstalks, impedance impairments, bridge taps, load coils, and AM radio interference). In the case of single-carrier modulation, any degradation of SNR at a given frequency has a negative impact on the whole system. This is not true with multicarrier modulation, subchannels being independent.

- In DMT, the symbol period T_0 used commonly by the N subchannels is much larger than the CAP symbol period T'_0. Indeed, the speed of a CAP modem is 1,024 kHz, whereas it is 4 kHz with DMT (this speed is

256 times lower than the speed of CAP modems). We have outlined that impulse noises have in their great majority a duration lower than 500 μs. In practice, the duration of impulse noise is in many cases low enough in comparison to T_0 to be neglected. This is not the case with T'_0.

- Flexibility is presented as the first advantage of DMT compared to CAP. A DMT modem evaluates almost permanently the quality of the transmission line. It is then adaptable to a wide range of subscriber lines. In the case of CAP, a lower quality line means a size reduction of the constellation. Some of the CAP-ADSL modem manufacturers have implemented multiple hardware in their equipment corresponding to different line conditions. The choice of the hardware version activated on the line is determined after an automatic line test. In DMT, capacity reduction of the system may be carried out with a granularity of 32 Kbps. Such a granularity is much thinner than the granularity achievable with CAP-ADSL modems on the market.

- Single-carrier modulation technologies are more mature than multicarrier modulation technologies. The electronic devices used for single-carrier modulation are simpler and less power-consuming than those used for multicarrier modulation. The problem of power consumption is more stringent in the ATU-C than in the ATU-R. Indeed, as presented in Chapter 8, ATU-Cs are grouped by tens or hundreds in a DSLAM at the EO.

- DMT uses FFT and IFFT. The computation delay required by such electronic circuits may be unacceptable in reference to the expected end-to-end delay for real-time applications such as digital telephony. Chapter 10 of this book is dedicated to *voice over DSL* (VoDSL) and to video over DSL.

- DMT-ADSL modems are less sensitive to the fluctuations of the physical channel than CAP-ADSL modems. Thus, DMT is better suited to fight against egress crosstalk due, for instance, to AM radio stations. It is sufficient in this case not to use the corresponding subchannels. Globally, the procedures inherent to constellation adjustment consume CPU and memory space but are more efficient than an isolated rate adaptation carried out with a single-carrier modulation.

- Line initialization is more complex in DMT than in CAP and is in general more time-consuming.

- EC in DMT-ADSL is more complex than in CAP-ADSL.

- In the case of CAP-ADSL modems, upstream and downstream channels may be separated by means of active analog filters. For these type of modems, one uses Tomlison-Harashima equalizers.

- In the case of DMT-ADSL modems, phase distortion may be penalizing for ICI whereas CAP-ADSL modems are insensitive to phase distortion because of the absence of the demodulation carrier on the line.

References

[1] Reusens, P., et al., "A Practical ADSL Technology Following a Decade of Effort," *IEEE Communications Magazine,* October 2001, pp. 145–151.

[2] ADSL Forum, http://www.adsl.com.

[3] *Asymmetric Digital Subscriber Line (ADSL) Metallic Interface,* American National Standard (ANSI) for Telecommunications T1.E1.413/97-007R4, 1995.

[4] *Network and CustomerInstallation Interfaces, Asymmetric Digital Subscriber Line (ADSL), Metallic Interface,* American National Standard (ANSI) for Telecommunications T1-413-1998, 1998.

[5] *Asymmetric Digital Subscriber Line (ADSL) Transceivers,* ITU-T Recommendation, G.922.1, June 1999.

[6] *Asymmetric Digital Subscriber Line (ADSL) European Particularities: ETR-328,* European Telecommunications Standard Institute, *ETSI TM6,* 1996.

[7] Maxwell, K., "Asymmetrical Digital Subscriber Line: Interim Technology for the Next 40 Years," *IEEE Communications Magazine,* October 1996.

[8] Teger, S., and D. J. Waks, "End-User Perspectives on Home Networking," *IEEE Communications Magazine,* April 2002, pp. 114–119.

[9] Ethernet in the First Mile Alliance (EFMA), http://www.efmalliance.org.

[10] Gagnaire, M., *Towards Optical Ethernet in Access and Metro Networks,* Business Briefing: Global Optical Communications, World Market Research Center, 2002.

[11] Cherubini, G., E. Eleftheriou, and S. Olçer, "Filtered Multitone Modulation for Very High-Speed Digitial Subscriber Lines," *IEEE Journal on Selected Areas in Telecommunications,* Vol. 20, No. 5, June 2002, pp. 1016–1028.

[12] Im, G- H., and J. J. Werner, "Bandwidth Efficient Digital Transmission over Unshielded Twisted-Pair Wiring," *IEEE Journal on Selected Areas in Communications,* Vol. 13, No. 9, December 1995.

[13] Mueller, K. H., and J. J. Werner, "A Hardware-Efficient Passband Equalizer Structure for Data Transmission," *IEEE Trans. on Communications,* Vol. COM-30, March 1982, pp. 538–541.

[14] Weste, N., and D. J. Skellern, "VLSI for OFDM," *IEEE Communications Magazine,* October 1998, pp. 127–131.

[15]   Lee, I., J. Chow, and J. M. Cioffi, "Performance Evaluation of a Fast Computation Algorithm for the DMT in High-Speed Subscriber Loop," *IEEE Journal on Selected in Communications,* Vol. 13, No. 9, December 1995, pp. 1564–1570.

[16]   Pollet, T., et al., "Equalization for DMT-Based Broadband Modems," *IEEE Communications Magazine,* Vol. 38, No. 5, May 2000, pp. 106–113.

[17]   Gappmair, W., "Claude E. Shannon: The 50th Anniversary of Information Theory," *IEEE Communications Magazine,* April 1999, pp. 102–105.

[18]   Ungerboek, G., "Trellis-Coded Modulation with Redundant Signal Sets: Part 1, Introduction," *IEEE Communications Magazine,* February 1997, pp. 5–21.

[19]   Proakis, J. G., *Digital Communications,* 3rd ed., New York: McGraw-Hill, 1995.

[20]   Bahl, L. R., et al., "Optimal Decoding of Linear Codes for Minimizing Symbol Rate," *IEEE Trans. on Information Theory,* Vol. IT-20, March 1974.

[21]   Saltzberg, B. R., "Comparison of Single-Carrier and Multitone Digital Modulation for ADSL Applications," *IEEE Communications Magazine,* November 1998, pp. 114–121.

[22]   Price, R., "Nonlinearly Feedback Equalized PAM Versus Capacity for Noisy Filter Channels," *Proc. of the International Communications Conference,* 1972, pp. 2212–2217.

[23]   Tomlinson, M., "New Automatic Equalizer Employing Modulo Arithmetic," *Electronic Letters,* Vol. 7, No. 5, March 1971, pp. 138–139.

[24]   Berlekamp, E. B., "Bit-Serial Reed-Solomon Encoders," *IEEE Transactions on Information Theory,* Vol. IT-28, November 1982.

[25]   Pollet, T., and M. Peeters, "Synchronization with DMT Modulation," *IEEE Communications Magazine,* April 1999, pp. 80–86.

Selected Bibliography

Weinstein, S. B., and P. M. Ebert, "Data Transmission by Frequency Division Multiplexing Using the Discrete Fourier Transform," *IEEE Trans. on Communications,* Vol. 19, No. 5, October 1971, pp. 628–634.

7

G.lite and VDSL

7.1 Introduction

Chapter 6 presented details of the characteristics of ADSL access systems. We focused mainly on physical layer aspects; higher layers referring to end-to-end protocol architectures are the subject of Chapter 9. This chapter discusses two variants of ADSL access systems: G.lite and VDSL. G.lite was developed to bypass the technological and administrative constraints inherent to POTS splitters. Whereas G.lite corresponds to a limited version of ADSL, VDSL aims at higher performance and extended services on existing copper wires in the local loop.

7.2 G.lite

G.lite technology is a light version of DMT-ADSL [1]. As outlined in the following, the performance and services offered by a G.lite access system are limited in comparison to those offered by ADSL access systems. Because of this characteristic, DMT-ADSL is sometimes thought of as "full-rate ADSL." G.lite was standardized in 1999 by ITU-T as a subset of the ADSL standard under the reference G.992.2. In an analogy to the G.lite standard, full-rate ADSL ITU-T standard is known as G.dmt. One of the main motivations that has driven the development of G.lite technology is related to end users' expectations. Indeed, end users want to benefit from high-speed, low-cost, and easy-to-use access systems. In-home cabling is an important aspect of the cost of an access system and of its complexity of installation. From this point of view, the ideal configuration for the CPE is a "plug-and-play" approach. To facilitate the rapid commercial deployment of xDSL technologies, end users can purchase xDSL modems at a specialized shop or to get it from their access provider. The installation of this modem to the *network interface device* (NID) should be sufficiently easy to be carried out by the users themselves. Similarly, the connection of multiple devices on the private modem should also be as convenient as possible. We will discuss this last aspect in more detail in Chapter 8. As mentioned in Chapter 6, POTS splitters in charge of separating the [0, 25 kHz] and [25 kHz, 1,104 kHz] bands in ADSL systems did not authorized plug-and-play ATU-R at the beginning of the 1990s.

7.2.1   POTS Splitter Limitations

Two types of constraints, technical and regulatory, have prevented the emergence of plug-and-play ATU-R.

7.2.1.1   Technical Constraints

A POTS splitter is not a simple cascade of a lowpass [0, 25 kHz] filter with a passband [25 kHz, 1104 kHz] filter but a multiband (or hybrid) filter. Independent from the operating mode of an ADSL modem (EC or FDM), a POTS splitter must select simultaneously two sub-bands, the [0, 4 kHz] band dedicated to analog telephony and the [25 kHz, 1,104 kHz] dedicated to Internet services. Figure 7.1 illustrates the frequency characteristic of a POTS splitter. We have represented in Figure 7.1 (see also Figure 6.2) an isolated tone between the analog and the digital bands. Such a tone is sometimes used by POTS carriers to inform in real time the end users of their billing.

Section 6.4.1.1 introduced the roll-off factor β of a filter (see also Figure 6.10) such that:

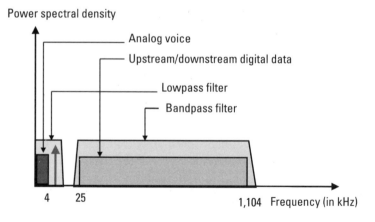

Figure 7.1   Frequency characteristic of a POTS splitter.

$$\beta = \frac{2 \cdot f_c - \frac{1}{T}}{\frac{1}{T}}, \text{with } 0 \le \beta \le 1 \qquad (7.1)$$

Because of the narrow spectral gap separating the upper bound of the low-pass filter from the lower bound of the bandpass filter, POTS splitters must be characterized by two roll-off factors (one around 10 kHz and the other one around 20 kHz) close to zero. Two techniques are possible to implement POTS splitters:

- *Passive POTS splitters:* A passive hybrid filter made only with passive devices such as resistances R, inductances L, conductances G, and capacitances C does not need any powering. Thus, a passive POTS splitter is always operational, even in case of lack of powering at the end user. In practice, it has been outlined that designing passive POTS splitters with a roll-off close to 0 is difficult and quite expensive.

- *Active POTS splitters:* Such splitters are made with electronic circuits that need local powering to be operational. Unlike passive POTS splitters, active POTS splitters are much easier to design. ADSL access is assumed to be "always on." This requires thus a permanent availability of the local powering at the end user and a good robustness of the electronic chips. To avoid losing the availability of the analog telephone line in the case of failure of local powering, active POTS splitters should be accompanied by a secured powering system (batteries, for instance). In addition to their inherent cost, such batteries must be regularly checked either by the end user or by a third party.[1]

In an unbundling context (see Chapter 3), the access provider is in charge of the supervision of the subscriber line; the provider of the analog telephone service and the Internet access provider may correspond to a single, or two or even three distinct administrative entities. Whatever the sharing of the responsibilities, a validation of the POTS splitter installed at the customer premises is highly desirable in order to guarantee the audio quality of the analog telephone service. We outline in the following sections the technical constraints in terms of cabling and of crosstalk imposed on POTS splitters by high ADSL frequencies.

7.2.1.2 Regulation Constraints

We have underlined in Chapter 3 that three regulation contexts are today possible in the local loop: interconnection, partial unbundling, and full unbundling.

---

1. We discuss the identity of this third party in Section 7.2.1.2.

Some of the major constraints inherent to the regulation context are listed as follows.

- *Case 1—Interconnection.* The interconnection context is the most simple to manage. The analog telephone service provider is also in charge of the provision of digital access to public data networks. This administrative entity corresponding to the ILEC has the responsibility to supervise and maintain the copper wire infrastructure. The ILEC prefers in general to provide the ATU-R and the ATU-C. This gives the ILEC the opportunity to install by itself the ATU-R at the end user and to validate in situ the conformance of the POTS splitter.

- *Case 2—Partial unbundling with ATU-R provisioning.* In a situation of partial unbundling, the provider of the analog telephone service is different from the provider of digital access to public data networks. In most cases, the former is the ILEC, the latter is the CLEC. Like in case 1, the CLEC may wish to provide the ATU-R and ATU-C to ensure their compatibility. One assumes here that the CLEC prefers to install by itself the ATU-R. The ILEC may want to validate the conformance of the POTS splitter coupled with the ATU-R. Such filters condition the audio quality of the analog telephone system. This situation is a bit risky for the CLEC. Indeed, the ILEC may choose not to validate the service offered by the CLEC if it judges that, due to local conditions, the end user's POTS splitter is not able to guarantee the expected audio quality to the analog telephone service. In practice, POTS splitter validation may give the opportunity to the ILEC to discover which kinds of services are offered to the customer by the CLEC. Another point of friction between the ILEC and the CLEC concerns the billing of POTS splitter validation. This cost must be regulated by official and independent regulation bodies[2] in order to guarantee a certain fairness of access to the market between the ILEC and the CLEC.

- *Case 3—Partial unbundling without ATU-R provisioning.* This situation is the same as the one considered in case 2 but the CLEC gives the liberty to the end user to choose, buy, and install by herself or himself an ATU-R. In addition to the risk represented by the validation of the end user's POTS splitter by the ILEC, a second risk is taken by the CLEC in that case. Indeed, the compatibility between the ATU-C provided and installed by the CLEC at the EO and the ATU-R is mandatory. This compatibility may be, for instance, easily guaranteed by requiring end users to choose their modem in a catalog from different vendors.

---

2. For example, these regulation bodies are the ART and the FCC in France and in the United States, respectively.

- *Case 4—Full unbundling.* Full unbundling means that the CLEC inherits the whole copper pair capacity. In general, the CLEC may decide to provide both the analog telephone service in addition to digital access to public data networks. Like in cases 2 and 3, the CLEC may or may not provide and install the ATU-R. Concerning the POTS splitter validation, this situation is more favorable to the CLEC than the situations considered in cases 2 and 3. Globally this fourth case enables the CLEC to have a full control of the installation cost.

7.2.1.3 In-Home Cabling with G.dmt

Figure 7.2 depicts the typical cabling of a full-rate ADSL installation at the customer premises. In Figure 7.2, one has represented the lowpass and the bandpass filters of the POTS splitter. The NID corresponds to a box that must be installed between the subscriber line coming from the public network and the private in-home cabling. The POTS splitter and the NID are in many cases in the same box.

The copper pair plugged on the output of the lowpass filter corresponds to the existing telephone cable. The other pair connected on the output of the bandpass filter is used to plug the ATU-R onto the NID. In most cases, this second branch of the "T" in-home cabling system represented in Figure 7.2 is not available and must be installed. The standard solution consists of providing a USB[3] interface on the NID in order to connect the ATU-R (one assumes that a single PC is connected to the NID). With full-rate ADSL (or G.dmt), a visit from a technician from either the CLEC or the ILEC is necessary in order to validate the conformance of the in-home cabling system. This technician is in charge of the following tasks:

- Installing the POTS splitter;

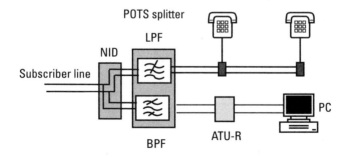

Figure 7.2 Private cabling system for full-rate ADSL.

---

3. USB stands for Universal Serial Bus.

- Eventually installing a new copper pair between the POTS splitter and the customer's PC;

- Validating the conformance of the POTS splitter.

7.2.2   In-Home Cabling with G.lite

Because of the technical and administrative constraints inherent to full-rate ADSL, a light version of ADSL systems has been developed under the name of G.lite. The main originality of G.lite versus G.dmt is that is does not require any POTS splitter at the customer premises. To bypass the problem of the design of passive POTS splitters, the basic idea of G.lite is very simple. By reducing considerably the target upstream and downstream data rates, one reduces considerably the amount of active DMT subchannels in the system. One uses then this capacity reduction for inserting a guard band between the analog band and the digital band of the ADSL system. A first approach of G.lite named *splitterless ADSL* has consisted of suppressing any filter within the private cabling system. We outline in the next section some of the problems that may still coexist with splitterless ADSL. This is the reason why a third version named *distributed splitter ADSL* has been developed.

7.2.2.1   Splitterless ADSL

Figure 7.3 illustrates a splitterless in-home cabling system. The benefit of this new configuration in comparison to the one presented in Figure 7.2 is twofold. First, one makes the economy of a visit of a technician to validate the conformance of the POTS splitter. Second, the existing in-home cabling can be reused ad hoc for connecting telephone sets and a PC.

The reader may notice a similarity between the upper branch of the passive "T" represented in Figure 7.3 and a bridge tap along a subscriber line (see Figure 4.6). Fortunately, no effect comparable to the strong reduction in the SNR is observed at the level of the NID for the ADSL frequency range and the length of the "T" branches. When all the handsets are on-hook (Figure 7.3), upstream

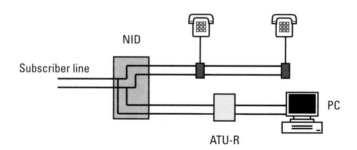

Figure 7.3 In-home cabling system for splitterless ADSL

and downstream data transit correctly between the EO to the end user's PC (Figure 7.4).

Nevertheless, when one or several handsets transit from the on-hook position to the off-hook position whereas a data connection is activated, strong degradations may be observed in SNR for data traffic. These degradations appear during short periods of time but induce bursts of errors. Let us analyze the reasons for this phenomenon. The input impedance of a telephone set varies in fact depending on the handset position. In the on-hook position, this impedance is equal to Z_{on}. Let Z_0 and Z_{mod} be the input impedance of the subscriber line and of the ATU-R, respectively. Figure 7.5 depicts the behavior of the system when suddenly one of the two handsets passes from the on-hook position to the off-hook position. Let Z_{off} be the input impedance of an off-hook handset.

During the changing of position of the handset, the input impedance of the subscriber line passes from Z_0 to Z_0'. Between these stable states, a fraction of the energy of the transmitted data signal transits through the upper branch and degrades the audio quality of the active handset. This phenomenon, although rather unpleasant for the user, remains acceptable because of its brevity. In the

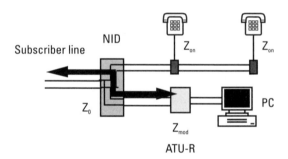

Figure 7.4 Correct behavior of a splitterless cabling system for data traffic.

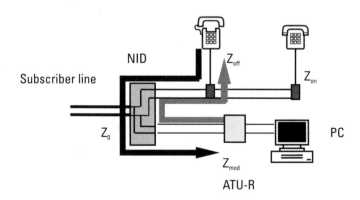

Figure 7.5 An example of nonlinearity at the NID.

reverse direction, a fraction of the energy of the audio signal enters into the lower branch and overlaps the upstream data signal. In other words, although analog telephony and digital data operate in distinct frequency domains, a shift of certain frequencies may occur at the level of the NID because of the transient electrical behavior of the cabling system. Like G.dmt, G.lite uses concatenated encoding, that is, an RS outer code and a convolutional inner code. It also uses an interleaver/deinterleaver between the outer and the inner encoders. In spite of these inherent error protection techniques, G.lite is not able to support the bursts of errors induced in splitterless configuration. In that case, error recovery must be carried out at the TCP level. One knows that TCP manages end-to-end connections and reacts then much slower than concatenated codes in case of packet loss. This means, for instance, that the end user may notice an abrupt degradation in the interactivity of its Internet access. At last, let us remark that the echo canceler of the ATU-R cannot fight against this type of parasite noise which is totally decorrelated from the transmitted data. Figure 7.6 presents the evolution of the real part of the ratio $Z_{on}/Z_{off} = (R_{on} + j \cdot \omega L_{on})/(R_{off} + j \cdot \omega L_{off})$ versus frequency f. One notices that in the best case, $\text{Re}[Z_{on}] = 8 \cdot \text{Re}[Z_{off}]$ for frequencies around 450 kHz. The disparity between $\text{Re}[Z_{on}]$ and $\text{Re}[Z_{off}]$ is more noticeable under low frequencies, below 100 kHz.

The amplitude of the noise induced on upstream data flow varies according to the telephone-set manufacturer. To prevent such disruptive effects, certain G.lite modem vendors have implemented an automatic and rapid detection technique of any state transition of a handset. In that case, the modem enters a "sleepy mode" during which a new data transfer cannot be initialized. Once the impedance discontinuity has disappeared, the G.lite modem reevaluates the subchannels' capacity and becomes again operational. In general, the sleeping mode

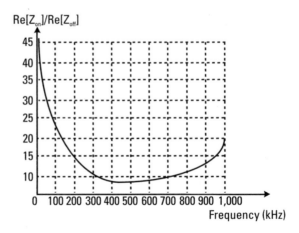

Figure 7.6 Evolution of $\text{Re}[Z_{on}]/\text{Re}[Z_{off}]$ with frequency f.

lasts 1.5 seconds. This reinitialization of the line is called the *fast retrain procedure.*

7.2.2.2 Distributed Splitter ADSL

Certain POTS service providers do not accept the degradations induced by non-linearities inherent to splitterless configuration. The reader may understand this attitude in the case of partial unbundling. This is the reason why another variant of in-home cabling called distributed splitter ADSL has been proposed. Figure 7.7 describes the configuration of an in-home cabling for distributed splitter ADSL. Instead of using a hybrid filter, one uses an isolated [0, 4 kHz] LPF for each telephone set and an isolated [20 kHz, 1,104 kHz] *bandpass filter* (BPF) at the input of the ATU-R. The LPFs are today available on the shelves of the shops. They are also sometimes directly integrated within the telephone sets. These LPFs or BPFs can easily be installed by the end user.

The price of a distributed filter is much lower than the price of a passive hybrid filter. The price of a passive hybrid filter may be on the order of an ATU-R, whereas the price of a distributed filter is the price of a simple connector.

7.2.3 G.lite Performance

Table 7.1 compares the performance of G.dmt and G.lite for an equivalent range (i.e., up to 5 km between the ATU-R and ATU-C). G.lite modulation relies on the same principles as G.dmt modulation. It uses IFFT and FFT in the Tx-modem and Rx-modem, respectively. The difference in terms of

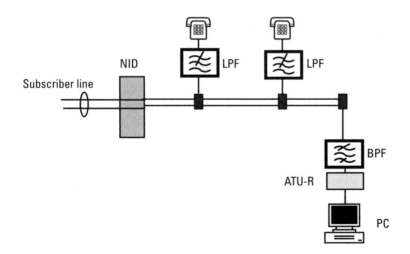

Figure 7.7 In-home cabling system for distributed splitter ADSL.

Table 7.1
Comparison Between EC-DMT-G.dmt and FDM-DMT-G.lite

Criteria	G.dmt (EC) [ITU-T G.992.1]	G.lite (FDM) [ITU-T G.922.2]
Upstream rate	800 Kbps	512 Kbps
Downstream rate	8 Mbps	1.5 Mbps
Number of subcarriers	256	128
Maximum capacity in bits per second per hertz	15	8

performance between the two systems is explained by a number of subchannels divided by two and by smaller QAM constellations. Whereas G.dmt may operate under two possible modes, either FDM or EC, a preference for FDM has been adopted for G.lite.

Figure 7.8 illustrates G.lite bandwidth utilization. One notices that the analog band and the digital band are clearly separated thanks to a frequency guard band. Similarly, upstream channels located in the [25 kHz, 130 kHz] band and downstream channels located in the [140 kHz, 552 kHz] band are separated by a 10 kHz guard band in order to limit the impact of nonlinearities.

7.2.4   Which Market for G.lite?

Unlike G.dmt, G.lite performance does not enable the provision of an MPEG-2 video channel that requires downstream bit rates between 2 and 4 Mbps. Although an MPEG-4 video channel requires only about 500 Kbps, the 512-Kbps downstream capacity of a G.lite system remains too limited. The main

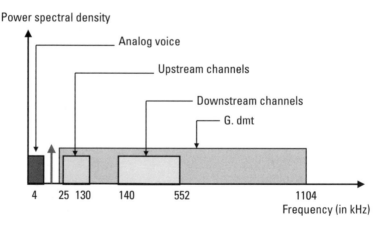

Figure 7.8 Bandwidth utilization in G.lite access systems.

objective of G.lite is to enable cost-effective high-speed access to the Internet in keeping the availability of the existing analog telephone line. The market dedicated to this type of service is made of residential users and *small office home offices* (SOHOs). A group of interests called *Universal ADSL Working Group* (UAWG) formed by major manufacturers such as HP-Compaq, Intel, and Microsoft in 1998 has developed a G.lite version for which the ATU-R is an additional card for PCs [2]. Some vendors should propose very soon bistandard V.90/G.lite *network interface cards* (NICs). Thus, a user may just connect his or her PC directly to a phone plug to benefit from high-speed access to the Internet, with telephone sets being connected on other plugs of the same private installation.[4] This plug-and-play version of ADSL access should appeal to a vast public.

7.3   VDSL

VDSL is an extended version of ADSL aiming to offer bit rates of a few tens of megabits per second on existing subscriber lines. We have investigated in Chapter 4 twisted copper pair characteristics. We have outlined the numerous physical impairments of the telephone loop plant, such as NEXT, FEXT, SNEXT, impulse noise, and bridge taps. Bit rates of several tens of megabits per second in the local loop are in fact achievable only on short distance limited to a few thousand feet[5] (around 1 km). Thus, VDSL cannot be considered as an autonomous solution for high-speed access to public data networks but as a complement to other techniques. During the last 10 years, VDSL has been considered to be the ideal complement of optical access networks. This is the reason why VDSL is also supported by the *full-service access network* (FSAN) initiative. The FSAN is a group of telephone companies from Europe, Australia, the United States, and Asia that work for a standard of optical access network called *passive optical access network* (PON). A PON is a point-to-multipoint access system based on a passive optical tree topology. Like VDSL, PON access systems standardization is elaborated in close collaboration with ITU-T, ANSI, and ETSI. Chapter 12 is fully dedicated to optical access networks. The high capacity of a VDSL system may be used to broadcast in parallel several digital TV channels and to provide bandwidth-consuming interactive services like high-definition medical images. Similar to ADSL or G.lite, VDSL makes it possible to keep simultaneously with these data services the existing analog telephone channel in the [0, 4 kHz] band. Because of its inherent capacity, a VDSL access system may also be configured to provide narrowband ISDN access (see Section 2.4.4) instead of POTS

---

4.   Of course, this user must have subscribed to an IAP.

5.   Let us recall that in urban areas, the average distance between customer premises and the nearest EO is about 2 km.

service. VDSL is an interim standard of the Committee T1E1.4 of ANSI. This standard has been elaborated in collaboration with Committee TM6 of ETSI [3, 4]. The ITU-T Study Group 15 considers the work from the ANSI project as a potential candidate for a new ITU-T VDSL international standard [5].

7.3.1 VDSL Access System Configuration

An end user located at a distance d of at most 1,500m from the nearest EO may be directly connected in VDSL by means of his existing subscriber line. Such distance concerns exclusively urban areas with a high population density. VDSL signals may occupy a very wide spectrum from 300 kHz to 10 MHz. Above a few megahertz, a copper wire behaves like an antenna. Thus, one of the main challenges of VDSL is its electromagnetic coexistence with other DSL systems or with voiceband modems. In most cases, d is larger than 1,500m. Two options are then possible: DLC access systems or PON access systems. The first option is based on the concept of the DLC introduced in Section 2.6. Let us briefly recall that a DLC system consists in a point-to-point digital link between a COT and an RT, as illustrated in Figure 7.9. Two types of mediums are considered for DLC, either a pair of contra-directional optical fibers or a binder cable with multiple copper wires between the COT and the RT. In the VDSL context, one only considers the case of optical fiber DLC. For such a configuration, the RT and the COT are called ONU and OLT, respectively. The last mile of the access system based on existing copper wires is then equipped with a pair of VDSL modems installed at the customer premises and at the RT, respectively. We mentioned in Chapter 2 that about 30% of the subscriber lines were connected

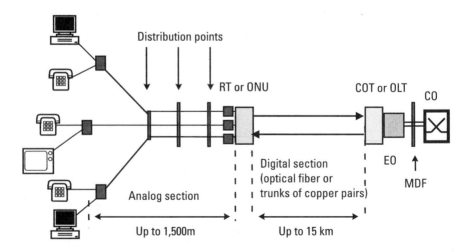

Figure 7.9 DLC+VDSL access system. RT: remote terminal, COT: central office terminal.

to a DLC system in the United States. Among these DLC systems, a fraction representing 15% of the American population is based on optical fibers [5].

Figure 7.10 illustrates the combination of a PON optical access system with several VDSL point-to-point access links. The optical modem used at the head of the optical tree is the OLT whereas the optical modems used at the leaves of the optical tree are the ONUs. The ONUs are installed in cabinets located at the level of a distribution point of the copper wire infrastructure. These cabinets must be able to power the ONU and a set of VDSL modems.

We underlined in Chapter 2 that the main role of an EO is to concentrate POTS traffic by means of a codec. In the case of DLC+VDSL access systems, no traffic concentration is carried out between the CPEs and the COT. This is not the case with PON+VDSL access systems where the ONUs concentrate upstream traffic on the same optical channel toward the OLT. This traffic concentration functionality is one of the major tasks dedicated to PON access systems that replaces then the EO. PON access systems have been standardized by the ITU-T for the transport of end-to-end ATM connections. This ATM-oriented version of PON access systems is also called APON. We mentioned in the introduction of this section that VDSL systems make it possible to keep the existing analog telephone line. Like ADSL, VDSL needs for that purpose the use of POTS splitters.

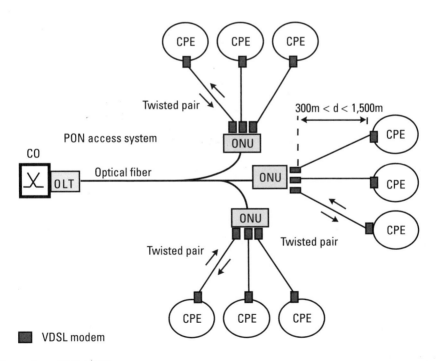

Figure 7.10 PON+VDSL access system.

7.3.2   Symmetrical or Asymmetrical Behavior

Whereas ADSL modems operate on an asymmetrical mode, downstream capacity being greater than upstream capacity, VDSL modems can be programmed to operate either symmetrically or asymmetrically. The choice of implementation between these two options is dictated by the targeted services (e.g., LAN interconnection and data transfers for the former and Web surfing and TV distribution for the latter). Table 7.2 summarizes the target bit rates and ranges for asymmetrical VDSL access systems. Asymmetrical services characterize in general the behavior of residential users.

Table 7.3 summarizes the target bit rates and ranges for symmetrical VDSL access systems. Symmetrical services characterize in general the behavior of professional users.

The two network configurations described in Figures 7.9 and 7.10 refer to public VDSL access systems. VDSL technology may also be used for private networking. A typical application of symmetrical VDSL is ATM or Ethernet/IP LAN interconnection at 10 Mbps or 25 Mbps. As an example, Figure 7.11 illustrates a campus network configuration. The main building of the campus (building *A*) is connected at high speed (several tens of megabits per second) to

Table 7.2
VDSL Performance Under Asymmetrical Mode

Upstream Bit Rate	Downstream Bit Rate	Range
6.4 Mbps	52 Mbps	300m, 984 ft
3.2 Mbps	26 Mbps	1,000m, 3,281 ft
1.6 Mbps	13 Mbps	1,500m, 4,921 ft
1.6 Mbps	6.5 Mbps	2,000m, 6,562 ft

Table 7.3
VDSL Performance Under Symmetrical Mode

Upstream/ Downstream Bit Rate	Range
25 Mbps	300m; 984 ft
13 Mbps	1,000m; 3,281 ft
6.5 Mbps	1,500m; 4,921 ft
4.3 Mbps	2,500m; 8,202 ft
2.3 Mbps	3,500m; 11,483 ft

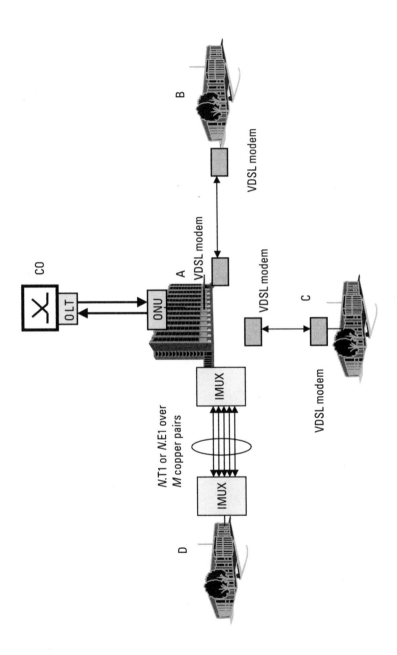

Figure 7.11 VDSL technology for private networks.

the CO by means of a point-to-point optical link. Within the campus, buildings *B*, *C*, and *D* are linked to building *A* by means of parallel copper wires. Over the years, the bit rate achievable on each of these pairs has been limited to E1/T1 rates. In order to benefit from higher bit rates on point-to-point links, inverse multiplexers have been frequently used in the past (example of the connection between buildings *D* and *A*). Today, inverse multiplexers are more expensive than VDSL modems for the provision of data pipes with a bit rate larger than E1/T1. High-speed Ethernet over copper wires (100-Mbps Ethernet, gigabit Ethernet) is a second field of application for VDSL technology [5]. Chapter 12 is partially dedicated to EFM systems. Variants of EFM systems rely on twisted copper pairs in the local loop and on VDSL technology.

7.3.3   Spectral Occupancy

Figure 7.12 illustrates bandwidth utilization of VDSL access systems. One notices that the frequency upper bound f_{max} in VDSL is between 10 MHz and 30 MHz (i.e., much higher than the frequency upper bound of 1.104 MHz of ADSL access systems). The [0, 120 kHz] band is dedicated either to the existing POTS service or to a narrowband ISDN access. The [300 kHz, 10 MHz] band represents the minimal bandwidth for digital data traffic in a VDSL access system. Several research laboratories[6] try to extend the upper bound of this second band up to 30 MHz. Let us recall that the higher the frequency, the higher the attenuation and the more difficult the equalization.

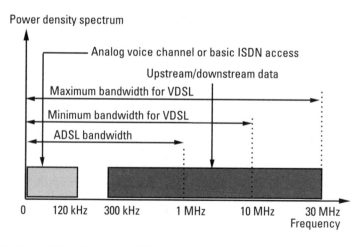

Figure 7.12   Bandwidth utilization in VDSL access systems.

---

6. These include Stanford University, IBM Zurich, Bell Labs, Alcatel Corporate Research center, and Texas Instruments.

7.3.3.1 FDM

Like in ADSL, two techniques are possible to provide full-duplex communication in VDSL access systems: FDM and EC. In FDM-VDSL, one considers two sub-bands for data traffic. The [300 kHz, 700 kHz] sub-band is dedicated to upstream traffic, whereas the [1 MHz, f_{max}] sub-band is dedicated to downstream traffic (see Figure 7.13). Let us recall that FDM avoids NEXT disturbance. Only FEXT interacts with the received signal. In ADSL systems, it is possible to reprogram the values of the two FDM sub-bands. This reconfiguration of the two FDM sub-band is not possible in VDSL due to the cost and complexity of the hardware. Thus, the main drawback of FDM-VDSL is its lack of flexibility, the degree of asymmetry being fixed within the hardware once for all.

7.3.3.2 EC

EC improves the performance of a VDSL access system by enabling an overlap between the bandwidths dedicated to upstream and to downstream traffic, respectively (see Figure 7.14). The SNR is strongly degraded under high frequencies by NEXT, FEXT being negligible. EC-VDSL performance also depends on SNEXT in the sub-band dedicated to both upstream and downstream traffic. We have outlined in Chapter 6 that SNEXT cancellation efficiency increases with the speed of electronics. According to the current state of the technology, this cancellation is at most applicable to a frequency band [300 kHz, 1 MHz]. Thus, echo cancellation cannot be adopted to provide the symmetrical bit rates mentioned in Table 7.3.

7.3.4 POTS Splitters in the VDSL Environment

VDSL access systems can be offered on existing copper wires used either for analog POTS services or for narrowband ISDN services. We have analyzed in

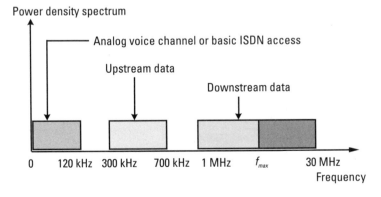

Figure 7.13 Bandwidth utilization in FDM-VDSL access systems.

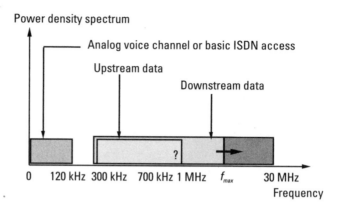

Figure 7.14 Bandwidth utilization in EC-VDSL access systems.

Section 7.2 the main constraints inherent to POTS splitters in charge of separating the analog sub-band and the digital sub-band in ADSL access systems. From this point of view, two different types of network configuration are considered for VDSL access systems.

7.3.4.1 VDSL Access System over POTS Subscriber Lines

Figure 7.15 depicts a VDSL access system reusing analog telephone lines. For such a configuration, VDSL splitters dedicated to each copper termination must be installed at the customer premises and in the cabinet located at the distribution point.

Like in ADSL, a VDSL splitter is a hybrid filter able to isolate the [0, 120 kHz] band and the [300 kHz, f_{max}] band. The [0, 120 kHz] band includes the [0, 4 kHz] band of the analog telephone line. The telephone sets installed at the

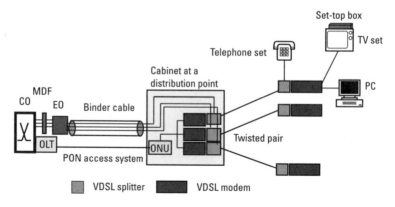

Figure 7.15 VDSL access system based on existing POTS subscriber lines.

customer premises must remain remotely powered by the EO. VDSL splitters installed at the distribution point are transparent to this remote powering of the telephone sets. Upstream voice calls arriving at the distribution point are redirected toward an existing binder cable. The upstream VDSL signal transported in the [300 kHz, f_{max}] band on each copper wire is demodulated at the cabinet to be regenerated on the PON access system. From the baseband signal obtained at the output of each VDSL modem, data packets are forwarded to the ONU located in the same cabinet. Let us recall that in APON access systems, these data packets correspond to ATM cells. The ONU carries out several operations for upstream traffic. First, data packets received from each copper wire are buffered into different queues corresponding to different classes of service. In PON access systems, upstream traffic from the various ONUs is concentrated on a single wavelength.[7] A MAC protocol is then mandatory in order to share efficiently and fairly this unique upstream wavelength between the different ONUs. The ATM cells arriving from different end users but assigned to the same ISP are forwarded toward a common ATM virtual path.[8] The physical and MAC layers of APON access systems are presented in Chapter 12.

Downstream data traffic arriving from the PON must be demodulated and demultiplexed at the ONUs. A routing function is necessary within the ONU to forward downstream data packets toward the right VDSL modems. One notices that, unlike ADSL access systems, VDSL access systems do not use the equivalent of a DSLAM close to the MDF. In fact, the set of VDSL modems located at the distribution point are the equivalent of a DSLAM (see Chapter 8). At the customer premises, a VDSL POTS splitter is associated to each VDSL modem.

7.3.4.2 VDSL Access System over Narrowband ISDN Subscriber Lines

We consider now the case of a VDSL access system including narrowband ISDN services in the [0, 80 kHz] band. In theory, ISDN services being digital, VDSL modems could be designed to multiplex ISDN traffic with pure VDSL data traffic by means of the same modulation technique. In practice, one only considers in this section the case of VDSL services offered on existing ISDN access lines. In other terms, the 2B1Q modulation adopted for existing ISDN services cannot be assimilated to the modulation techniques considered for VDSL services. This is the reason why VDSL splitters are necessary at the distribution points and at the customer premises. Figure 7.16 depicts a VDSL access network configuration using existing narrowband ISDN access lines. In reference to Figure 2.11 describing a narrowband ISDN subscriber line, the two end modems (NT1 and LT) of an ISDN access system must be separated for at most 3,600m.

---

7. All the ONUs are equipped with the same laser diode.

8. Protocol aspects of xDSL access systems are the topic of Chapter 9.

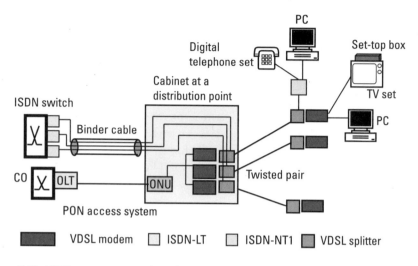

Figure 7.16 VDSL access system based on narrowband ISDN subscriber lines.

7.3.4.3   Fast and Slow VDSL Channels

We underlined in Chapter 6 the disruptive impact of impulse noise on ADSL access systems. Particular attention has also been dedicated to the protection against impulse noise in VDSL. Like in ADSL, FEC and interleaving/deinterleaving techniques have been developed for VDSL. Like in ADSL, one considers in VDSL two types of VDSL channels: slow and fast. Slow VDSL channels operate correctly for impulse noise with a duration under 500 μs.[9] They include FEC and interleaving/deinterleaving. An admissible upper bound of 20 ms for the one-way propagation delay has been specified for slow VDSL channels. Fast VDSL channels are more sensitive to impulse noise because they do not benefit of FEC and interleaving. They enable a one-way propagation delay under 1.2 ms and are then well suited to interactive applications like digital telephony.

7.3.5   VDSL Standardization

Various standardization bodies and groups of interest are working in cooperation for a standardization of the various possible implementations of VDSL access systems. They are described as follows:

- *ANSI T1E1.4 committee:* ANSI studies were initiated in 1995. By the end of 1998, a first document titled "VDSL System Requirements Document" was edited [6]. This document specifies the reference

---

9. Chapter 6 explains that impulse noise duration is in most cases below 500 μs.

configuration of a VDSL access system (possible data rates, requirements in terms of such factors as interoperability, services, and OAM
functionalities).

- *ETSI TM6 committee:* ETSI investigations started in June 1996. A first
document dating from April 1998 specifies the functionalities of the
various modules of a VDSL transmission channel. ANSI and ETSI have
specified a common characterization of the various noise sources in
VDSL access systems.

- *ITU-T study group 15:* ANSI, ETSI, and ITU-T cooperate on the same
basis principles (choice of the modulation and multiplexing techniques).

- *FSAN initiative:* The FSAN initiative is involved in the study of VDSL
deployment scenarios with a particular interest in electromagnetic compatibility between VDSL and other xDSL techniques.

- *VDSL Forum:* The VDSL Forum was created in 1997. Like the ADSL
Forum, the VDSL Forum is a group of modem vendors, electronic chip
manufacturers, and carriers involved in the development of VDSL
equipment. VDSL Forum and FSAN are not in charge of the standardization of the VDSL technique. The result of their investigations are
communicated to ETSI and ANSI for recommendation.

- *DAVIC:* DAVIC is an association of various actors in the field of video
broadcasting in the United States. Historically, DAVIC investigations
focus on point-to-multipoint access network configurations. Cable network operators are the main actors within DAVIC (see Section 3.7).

7.3.6   Which Modulation for VDSL?

Several modulation techniques have been investigated for VDSL. Until the
mid-1990s, the same types of modulation techniques were considered for VDSL
as for ADSL: CAP and DMT. CAP is a single-carrier modulation technique (see
Section 6.4.2). Let us simply recall that in order to avoid phase distortion, one
adopts in CAP digital filters that enable a demodulation at the receiver without
requiring the modulation carrier. CAP modulation is characterized by a high
spectral efficiency. Meanwhile, according to the state of the hardware technology, with CAP, it is not possible to achieve the performance objectives mentioned in Tables 7.2 and 7.3. Unlike CAP, DMT is a multicarrier modulation.
It is presented in Section 6.4.3. In reference to the VDSL bit rates mentioned in
Tables 7.2 and 7.3, it appears that the main limitation of DMT is its incapabil-
ity to prevent ICI with large-size QAM constellations. At the end of the 1990s,
two other multicarrier modulation techniques were proposed. The first one
called *discrete wavelet multitone* (DWMT) has been proposed to efficiently

prevent ICI [7]. Like CAP-VDSL, DWMT-VDSL remains difficult to implement according to the state of the technology. The second modulation technique proposed for VDSL is called *synchronized discrete multitone* (SDMT) [8, 9].The most recent investigations in the field of VDSL modulation techniques are *filtered multitone* (FMT) [10] and Turbo multiuser detection for coded DMT-VDSL systems [11].

7.3.6.1   From DMT to DWMT

In the mid-1990s, extensions of DMT modulation were investigated for VDSL. Like in ADSL, the [300 kHz, f_{max}] VDSL band is divided into up to 4,096 sub-channels of 4.3125 kHz each. Thanks to their very limited bandwidth, the impact of amplitude and phase distortions within each of these subchannels may be neglected at the receiver. A QAM is used in each subchannel. Similarly to ADSL, VDSL needs a line initialization to evaluate the value of the SNR in each subchannel. From these values, one determines the optimum size of the QAM constellation to adopt in each subchannel. In [10], numerical simulations of DMT transmission systems operating at the VDSL bit rates are presented. Figure 7.17 illustrates some of these results for a modulator with 64 subchannels and the same bit rate per subchannel. One notices from Figure 7.17 that the amplitudes of the main lobes inherent to adjacent subchannels cross at −3 dB. The maximum amplitudes of the first right and left lobes of each subchannel are 13 dB below the maximum amplitude of the main lobe of the same subchannels. Considering the size of the constellations required to achieve VDSL performance, it appears that such an overlap between adjacent subchannels is unacceptable at the receiver.

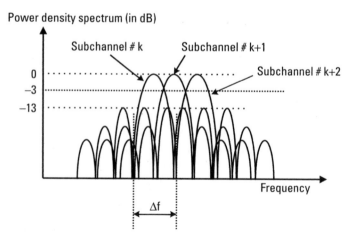

Figure 7.17 Subchannel frequency response in DMT-VDSL. (*From:* [12]. © 1995 IEEE. Reprinted with permission.)

To reduce ICI, another multicarrier modulation technique has been proposed in 1998. This technique known as DWMT is very similar in its principle to DMT. The IFFT and FFT used in the Tx-DMT-VDSL and in the Rx-DMT-VDSL, respectively, are based on a Fourier series expansion obtained from a set of complex terms such that:

$$g(t) = \sum_{n=-\infty}^{\infty} \left[Dn \cdot e^{jn\frac{2\pi}{T_0}\cdot t} \right] \qquad (7.2)$$

The exponential functions (harmonics) of the form $\exp(+j \cdot n\omega_0 t)$ considered in the Fourier series expansion of function $g(t)$ have an infinite duration. These harmonics remain with an infinite duration in the case of a function $g(t)$ with a finite duration. In other terms, the traditional Fourier transform is unsuited to describe functions with finite duration. The concept of wavelet used in DWMT corresponds to a family of functions with a finite duration from which it is possible to describe any signal, especially if this signal is itself with a finite duration. Equation (7.3) gives the general expression of a wavelet:

$$\psi_{a,b}(t) = \frac{1}{\sqrt{a}} \cdot \psi\left(\frac{t-b}{a}\right) \qquad (7.3)$$

where a is a time dilatation factor and b is a time shift factor. The expression of a continuous wavelet transform is given in (7.4), this expression can be compared with the expression of a continuous Fourier transform:

$$\omega_{a,b}(t) = \frac{1}{\sqrt{a}} \cdot \int_{-\infty}^{+\infty} g(t)\psi * \left(\frac{t-a}{b}\right) \cdot dt \qquad (7.4)$$

The Fourier transform corresponds to a representation in the frequency domain of a signal described in the time domain. In comparison, the wavelet transform corresponds to a two-dimensional representation of a signal described in the time domain. The two dimensions correspond to the values of a and b parameters. Parameters a and b are used to adjust the frequency and the time positioning of a wavelet, respectively. From a specific "mother wavelet," it is possible to obtain various variants of wavelets corresponding either to a time contraction or to a time dilatation of this mother wavelet. Equation (7.5) gives an example of mother wavelet known as the Morlet wavelet:

$$\psi(t) = e^{+j\omega t} \cdot e^{-t^2/2} \qquad (7.5)$$

Wavelet theory dating from the 1970s is essentially known for its applications to the domain of image coding and in signal processing. Wavelet transform qualifies as multiresolution in the sense it analyzes high and low frequencies of a signal with different time window sizes. The lower the frequency, the larger the size of the time window. As a comparison, the short-term Fourier transform analyzes a signal on a fixed time window size whatever the frequency of this signal. The usage of discrete wavelet transform in the field of digital communications is much more recent. Without detailing the principle of DWMT, which is discussed in [12], we simply want to outline the interesting mathematical properties of wavelets for reducing ICI in multicarrier modulation. Figure 7.18 describes the frequency response (in decibels) of adjacent subchannels obtained at the output of a DWMT modulator by means of a numerical simulation. One considers like in Figure 7.17 a modulator with 64 subchannels and the same bit rate per subchannel [10].

On one hand, one notices in comparison to Figure 7.17 that the amplitudes of the main lobes of adjacent DWMT subchannels still cross at –3 dB. On the other hand, the maximum amplitudes of the first right and left lobes of each DWMT subchannel are 60 dB below the maximum amplitude of the main lobe of the subchannels. One also notices that the main lobe of each DWMT subchannel occupies half the bandwidth of the main lobe of the corresponding to DMT subchannel. In general, because of amplitude and phase distortions induced by the transmission medium, orthogonality between adjacent subchannels does not exist any longer at the receiver. We have seen in Chapter 6 that a solution to this problem may consist of introducing cyclic extensions under the form of a cyclic prefix or of a cyclic suffix.

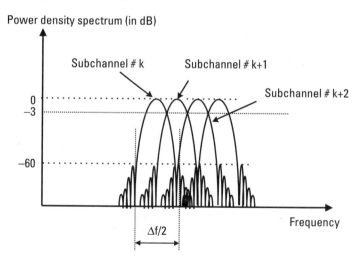

Figure 7.18 Subchannel frequency response in DWMT-VDSL. (*From:* [12]. © 1995 IEEE. Reprinted with permission.)

7.3.6.2 SDMT Modulation

SDMT was developed at the end of the 1990s. Like DMT-VDSL, SDMT-VDSL is a multicarrier modulation based on DM7T. The main originality of SDMT consists in the use of a ping-pong *time division duplexing* (TDD) technique to isolate upstream and downstream data flows. Unlike FDM and EC techniques, TDD prevents a VDSL modem to send and receive data simultaneously. In this way, SDMT-VDSL access systems avoid NEXT or SNEXT. In SDMT-VDSL, FEXT is then the only noticeable crosstalk.[10] When a SDMT-VDSL modem is authorized to transmit, it may use the whole VDSL band [300 kHz, f_{max}] if one knows that no HDSL or ADSL access system shares the same cable binder with this VDSL line. In practice, one defines alternatively common periods of transmission and common periods of reception for the set of the VDSL modems depending on the same ONU. Figure 7.19 illustrates this set of synchronized modems located either at the distribution point or at the customer premises.

To provide this synchronization, one defines successive superframes used by all the modems. These superframes specify downstream and upstream transmission periods separated by guard times. Thanks to TDD, a SDMT-VDSL modem can either transmit or receive during distinct periods. In other words, the same band is used for upstream and downstream traffic. This characteristic makes it possible to reduce the hardware complexity of VDSL modems. Indeed,

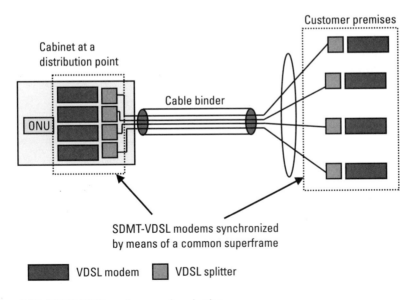

Figure 7.19 SDMT-VDSL modem synchronization.

---

10. Let us recall that FEXT is much less penalizing than NEXT or SNEXT.

the same hardware is used to compute only one DFT for both upstream and downstream traffic. The guard times considered in the superframes are dimensioned according to three parameters:

- The maximum propagation delay between two end-VDSL modems;

- The delay required for the evanescing of the parasite echoes following each transmission;

- The switchover time between the transmit and receive states of the modems.

Figure 7.20 depicts an example of SDMT superframe configuration. One notices that in fact two superframe formats are considered at the distribution point and at the customer premises, respectively. These formats are common to all the subchannels. At the distribution point, a superframe corresponds to a transmission period of, let us say, A DMT symbols followed by a guard time of duration Q_1 itself followed by a reception period of, let us say, B DMT symbols. Finally, a superframe at the distribution point ends by a guard time of duration Q_2. At the customer premises, a superframe corresponds to a reception period of A DMT symbols followed by a guard time of duration T_1 itself followed by a transmission period of B DMT symbols. Finally, a superframe at the customer premises ends by a guard time of duration T_2.

Although the values of the guard times are different at the distribution points and at the customer premises, one has:

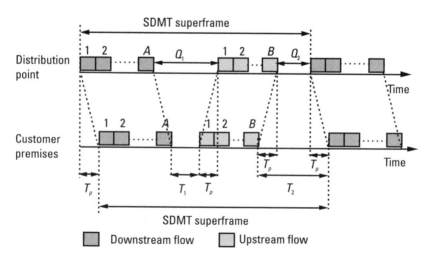

Figure 7.20 Example of SDMT-VDSL superframe.

$$T_2 + T_2 = Q_1 + Q_2 \tag{7.6}$$

If T stands for the symbol duration, the efficiency η of the SDMT-VDSL access system may be defined as:

$$
\begin{aligned}
\eta &= \frac{(A+B)\cdot T}{\left\{(A+B)\cdot T + Q_1 + Q_2\right\}} = \frac{(A+B)\cdot T}{\left\{(A+B)\cdot T + T_1 + T_2\right\}} \\
&= \frac{1}{1 + \dfrac{Q_1 + Q_2}{(A+B)\cdot T}}
\end{aligned}
\tag{7.7}
$$

One tries then to maximize $A + B$ and to minimize $Q_1 + Q_2$. Let us recall that meanwhile, Q_1 and Q_2 must remain greater than the summation of the three parameters mentioned earlier. SDMT-VDSL access systems may operate in a symmetrical or in an asymmetrical mode with $A = B$ or $A \neq B$, respectively. The structure of the two SDMT superframes are software-programmable, depending on the type of service (symmetrical or asymmetrical) offered by the carrier. In the SDMT-VDSL draft standard [13], typical superframes configurations are proposed for a SDMT-VDSL access system with 256 subchannels with a width of about 40 kHz corresponding to $f_{max} = 10.24$ MHz. The superframe duration of 500 μs at the customer premises corresponds to 20 DMT symbols of duration $T = 25$ μs. In the case of symmetrical configuration, the draft standard recommends $A = B = 9$ and $T_1 = T_2 = 1$. In the case of asymmetrical configuration, the draft standard recommends $A = 16$, $B = 2$, and $T_1 = T_2 = 1$.

Figure 7.21 illustrates the risk of NEXT interference when the modems of multiple end users depending from the same ONU are not correctly synchronized. By simplification, one assumes that the two end modems considered in Figure 7.21 are equidistant from the same distribution point. In spite of the same superframe configuration at the two end modems, the time shift between these frames induces NEXT interference on a certain amount of received symbols. In practice, electronic delay buffers correctly dimensioned may compensate distance disparities between the various end modems and the distribution point. Two techniques are considered for a correct synchronization between VDSL modems:

- An 8-kHz frequency carrier implemented at the distribution point may be broadcast toward all the end modems depending on the same ONU.
- A common clock reference can be obtained by all the modems by means of *global positioning system* (GPS) satellite signals.

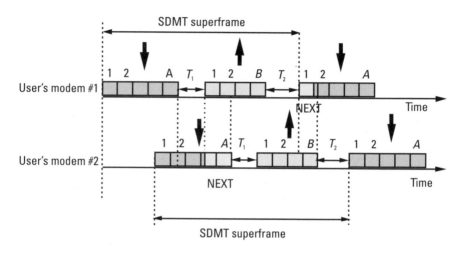

Figure 7.21 NEXT induced by a lack of synchronization in a SDMT-VDSL access system.

One of the main drawbacks of SDMT-VDSL access systems is their lack of flexibility in terms of service personalization. Indeed, the same degree of asymmetry is imposed to all the end users depending on the same ONU. Figure 7.22 depicts NEXT perturbation induced by a service disparity between the various end users depending on the same ONU. Several algorithms have been investigated in order to define a judicious assignment of the time slots to each symbol duration within SDMT superframes in order to minimize NEXT [12]. The Zipper technique presented in the following section has strongly contributed to this objective.

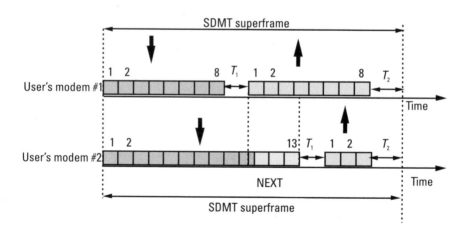

Figure 7.22 NEXT induced by service disparity in a SDMT-VDSL access system.

7.3.6.3   Zipper

Thanks to the Zipper technique,[11] a more flexible allocation of the subchannels between upstream and downstream flows is possible. Thus, one does not need like in DMT-VDSL to have upstream and downstream subchannels grouped into dedicated sub-bands as described in Figures 7.13 and 7.14. The position of an upstream or downstream subchannel may vary with time according to the dynamic transmission requirements of the end users. The Zipper technique prevents the use of traditional analog filters for subchannel selection at the receiver. For that purpose, the Zipper technique consists of inserting at the transmitter a cyclic suffix at the end of each DMT symbol. The cyclic suffix of current symbol k is obtained by taking a portion of the last bits of symbols $k - n$, $k - (n - 1)$, $k - (n - 2)$, ..., $k - 1$. This suffix introduces a guard time between successive symbols and reduces then the ISI at the receiver. At the receiver, this cyclic suffix is withdrawn before the received signal is demodulated by means of FFT. By their composition, cyclic suffixes also facilitate a reduction in ICI by maintaining a certain orthogonality between subchannels at the receiver. It is observed that DMT-VDSL access systems perform best if DMT symbols arrive simultaneously at destination. Cyclic suffix enables this simultaneity at the price of a reduced bandwidth utilization ratio. The number of subchannels in a VDSL system using the Zipper technique may be increased in order to compensate this performance degradation. The reader may notice a similitude between the Zipper and the cyclic prefix (Section 6.6.3) used in ADSL: The value of m is dynamic for the former, whereas it is static for the latter.

7.3.6.4   FMT Modulation

FMT is the most recent multicarrier modulation technique proposed for VDSL. In comparison to other modulation techniques such as DMT and DWMT, FMT is the most efficient in reducing ICI. DMT-VDSL and a FMT-VDSL access system adopt different approaches in order to provide orthogonality between adjacent subchannels. We have seen that the former uses cyclic suffix whereas the latter exploits very efficient baseband filters for each subchannel. A more detailed description of the realization of FMT-VDSL filters is given in [14]. In order to illustrate the efficiency of FMT-VDSL facing DMT-VDSL and DWMT-VDSL, let us consider again the same multicarrier modulation system with $M = 64$ subchannels as in Section 7.3.6.1.

Let us compare the performance of DMT, DWMT, and FMT modulation by means of the curves presented in Figures 7.17, 7.18, and 7.23. The maximum amplitudes of the first right and left lobes of each FMT subchannel are 65 dB below the maximum amplitude of the main lobe of the same subchannels. In comparison, this gap is equal to 60 dB for DWMT and 13 dB for

---

11. The Zipper technique is also referred to as digital duplexing.

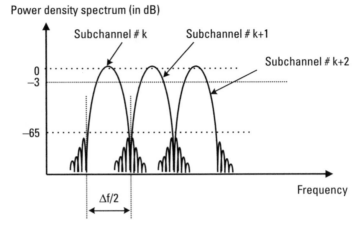

Power density spectrum (in dB)

Figure 7.23 Subchannel frequency response in FMT-VDSL. (*From:* [10]. © 2002 IEEE. Reprinted with permission.)

DMT. Whereas the amplitudes of the first lobes of adjacent subchannels cross at −3 dB with DMT and with DWMT, these amplitudes cross at −65 dB with FMT. For the same bit rate per subchannel, the width of the main lobes in DWMT and DMT is half the width of the main lobes in FMT. All these results outline the suitability of FMT modulation to avoid ICI. The basic principles of FMT have been added as an informative annex to the most recent version of the VDSL ANSI draft standard [3].

7.3.7 Electromagnetic Compatibility

Particular attention is dedicated to the problem of *electromagnetic compatibility* (EMC) between VDSL access systems and other transmission techniques (xDSL and voiceband access systems) that could share the same binder cables. EMC also concerns the interaction between VDSL access systems with their external environment. Two main types of electromagnetic interference are then considered:

- *Ingress interference:* A VDSL copper pair is subject to *radio frequency* (RF) ingress noise at the receiver. The main source of ingress *RF interference* (RFI) are AM radio transmitters and amateur radio operators. AM radio transmitters occupy on average 10-kHz bands located between 560 kHz and 1,600 kHz. The level of interference between AM radio transmitters and VDSL systems may reach −80 dBm/Hz. Amateur radio transmitters use 2.5-kHz frequency bands around different discrete frequencies between 2 MHz and 10 MHz. The exact values of these discrete bands depend on the national regulation bodies. In the worst cases, amateur radio interferers may represent a signal with a power of 1 mW, which is nine times greater than the power of the NEXT or impulse noise observed usually on subscriber lines [5]. Figure 7.24 gives the spectral position of the frequencies used by amateur radio operators in the United Kingdom [15].

* *Egress interference:* Above 10 MHz, a VDSL phone line may itself induce a noise into amateur radio receivers. This egress noise is particularly noticeable in the case of elevated phone lines. Noise level induced by egress interference on amateur radio receivers is always much lower than noise level due to ingress interference on VDSL receivers because of the much lower transmission power used for the former.

British Telecom research laboratories have been in charge within the VDSL Forum of investigations about the characterization and impact of ingress and egress RFI. We represented in Figure 7.25 an example of ingress RFI. It has been observed experimentally that external electromagnetic fields with a frequency of several megahertz induce parasite voltage V_1 and V_2 between each of the two wires of a phone line and the ground. One qualifies such voltage signals as longitudinal mode. The fluctuations of V_1 and V_2 may have a strong impact on the measured signal $V(t)$ at the receiver. Let us recall that $V(t)$ considered at the input of Rx-modem is evaluated between the two wires of the line. Such a voltage qualifies as transversal mode.

A first approach to reduce RFI is to use an adaptive analog filter that estimates the *differential mode* (DM) component of the RFI signal by using the CM component of the RFI as an input [5]. Such a filter is updated dynamically in order to evaluate the impact of ingress RFI on $V(t)$. Two main techniques

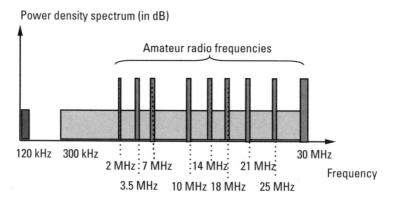

Figure 7.24 Ingress noise inherent to amateur radio transmitters/receivers.

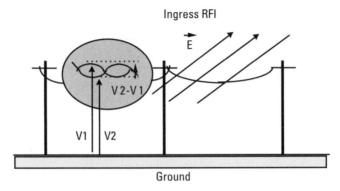

Figure 7.25   Longitudinal and transversal modes over a twisted copper pair.

known as *time equalization* (TEQ) and windowing can be used to reduce ICI. TEQ reduces both ICI and ISI by limiting the duration of the impulse response of the physical channel (the subscriber line) to the duration of the Zipper guard time. Windowing can be assimilated to a raised cosine filter used at the receiver.

7.4   Conclusion and Perspectives

This chapter has presented two variants of ADSL multicarrier access systems: G.lite and VDSL. On one hand, the G.lite technique has been developed because of the difficulty to implement robust and efficient POTS splitters in ADSL access systems. G.lite modems are today widely deployed by IAPs. On the other hand, the VDSL technique aims to enable very high bit rates (up to several tens of megabits per second) over existing subscriber lines either for symmetrical or asymmetrical communications. We have seen that such bit rates could be achieved only on very short-range (up to 1 km) copper pairs. This is why VDSL access systems are considered by many carriers as the ideal complement to optical access systems. Since the very first VDSL investigations at the beginning of the 1990s, several modulation techniques have been proposed: DMT, DWMT, SDMT, Zipper, and FMT. Currently, SDMT and Zipper techniques are good candidates for VDSL standardization. FMT looks very promising but its implementation in real equipment has not yet been demonstrated. Chapter 6 emphasized the need to use concatenated coding[12] in order to

---

12. We have seen that concatenated coding associates an outer RS code with an inner convolutional code.

improve the SNR at the receiver. Since 2000, several investigations concerning an adaptation of the Turbo-code techniques to VDSL systems have been carried out. It has been shown in [11] that a gain of 7 dB on SNR could be achieved thanks to turbo multiuser detection in comparison to traditional DMT-VDSL for a BER equal to 10^{-7} under severe ISI and AWGN. Applications of the VDSL technique are not limited to the local loop. High-speed in-home networking should be an important field of utilization of VDSL in the very next years, for instance, for gigabit or 10-Gb Ethernet. Chapter 8 is partially dedicated to in-home networking.

References

[1]   Henkel, W., and S. Olçer, guest editorial: "Twisted Pair Transmission—Ever Increasing Performances on Ancient Telephone Lines," *IEEE Journal on Selected Areas in Communications,* Vol. 20, No. 5, June 2002, pp. 877–879.

[2]   UAWG, http://www.uawg.org.

[3]   "Very High-Speed Digital Subscriber Line (VDSL) Metallic Interface, Parts 1 and 3," Committee T1, Working Group T1E1.4, ITU-T Contribution T1E1.4/2000-009R3, 011-013R4, February 2001.

[4]   ETSI technical specifications TS 101 270-1, "Transmission and Multiplexing: Access Transmission Systems on Metallic Access Cables; Very High-Speed Digital Subscriber Line (VDSL); Part 1: Functional Requirements," ETSI publications, April 1998.

[5]   Cioffi, J. M., et al., "Very High-Seed Digital Subscriber Line," *IEEE Communications Magazine,* April 1999, pp. 72–79.

[6]   VDSL Coalition (V. Oksman, ed.), *VDSL Draft specification,* ANSI cont. T1E1.4/98-045R1, June 1998.

[7]   Akansu, A. N., et al., "Wavelet and Subband Transforms: Fundamentals and Communications Applications," *IEEE Communications Magazine,* December 1997.

[8]   Pollet, T., and M. Peeters, "Synchronization Techniques with DMT Modulation," *IEEE Communications Magazine,* April 1999, pp. 80–86.

[9]   Jacobsen, K. S., "Synchronized DMT (SDMT) for Very High-Speed Digital Subscriber Line (VDSL) Transmission," *Proc. of IEEE Globecom'98 Conference,* Sydney, Australia, 1998.

[10]   Cherubini, G., E. Eledtheriou, and S. Olçer, "Filtered Multitone Modulation for Very High-Speed Digital Subscriber Lines," *IEEE Journal on Selected Areas in Communications,* Vol. 20, No. 5, June 2002, pp. 1016–1027.

[11]   Dai, H., and H. V. Poor, "Turbo Multiuser Detection for Coded DMT VDSL Systems," *IEEE Journal on Selected Areas in Communications,* Vol. 20, No. 5, February 2002, pp. 1016–1027.

[12]   Sandberg, S. D., and M. A. Tzannes, "Overlapped Discrete Multitone Modulation for High-Speed Copper Wire Communications," *IEEE Journal on Selected Areas in Communications,* Vol. 13, No. 9, December 1995, pp. 1571–1585.

[13]   VDSL Alliance (K. Jacobsen), *VDSL Alliance SDMT VDSL draft standard proposal,* ANSI cont. T1E1.4/98-265, August 1998.

[14]   Cherubini, G., et al., "Filter Bank Modulation Techniques for Very High-Speed Digital Subscriber Lines," *IEEE Communications Magazine,* Vol. 38, May 2000, pp. 98–104.

[15]   Foster, K., "EMC and VDSL: An Overview from a Telco Perspective," *Proc. of IEEE Globecom'98 Conference,* VDSL Workshop W04, 1998.

8

DSLAM and Home Network

8.1 Introduction

Chapters 5–7 presented the various versions of xDSL modems (HDSL, ADSL, G.lite, and VDSL) with their variants. Our approach has essentially been focused on the description of these modems from an end user's perspective. This chapter is divided into two parts dedicated to DSLAM and to home networking, respectively. A DSLAM may be seen as the complementary of an xDSL end user's modem, at the CO side. We have already introduced in Chapter 3 the concept of DSLAM as the set of xDSL modems installed in the carrier premises. For each xDSL modem installed at the end user, another xDSL modem is installed by the carrier between the MDF and the POP of a packet-switching network carrier (see Figure 3.6). We detailed in Chapter 3 unbundling and the technical and regulation aspects related to the management of the MDF and of the DSLAM. We describe in the first part of this chapter the role and interfaces of a DSLAM. The second part is dedicated to home networking. In reference to the ITU-T terminology, we have underlined that an ATU-R and an ATU-C both correspond to a *network termination of type 1* (NT-1) and to a *line termination* (LT). When we have presented the reference points of an ADSL access system (see Figure 6.5), we have introduced the concept of network terminations of type 2 (NT-2). In the ITU-T terminology, an NT-2 is an optional piece of equipment that a carrier may provide to the customer to complement his or her modem (NT-1). The main objective of an NT-2 is to facilitate a fair share between various pieces of equipment under the responsibility of the end user of a unique physical access to a data network via the NT-1. In general, an NT-2 consists of the provision of a LAN service also called a home network to the end user.

8.2   DSLAM

In a multioperator environment, DSLAMs have an important impact on the design of the global architecture of multiservice networks. Indeed, a DSLAM is placed at the boarder between local loops and public networks on which are connected service providers. These public networks may be of different types, either circuit-switching–oriented (PSTN managed by an ILEC) or packet-switching–oriented (ATM, IP, Ethernet-based or frame relay networks managed by a CLEC). A DSLAM is not only a set of xDSL modems used at the end of each subscriber line. Located either at the EOs or at CO,[1] it serves as a protocol adapter and as a traffic concentrator. We have also underlined in Chapter 3 the key role of a DSLAM in the unbundling context.

8.2.1   DSLAM Functionalities

A DSLAM is a piece of equipment that assembles multiple xDSL modems. Current DSLAMs available on the market may include up to a few hundreds of xDSL modem-cards.[2] Knowing that an EO concentrates up to 5,000 subscriber lines, several tens of DSLAMs are in fact installed close to the MDF. By simplification, we assimilate in the following the set of these DSLAMs to a single piece of equipment. Figure 8.1 illustrates the location of a DSLAM in its typical environment. A DSLAM is a multitechnology piece of equipment, its modem cards being in general of different types (HDSL, HDSL2, SHDSL, G.dmt, and G.lite). One distinguishes in Figure 8.1 five types of access lines:

- *Case 1:* The lines that only transport traditional analog voice channels do not use any modem or POTS splitter at the DSLAM. Such lines are directly redirected at the MDF to the inputs of the EO.

- *Case 2:* The lines used for ADSL and G.lite services transport both analog voice channels and data traffic. Each of these lines needs a modem and a POTS splitter within the DSLAM. Let us remark that although a G.lite modem does not need a POTS splitter at the customer premises, it is not the case at the EO or at the CO.

- *Case 3:* IDSL lines that are specifically designed for ISDN services and HDSL lines that are used for T1/E1 leased lines services need a dedicated modem at the DSLAM. On the other hand, no POTS splitter is required at the CO/EO side. In addition, as ISDN and T1/E1 LL are

---

1. Chapter 2 explained that the proximity between an EO and its CO depends on the end users' density.

2. The size of a DSLAM is comparable to the size of an IP router.

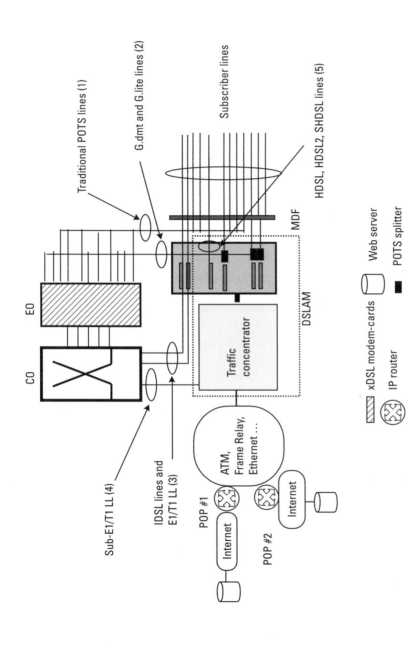

Figure 8.1 Location of a DSLAM in its typical environment.

fixed bit rate services, they to not need to use a traffic concentrator at the DSLAM.

- *Case 4:* HDSL/SHDSL modems may be used to provide circuit-oriented LL services with a fixed bit rate lower than E1/T1. In that case, one may use the services of a traffic concentrator in fact quite similar to an EO in order to concentrate the traffic of various analog lines on dedicated E1/T1 links.

- *Case 5:* The HDSL, HDSL2 and SHDSL lines may also transport packet-oriented data traffic. These packets are forwarded by the DSLAM toward packet-switching networks.

8.2.2 Reference Model for xDSL Services

In the context of client-server applications like remote access to Web servers, a DSLAM routes the multiple requests sent by the end users to such or such POP of an NSP. Figure 8.1 illustrates such a situation for which the NSPs cannot afford the installation of a POP close to every DSLAM. In other terms, IP networks on which are connected the Web servers are not accessible directly from the EO/CO but need the crossing of a data network, which is either a long-distance core network or a MAN. The choice of such or such a POP by the DSLAM is related to the fact that this POP makes it possible to reach the remote IP network on which is connected the requested server. The transport network between the DSLAM and the POPs may rely on various technologies and protocols (ATM, frame relay, and probably Ethernet in the next few years). This is the reason why a DSLAM must in general proceed to protocol conversion. Chapter 9 is totally dedicated to the description of end-to-end protocol architectures. Currently, ATM or frame relay networks are frequently used between the DSLAMs and the POPs. Instead of opening as many virtual circuits from the DSLAM to the POPs as the amount of simultaneous requests received from the end users, a DSLAM uses a single permanent virtual circuit to be connected to each POP. Thus, in our example of Figure 8.1, two permanent virtual circuits link the DSLAM to the two POPs. On the basis of the requested URL, IP packets received from the end users are forwarded toward such a permanent virtual circuit.

Several technologies are possible to proceed to an upstream traffic concentrator at the DSLAM The choice of technology is a key point in the cost, the efficiency, and the complexity of the offered services. Current implementations are in their majority based on ATM switches. Since 2000, DSLAM vendors consider with an increasing interest the benefits of gigabit Ethernet switches for concentrating upstream traffic. We have already underlined this evolution from ATM to Ethernet in the local loop in Chapter 6 when we introduced the concept of EFM. For economy of scale, the modem cards installed within a

DSLAM are software-programmable. Thus, these cards can be dynamically reconfigured in order to operate under different modes, including G.dmt, G.lite, HDSL, HDSL2, and SHDSL. This flexibility assumes that several line codes and modulation techniques are possible from a same hardware. This is the reason why an xDSL modem at the CO side is in fact much more complex and expensive than an xDSL modem at user's side.

8.2.3   DSLAM Supervision

Because of their multiple functions and of their key role in Internet access provision, efficient tools are necessary to manage the proper working of DSLAMs. DSLAMs available on the market are equipped with network management tools like *simple network management protocol* (SNMP). As an example, such tools are able to provide statistics on subscriber lines quality in terms of SNR or in terms of utilization ratio per subcarrier in DMT systems. As mentioned above, a DSLAM may be under the responsibility of a CLEC. If colocation is possible, it may be located within the CO or EO building of an ILEC. The IAP installs in its own room (see Figure 3.8) a work station on which is implemented the DSLAM management software.

8.2.4   Remote Access Multiplexer (RAM)

Telcordia laboratories (formerly Bellcore) proposed a new version of DSLAM called RAM, which makes it possible to provide xDSL services to users located further than 4 km from the EO. Typically, a RAM is a small-size version of a DSLAM able to manage up to about 10 subscriber lines. A T1/E1 LL is used between the CO and the RAM. Figure 8.2 illustrates how a RAM may be attached to a DSLAM in order to serve the most distant users from the CO. The LTs of these digital LL are directly connected to the internal bus of the DSLAM.

8.3   Home Networking

8.3.1   The Emergence of Home Networking

Typically, the families using more than one PC are those for which the parents bring back their own laptop at home at the end of the day and where one or several PCs owned by the family are used for private purposes. The homes of these families are usually connected to the Internet either by means of a cable modem or of a voiceband or xDSL modem. Let us recall that a modem is called an NT-1 in the ITU-T terminology. An NT-1 is coupled to the telephone line. In their traditional configuration, these modems enable the connection of a single end system at a time. The specification of the reference points of an ADSL modem

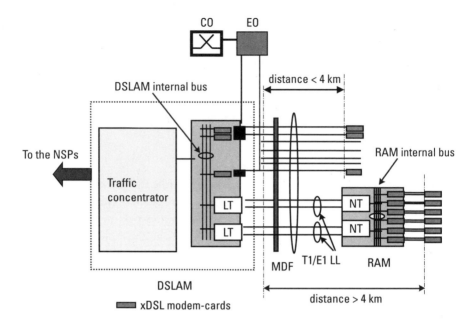

Figure 8.2 Example of RAM in an access network.

outlines the role of and NT-2 (see Figure 6.5). An NT-2 is an optional piece of equipment that can be provided by the access provider or that can be purchased by the end user himself or herself in order to have the possibility to connect several end systems on the NT-1. The role of an NT-2 is twofold. First it enables simultaneous access to public networks via a same NT-1 by multiple PCs. Second, an NT-2 is the equivalent of a LAN connected to the NT-1 via a standardized interface. Whereas a LAN is used for professional purposes, an NT-2 is used in general for residential purposes. Such a private LAN is called a *home network*. We have given an example in Chapter 2 of an NT-1 and NT-2 combination in reference to narrowband ISDN access (see Figure 2.11). Two major functions are dedicated to an NT-1 when it is connected to a home network: routing and access control. These two functions are carried out by means of an IP router and of a firewall, respectively.

Knowing that by the end of 2001, about 10 million xDSL lines (essentially G.dmt and G.lite) were in service in the world, home networking represents a huge market for electronic chip providers and electronic equipment vendors. Figure 8.3 (from the ADSL Forum) illustrates the main functions that should be carried out for home networking. As underlined in Chapter 6, in its current implementation, ADSL assumes an ATM cell-based transmission format. The *broadband NT-1* (BNT-1) is in charge of the physical layer functions mentioned in Chapter 6 (modulation, error detection and correction, synchronization).

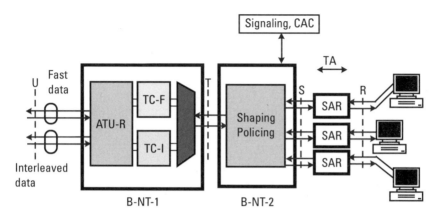

Figure 8.3 ADSL Forum reference interfaces for home networking.

The BNT-2 is in charge of ATM signaling for the supervision of ATM connections (OAM functions) and of CAC. The decision of acceptance or of rejection of a new ATM connection is not decided by the BNT-2 but by the nearest ATM switch in the carriers' network. The *terminal adaptors* (TAs) are in charge of protocol adaptation between the ATM-oriented BNT-1 and the IP/Ethernet-oriented end-equipment. A TA proceeds to *segmentation and reassembly* (SAR) of the variable size packets generated by users' applications. The function of terminal adaptation is either distributed onto the various end equipment or is carried out after traffic concentration between an NT-2 and a BNT-1. Several groups of interest are working on a standardization of home networking: ADSL Forum (TR-011/012 Group), ETSI (TM6 ANA Group), ATM Forum (Residential Broadband or RBB Group), DAVIC, ITU-T (SG 15/4) and finally IEEE.

Today, many homes are equipped with multiple PCs in North-America, Europe, and a few countries in Asia. Facing this situation, the most recent versions of cable modems or of xDSL modems include a router and multiple ports enabling the connection of multiple pieces of end equipment. Depending on the vendors, the first forms of built-in home networks have appeared on the market since 2000. In most cases, these first versions of NT-2 enable the connection of a very limited range of end equipment such as PCs, printers, and analog or digital telephone sets to the NT-1.

8.3.2 End Users' Expectations

These recent years, several polls have been carried out in order to determine consumers' expectations in terms of home networking [1]. The first demand from the end users is to benefit from real in-home networks that enable them to spread out the end equipment in the different rooms of the home. Via a single NT-1 a few pieces or all of this equipment should benefit concurrently from always-on connections for high-speed access to the Internet. Like a traditional LAN, the basic role of a home network is to facilitate file and peripheral sharing within the home. Data applications accessible via PCs such as Web access, e-mail, and printing represent a global bit rate of a few megabits per second. Voice applications requirements are low in terms of bandwidth (a few tens of kilobits per second). Thanks to compression techniques like MPEG-4, video over PC services are today possible at data rates under 500 Kbps. Traditional video over TV set requires higher bit rates up to 4 Mbps.

Today, the concept of home networking is enlarged to a great variety of end equipment and applications. Thus, consumers would like to be able to integrate their digital mobile phone to their home network in order to use it as a cordless system. The definition of a cordless system is given in Chapter 11. Let us state precisely that a cordless system assumes the reduced mobility of the end user. In this way, the cost of a cordless call is much lower than the cost of a regular cellular call. Indeed, operations like end-user localization, handover, or power control are much easier to achieve in cordless systems than in cellular systems. Consumers also would like to be able to transfer easily information from equipment such as their *personal digital assistants* (PDAs), digital cameras, and *digital video recorders* (DVRs) onto their home network. They could in this way share digital documents locally or with distant persons. Playing Internet radio stations or MP3 files on their home stereo systems is another example or consumers' demand. At last, a whole range of services deals with the remote control and telemetry of electrical in-home devices. Such remote controls should make it possible to regulate the temperature or the lightening within the home (concept of smart home). They could also be used for security purposes by means of video cameras. Considered one by one, these new services do not represent high bit rates but some of them (i.e., telephony, video, and conferencing) require stringent QoS constraints in terms of end-to-end delay and packet loss ratio.

8.3.3   Which Medium for Home Networking?

The previous section listed a wide range of applications based on various pieces of equipment. This equipment is usually incompatible in terms of networking because of reliance on different cabling or wireless systems. During the last 5 years, the multiplicity of commercial offers for private devices susceptible to be connected to a home network has been such that standardization will be necessary in the next years. Schematically, three types of cabling approaches are possible: structured wiring, existing wiring, and wireless systems.

8.3.3.1 Structured Wiring

Structured wiring assumes the installation of brand new wiring in the home based, for instance, on UTP copper pairs, coaxial cable, or optical fibers. Economically, a single type of medium is recommended. Today, 10- or 100-Mbps Ethernet switches or hubs may operate on UTP pairs. Coaxial cable is certainly too expensive and lacks the mechanical flexibility to be a considered as a future-proof alternative. Finally, optical fibers could be the solution for installations requiring the highest bit rates. The great default of structured wiring is that it is in practice applicable only to the case of brand-new apartments or houses, with recabling an existing house or apartment appearing less realistic.

8.3.3.2 Existing Wiring

The following are examples of existing wiring.

- *Home Phoneline Networking Alliance (HomePNA):* In the case of existing buildings or houses, a reuse of the installed wiring seems economically reasonable. Today, three types of wired mediums are installed at the customer premises. UTP copper pairs traditionally used for analog telephony are widely deployed in most of the rooms of the house or of the apartment. A consortium known as the HomePNA has already published a set of specifications enabling the constitution of home networks from in-home existing copper pairs [2, 3].

  Figure 8.4 depicts an example of home network configuration using the HomePNA standard. An ADSL modem (the BNT-1) is connected to a 10BaseT Ethernet hub or switch. Two PCs (1 and 2) equipped with an Ethernet card are connected on two ports of this hub/switch. An analog telephone set is connected to the POTS splitter, which is itself connected to the existing copper phone wiring. Another telephone set connected to a PC (PC 2) is used for digital voice (for instance, a VoIP phone). A HomePNA bridge makes it possible to extend the range of the home network based on the Ethernet hub/switch by making it possible to connect PCs directly into the existing phone wiring. Thus, PCs 3 and 4 must be equipped with specific HomePNA adaptors linking their USB port to a RJ45 phone connector. For such a home network configuration, the NT-2 corresponds to the set of the equipment included within the dotted line box represented on Figure 8.4. A TA is necessary between the Ethernet hub/switch and the ATU-R (i.e., the B-NT1) in order to carry out Ethernet-to-ATM conversion (the set of the NT-2 and TA corresponds logically to the BNT-2). Multiple IP addresses are dynamically assigned to the ADSL modem by the ISP, each of these addresses serving a given

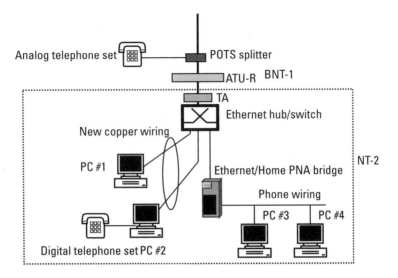

Figure 8.4 An example of a home network based on the HomePNA standard.

piece of end equipment. In our example, up to five IP addresses may be necessary, one for each of the four PCs and one for the voice-over-IP phone.

- *HomePlug:* In-house electrical wiring is another type of medium already installed at the customer premises. Section 3.8 introduced the HomePlug Powerline Alliance [4], a consortium of vendors interested in the definition and development of interfaces enabling to use electrical wiring as a home network. The MAC protocol specified in the HomePlug 1.0 reference document is based on OFDM. Compared to HomePNA networks, home networks based on PLC technology are characterized by a more limited capacity. According to the state of the technology, the capacity of such home networks is about 1 Mbps instead of several megabits per second to share between multiple pieces of end equipment for HomePNA. Current investigations carried out within the Home-Plug Powerline Alliance aim to enable a global capacity on the electrical home wiring up to 10 Mbps. Let us remark that in general, electrical wiring is installed in every room; that is not the case for phone wiring. Electrical wiring is equipped with a larger number of outlets than phone wiring.

- *Coaxial wiring:* Many homes are connected to coaxial cable networks all over the world. Nevertheless, a great disparity exists, especially in Europe where cable modems are, for instance, widely deployed in

Belgium but much less popular in France. We mentioned the limitations of coaxial cable in the section dedicated to structured wiring.

8.3.3.3 Wireless Systems

Compared to wired systems, wireless home networks have the advantage of not needing any recabling at the customer premises. This property will be certainly an important factor in the competitive context of home networking. Unlike HomePNA solutions, wireless technologies are better suited to the connection of mobile or semimobile radio equipment such as GSM cellular phones, laptops, PDAs or cordless phones to a home network. Numerous vendors proposing different technologies are in competition, each of them expecting their solution to become the de facto standard by occupying the fastest as possible this market segment. The IEEE 802.11b and Bluetooth technologies are currently the most popular approaches for wireless home networking. Other technologies like home RF, HiperLAN or ultra wideband are also among the candidates for wireless home network. These different techniques are presented in the following sections.

8.3.4 IEEE 802.11 and IEEE 802.11b

Historically, the first IEEE 802.11 standard, also known as wireless Ethernet, dates from 1997. This first version operating at 1 or 2 Mbps was replaced by an enhanced version operating either at 5.5 Mbps or 11 Mbps and known as the IEEE 802.11b standard in 1999. The two data rates considered for both IEEE 802.11 and IEEE 802.11b correspond to a noisy and to a favorable environment, respectively. Currently, investigations aim to increase the on-line bit rate up to 20 Mbps for the new IEEE 802.11g standard.

The IEEE 802.11b wireless LAN, also known as wireless fidelity or WiFi, operates at 2.4 GHz. As it is intended for private usage, no license from the regulation bodies is necessary to operate a WiFi network. The transmission power is of 20 dBm for a range of about 50m. Two configuration modes are possible for WiFi: the infrastructure mode and the ad hoc mode. An ad hoc IEEE 802.11b network consists of a set of terminal equipment (typically PCs and printers, for example) able to communicate directly between each other without any master node or base station. This set of terminals forms an *independent basic set service* (IBSS). By its nature, the services offered by the ad hoc mode are limited to best effort asynchronous data transfers. The other configuration mode, called the infrastructure mode, makes it possible to provide specific services to a terminal over a certain area. Infrastructure networks are based on the usage of base stations called *access points* (APs). A radio cell defined by an AP is called a *basic set service* (BSS). The AP are linked together by means of a traditional Ethernet bus called a *distribution system* (DS). Figure 8.5 illustrates an example of IEEE 802.11b network configuration associating three BSSs and one IBSS. One notices that a terminal may belong to an IBSS and to a BSS simultaneously. The most popular versions of IEEE 802.11b are based on the infrastructure mode.

Three different physical layers are possible for an IEEE 802.11 network: *frequency hopping spread spectrum* (FHSS), *direct sequence spread spectrum* (DSSS), and infrared. FHSS consists of splitting the 75-MHz capacity of a cell into 75 channels of 1 MHz each. An active terminal (transmitting or listening) scans permanently these 75 channels according to a procedure that is independent from one cell (IBSS or BSS) to another one. The objective of frequency hopping is to reduce considerably the risk of interference between communication channels. The main drawback of FHSS is that it reduces considerably the effective capacity of the system due to the switchover time inherent to frequency hopping. DSSS is better suited to high-speed data transfers than FHSS. The baseband signal is spread out in the frequency domain in order to reduce the impact of interference. In each transmitter, information bits are added up with a binary sequence in the Gallois field. The binary sequence corresponds to a set of "bits" called chips. The duration of a chip is an integer fraction of a bit duration. The sequence of chips used in IEEE 802.11 is called a Barker sequence with 11 chips. Figure 8.6 depicts the principle of DSSS. Both IEEE 802.11 and IEEE 802.11b operates with DSSS.

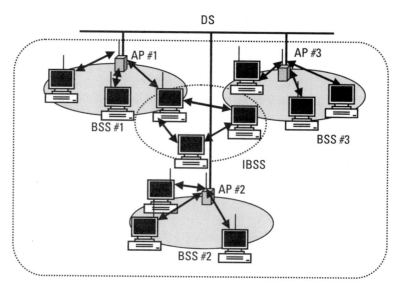

Figure 8.5 Example of IEEE 802.11b network.

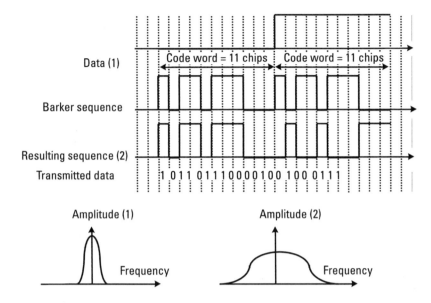

Figure 8.6 Principle of DSSS in IEEE 802.11.

A received signal during a bit duration with six chips equal to 0 or 6 chips equal to 1 corresponds to the bit "1" or "0," respectively. The Barker sequences imply a frequency spreading factor equal to 11. DSSS uses 14 channels of 5 MHz each. In IEEE 802.11b, a modulation technique inspired by QPSK called *complementary code keying* (CCK) associates to each suite of 8 bits a given phase according to the value of the first 2 bits and an 8-bit code word according to the value of the last 6 bits. At the MAC layer, the *carrier sense multiple access with collision avoidance* (CSMA-CA) protocol has been adopted. The protocol efficiency of this MAC protocol limits the effective capacity of the IEEE 802.11b to about 6 Mbps. A particular interest has been dedicated to security aspects in order to prevent a malicious user to access or to listen illegally to the data transferred on the network. For that purpose, the MAC layer includes an encryption sublayer called *wired equivalent privacy* (WEP). A detailed description of the IEEE 802.11b MAC protocol and of CCK is out of the scope of this book. The interested reader may consult [5, 6].

8.3.5 Bluetooth

Bluetooth is another good candidate for wireless home networking. At the difference of IEEE 802.11b, Bluetooth was conceived initially for applications requiring low bit rates over a few meters range (between 5m and 10m). Networks with such a very short range qualify as *personal area networks* (PANs). Like

IEEE 802.11b, Bluetooth operates around 2.4 GHz. It uses a bandwidth of 83 MHz in the [2.4 GHz, 2.4835 GHz] band. The transmission power of Bluetooth is on average of a few dBm, much lower than the 20 dBm required by IEEE 802.11b. Bluetooth technology is based on VLSI electronic chips that can be placed on any kind of electrical device susceptible to be remote-controlled (e.g., washing machine, heating system, air conditioning system, home stereo system, and PDAs) or to communicate with other equipment. Bluetooth is an ad hoc network. This means that the connection of an equipment to a "piconet" (a cell with 10m radius) is carried out automatically without any manual configuration. Several piconets may be linked together automatically to form a "scatternet." The first piece of equipment to be activated plays automatically the role of master. The other machines activated in the same cell after the master are qualified as slaves under the control of this master. Up to seven slaves may depend on a same master to form a piconet. In other words, an eighth station is considered automatically as a new master that will control up to seven new slaves and so on. The masters communicate between each others to form a scatternet. The FHSS transmission technique is adopted at the physical layer. The 83-MHz band mentioned above is thus divided into 70 sub-bands; an active piece of end equipment scanning permanently on these 70 channels. Master stations have the responsibility to assign randomly the sequence of frequency hops that two machines have to adopt during their dialogue. About 1,600 frequency hops per second are thus applied during communications. A detailed description of Bluetooth systems is given in [7]. Today, Bluetooth chips are implemented in such devices as cellular phones (Ericsson), PC cards (Anycom, 3Com), PDAs (Palm), and electronic pencils.

8.3.6   Home RF, HiperLAN, and Ultra Wideband

Other wireless techniques are also considered to be possible candidates for wireless home networking. Home RF and HiperLAN operate at 2.4 GHz whereas ultra wideband operates between 3 and 6 GHz. Meanwhile, the development of these techniques is less advanced than for the two previous ones. This is the reason why we do not detail their respective announced performance.

8.3.7   Toward a Universal Residential Gateway

8.3.7.1   Residential Gateway

We have listed in Section 3.2 a large variety of possible services that could be offered on next generation home networks. The most innovative among these services are those referring to the concept of smart house. We have also underlined that next generation home networks will probably be based for economical and technical purposes on a hybrid combination of multiple subnetworks using various types of mediums. Some of these subnetworks will reuse existing cabling systems by enlarging their field of utilization (HomePNA, PLC). Figure 8.7 illustrates an example of a home network combining existing phone wiring, Ethernet wiring, and a wireless system. In Figure 8.7, one has represented a residential gateway coupled with a remote modem in charge of interconnecting the various in-home networks onto a same user's modem. For downstream traffic, a residential gateway operates as an IP router to dispatch incoming IP packets to the right terminal. Locally, the residential gateway must be able to proceed to protocol conversion in order, for instance, to enable a PDA equipped with a Bluetooth chip to communicate with a HomePNA PC. In order to facilitate such local communications without the intervention of a carrier or ISP, the residential gateway must be able to assign dynamically IP addresses to any active piece of end equipment. Finally, the residential gateway must be able to carry out synchronous/asynchronous packet multiplexing in order to share the xDSL link between multiple egress connections. We have also mentioned above the necessity to protect the private installation from undesirable incoming traffic; this service is offered by means of a firewall included in the residential gateway.

8.3.7.2 Service Gateway

Because of the multiplicity of the actors (vendors, service providers, carriers) involved in the market of home networking, the emergence of a standard is necessary. The *open services gateway initiative* (OSGi) is the most advanced approach in this matter [8, 9]. The OSGi introduces the concept of a network

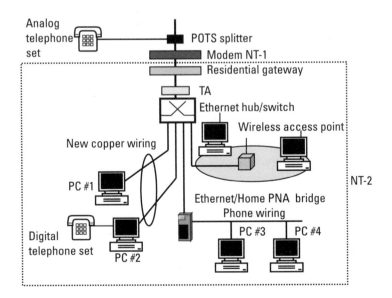

Figure 8.7 An example of hybrid home network.

service gateway that consists of providing each home a unique "virtual service gateway" based on the OSGi specifications. Thus, one enlarges the concept of residential gateway to a form of universal residential gateway made of a network terminal and of a service gateway. The network terminal provides IP bearer services by means of an NT-1 and of an NT-2. The service gateway is an application server using Java software and located at the customer premises. A piece of terminal equipment may dynamically download from the service gateway service applications called *bundles*. If these bundles necessary for the provision of the required service are not available in the service gateway, the service gateway may itself download them from dedicated servers located at the CO. In existing networks, the services offered to the end users are fixed and rely on specific terminal equipment. With OSGi, the idea is then to enable the end users to adapt partially the offered services to their own requirements, independently from the provider of the basic service.

8.4   Conclusion and Perspectives

The first part of this chapter was dedicated to the description of a DSLAM. We have seen that in addition to terminating at the EO or CO xDSL links, a DSLAM serves as a traffic concentrator and as protocol converter. The second part of this chapter was dedicated to home networking. Ad hoc wireless LANs may be considered good candidates for the coming home network standard because they are the most simple to implement. We have seen that IEEE 802.11b and Bluetooth are the most two important competitors for this market segment. In terms of range and power consumption, Bluetooth seems better suited to an in-home environment. On the other hand, some estimate that it should be easy to reduce the power of transmission in IEEE 802.11b in order to adapt the range to home networking, the reverse being not possible for Bluetooth, which is PAN-oriented. Efforts are currently being carried out to adapt both systems to the residential environment. Ethernet switches or hubs are already widely used as an NT-2 coupled to an ADSL modem. In-home networking will probably be based on hybrid technologies. It has been observed experimentally on real installations that Bluetooth, Home RF, and IEEE 802.11b systems could interfere with microwave systems. Because of the inevitable multiplicity and permanent evolution of the services inherent to the smart network concept, the definition of a universal residential gateway seems necessary. This is the reason why we have briefly introduced the OSGi standard. An OSGi that promotes an evolution toward universal home networks should motivate, for instance, the development of bistandard phones (cellular and cordless) by the manufacturers.

References

[1]   Teger, S., and D. J. Waks, "End-User Perspectives on Home Networking," *IEEE Communications Magazine,* April 2002, pp. 114–119.

[2]   Bisaglia, P., R. Castle, and S. H. Baynham, "Channel Modeling and System Performance for HomePNA 2.0," *IEEE Journal on Selected Areas in Communications,* Vol. 20, No. 5, June 2002, pp. 913–922.

[3]   HomePNA, http://www.homepna.org.

[4]   Homeplug Powerline Alliance, http://www.homeplug.org.

[5]   Ala-Laurila, J., et al., "Wireless LAN Access Network Architecture for Mobile Operators," *IEEE Communications Magazine,* November 2001, pp. 82–89.

[6]   Heegard, C., "High-Performance Wireless Ethernet," *IEEE Communications Magazine,* November 2001, pp. 64–73.

[7]   Schneiderman, R., "Bluetooth Slow Dawn," *IEEE Spectrum Magazine,* November 2000.

[8]   Marples, D., and P. Kriens, "The Open Services Gateway Initiative: An Introductory Overview," *IEEE Communications Magazine,* December 2001, pp. 110–114.

[9]   OSGi, http://www.osgi.org.

Selected Bibliography

Horowitz, B., N. Magnusson, and N. Klack, "Telia's Service Delivery Solution for the Home," *IEEE Communications Magazine,* April 2002, pp. 120–125.

Muñoz, L., et al., "Optimizing Internet Flows over IEEE 802.11b Wireless LANs: A Performance-Enhancing Proxy Based on Forward Error Correction," *IEEE Communications Magazine,* December 2001, pp. 60–67.

Paradyne Corporation, *The DSL Source Book: Plain Answers About Digital Subscriber Line Opportunities,* 1997.

Valtchev, D., and I. Frankov, "Service Gateway Architecture for a Smart Home," *IEEE Communications Magazine,* April 2002, pp. 126–132.

9

Protocol Architecture

9.1 Introduction

Previous chapters described the main characteristics of xDSL access systems (including HDSL, SHDSL, ADSL, G.lite, and VDSL). Throughout these chapters, we have mainly focused on the physical layer aspects at the customer premises and at the DSLAM. This chapter considers in an end-to-end perspective the higher protocol layers adopted in xDSL access systems. Our analysis applies essentially to ADSL, G.lite, and VDSL access systems. It can be partially generalized to wireless and optical access systems described in Chapters 11 and 12, respectively. Two modes of transmission have been considered for ADSL, either synchronous-oriented or *synchronous time multiplexing* (STM) or ATM cell-oriented. Section 9.2 explains why most vendors have implemented the ATM mode in their equipment. Section 9.3 describes data formatting at the physical layer for both the STM and ATM transmission mode. Sections 9.4–9.8 consider the various possible protocol architectures in the upper layers (2 and 3). Finally, Section 9.9 investigates the problem of the dimensioning of the TCP windowing mechanism, TCP being applied to end-to-end connections.

9.2 Why the ATM Technique in the Local Loop?

With the first commercial offers of ADSL access systems at the beginning of the 1990s, service providers and modem vendors considered ATM the best suited technique for the local loop. On one hand, in spite of its inherent complexity and cost, ATM was the only technique recognized as an international standard widely deployed by major carriers able to provide both real-time and nonreal-time services. On the other hand, the first advances for QoS provisioning in IP networks mentioned in Chapter 1 (IntServ, DiffServ) were not mature at that time. Thanks to the ATCs [1] described in Section 1.3.1, the ATM technique makes it possible to provide various types of QoS to the end users.

9.3 Physical Layer ADSL Data Formatting

We describe in the following sections the data formatting at the level of each of the interfaces of the ADSL reference model mentioned in Figure 6.5.

9.3.1 Reference Model of the ATU-C

Seven data channels, referred to as AS_0, AS_1, AS_2, and AS_3 and LS_0, LS_1, and LS_2, are specified at the V_c interface between the ATU-C and a public packet switching data network. The AS_x channels are unidirectional and enable simplex[1] communications whereas LS_x channels enable duplex[2] communications. The AS_x simplex channels are dedicated to downstream flows from the network to the CPE. The ATU-C carries out several operations, described as follows, to adapt these downstream flows to the U_{c2} interface specifications:

- Transfer of downstream data onto the various DMT channels, each channel using a specific line code (see Section 6.11 about the initialization of ADSL modems);

- Definition and application of FEC to the active data flows;

- Insertion of downstream data onto ADSL frames and superframes;

- Transmission of the data on the subscriber line.

We have seen in Chapter 8 that two types of transmission modes were possible in ADSL systems: fast buffer (TC-F) and interleaved buffer (TC-I). Real-time traffic uses the TC-F mode whereas asynchronous traffic uses the TC-I mode. The main difference between these two modes is the usage of an interleaver in the TC-I mode that does not exist in the TC-F mode. The data flows relying either on the TC-F or the TC-I modes are both transmitted in the same physical layer frames. The data using the TC-F mode are placed at the beginning of the payload of successive ADSL frames of 250 μs. In each ADSL frame, these data are followed by the data using the TC-I mode.

---

1. Simplex communication means in a single direction.

9.3.2 Reference Modem of the ATU-R

The T_r interface is specified at the ATU-R between the NT-1 and the NT-2 (see Figure 6.5). The AS_x channels are used only for the reception of downstream data. The bit rates generated by an ATU-R in the upstream direction are in general lower than the bit rates in the downstream direction received from a DSLAM. Only the three channels LS_0, LS_1, and LS_2 are used for upstream data. These three channels are multiplexed into upstream physical layer ADSL frames.

9.3.3 STM Mode and ATM Mode

Two modes of transmission are specified in the ANSI T1E1.4 standards, either synchronous-oriented (also called STM-based) or ATM-based. Figures 9.1 and 9.2 illustrate the data formatting reference points in an ATU-C for the STM transmission mode at the transmitter side (Tx) and at the receiver side (Rx), respectively. Several logical channels may be activated simultaneously on an ADSL link by programming the ATU-R and ATU-C. This is typically the case when an end user activates multiple pieces of terminal equipment connected on its ATU-R.

In Figures 9.1 and 9.2, we can see that three reference points (*A*, *B*, and *C*) are specified for data formatting. A multiplexer is in charge of building the physical layer ADSL frames from the logical channels submitted at the input of the modem. These frames are directed either toward the TC-I branch or the TC-F branch of the modem. Reference point *A* specifies data formatting on

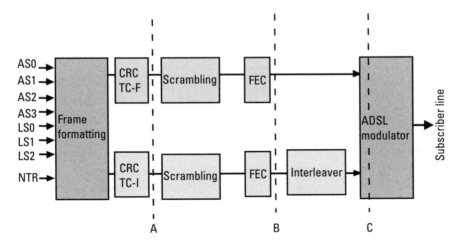

Figure 9.1 Reference points in an Tx-ATU-C for the STM mode.

---

2. Duplex communication means alternate communication in both directions.

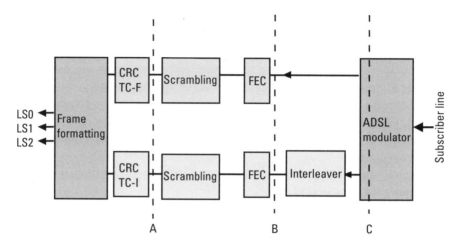

Figure 9.2 Reference points in an Rx-ATU-C for the STM mode.

each of these two branches after the calculation and the addition of a specific *cyclic redundancy checksum* (CRC). Reference point *B* refers to data formatting at the output (Figure 9.1) or at the input (Figure 9.2) of the FEC encoder, respectively. At point *B*, the frames taking into account the overload due to the FEC bits are generated at the DMT symbol rate. Finally, the reference point *C* specifies data formatting at a specific point within the ADSL modulator (Figure 9.1) or demodulator (Figure 9.2). The position of the dotted line associated to this point of reference in both Figures 9.1 and 9.2 corresponds more precisely to the interface between the circuit in charge of tone ordering and the circuit in charge of constellation encoding and gain scaling (see Chapter 6). A time reference of 8 kHz is used within the ATU-C (see Figure 9.1) by means of a port name NTR (network timing reference).

The bit rates of the LS_x duplex channels may differ depending on their directivity (upstream or downstream). These bit rates are programmable either by the end user or by the IAP by steps of 32 Kbps. The possible values for the AS_x and LS_x channels are given in Table 9.1.

In practice, most of the end users do not program by themselves the bit rates of the AS_x and LS_x channels and use a single logical channel. In that case, the AS_0 and LS_0 channels are used for downstream and upstream data, respectively. The on-demand provision of MPEG-2 video sequences from a central server located in an IP network to a residential user[3] is an example where a single logical channel is used. The AS_0 channel is used at 4 Mbps for the transmission of downstream MPEG-2 frames. The LS_0 channel is used by the end user at 32

---

3. This service is called VoD.

Table 9.1

Range of the Possible Bit Rates in Each AS_x and LS_x Channel

Channel	Bit Rate
AS_0	0–8.192 Kbps
AS_1	0–4.608 Kbps
AS_2	0–3,072 Kbps
AS_3	0–1,536 Kbps
LS_0	0–640 Kbps
LS_1	0–640 Kbps
LS_2	0–640 Kbps

Kbps to send command signals to the video server via a set-top box and to receive signaling messages from the server. We describe in detail in Chapter 10 the protocol architectures considered for *video-on-demand* (VoD) services.

As mentioned earlier, another transmission mode based on ATM cells may also be used in ADSL access systems. Unlike the STM mode, one does not consider in the ATM mode the multiplexing of various logical channels at the input of the modem. However, one assumes in that case that this multiplexing is carried out in the ATM-compatible end-equipment on the customer site. The downstream and upstream flows of ATM cells use the AS_0 and LS_0 channels, respectively. Although both the STM mode and the ATM mode have similar reference points *A*, *B*, and *C*, data formatting differ between these two transmission modes. Figure 9.3 illustrates the location of reference points *A*, *B*, and *C* in

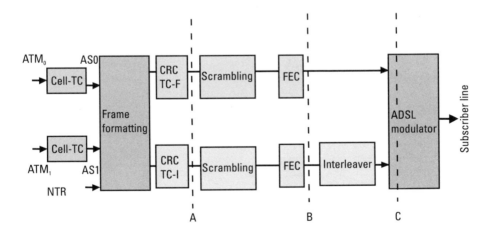

Figure 9.3 Reference points in an Tx-ATU-C for the ATM mode.

the case of a Tx-ATU-C with the ATM transmission mode (we do not describe the Rx-ATU-C which is similar to the Tx-ATU-C but that operates in the reverse direction).

9.3.4   ADSL Frame and Superframe Formats

An ADSL superframe is made of 68 ADSL frames numbered from 0 to 67. Each frame is coded by means of a DMT encoder and then modulated. An ADSL superframe is ended with the addition of a synchronization symbol with the same duration as a the duration of data frame. The global duration of an ADSL superframe is 17 ms. The duration of a DMT symbol is 250 μs. Because it is necessary to insert the synchronization symbol, the duration of a DMT symbol must be reduced with a 68/69 ratio. In other words, a DMT symbol duration is then 246 μs. In an ADSL modem, 4,000 ADSL frames are then generated every second. Figure 9.4 illustrates the format of an ADSL superframe.

Each frame within a superframe may transport real-time data for which interleaving is not applied (these data are referred to as *fast-buffer data*) and nonreal-time data for which interleaving is applied (these data are referred to as *interleaved-buffer data*). Each of the 68 frames of a superframe contains a certain amount of bytes belonging to each of the active logical channels (TC-F and TC-I). The first frame of the superframe (frame # 0) is made of the CRC

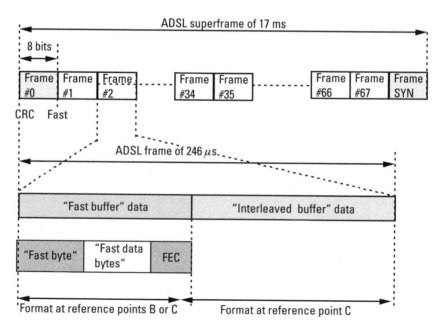

Figure 9.4 Format of an ADSL superframe.

associated to fast data. Frames # 1, # 34, and # 35 are used for the transport of operation and maintenance information relative to fast data.

9.3.4.1 TC-F Frame Format

Figure 9.5 illustrates in more detail the format of the fast-buffer section of an ADSL frame[4] at reference point *B*. The only difference between reference points *A* and *B* is the presence or not of an FEC field at the end of the frame. On the upper and lower parts of this figure, we have represented the format of the data at the ATU-C and the ATU-R, respectively.

The amount of bytes allocated to each downstream channel (AS_x) and to each upstream channel (LS_x) is not specified in the standard. Although ADSL frames are of fixed duration, the amount of bits transported in each of them depends on the capacity of the considered DMT subchannel. The first byte of a frame, named *fast byte*, is used for synchronization purposes. During silence periods (absence of data waiting for transmission), only the fast byte is transmitted. The last 2 bytes of a frame AEX and LEX are also used for synchronization purposes. Useful bytes AS_x and LS_x are encrypted for confidentiality purposes. The size in bytes induced by FEC may vary from 0 to 16 by step of 2 according to the capacity allocated to the DMT subchannels at line initialization.

9.3.4.2 TC-I Frame Format

Figure 9.6 depicts the format of interleaved-buffer frames in upstream and downstream directions at reference point *B*. Like in Figure 9.5, the only difference between reference points *A* and *B* is the presence or not of an FEC field at

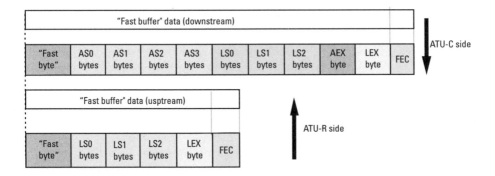

Figure 9.5 Format of the fast buffer section of an ADSL frame at reference point *B*.

---

4. Note that this format is not used for frames 0, 1, 34, 35, and 68.

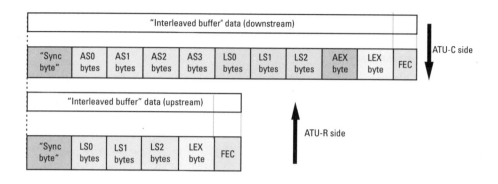

Figure 9.6 Format of the interleaved buffer ADSL frame at reference point *B*.

the end of the frame. One notices that the fast byte used in TC-F frames is replaced in TC-I frames by a *sync byte*. The description of the role of the fast byte and of the sync byte is out of the scope of this book; the interested reader may refer to [2].

Unlike downstream fast-buffer frames (see Figure 9.5) at reference points *A* and *B* for which an FEC block is calculated for each ADSL frame, a single FEC field is calculated by the RS encoder for a set of *S* frames in the case of interleaved-buffer sections. The value of the parameter *S* is determined at line initialization.

9.4   The PPP Protocol

Either a single or several data terminals (typically one or several PCs) may be connected to the same modem at the customer premises. A terminal that requires an Internet access must obtain a temporary IP address from a *Dynamic Host Configuration Protocol* (DHCP) server implemented in an access server located in the public network. An access server is called either *network access server* (NAS) or *broadband access server* (BAS) depending on the type of the user's modem. A NAS refers to voiceband dial-up modems whereas a BAS refers to xDSL modems. A NAS and a BAS serve as a gateway to an IP network. Thus, a NAS is located at the border between PSTN and an IP autonomous system. It is connected to an ISO. A BAS is located at the border between an ATM network and an IP autonomous system. It is connected to an ATM switch. An access server is in charge of controlling the identity of an end user before to assign him an IP address. Two reasons motivate this control. First, as it has been mentioned in Chapter 1, the lack of IP addresses in IPv4 prevents the assignment of a permanent IP address to each user. Second, by changing at each Internet access the IP address of an end user, one prevents any connection to the CPE from malicious end user. If several PCs share the same modem at the customer premises (e.g., in the case of a home network), two additional functionalities must be implemented in the user's modem: routing of IP packets and IP packets traffic management for QoS purposes. We have seen in the previous section that two different access modes are considered in ADSL modems, either STM-based or ATM-based. Whatever the adopted access mode, an end user must always connect to a NAS or a BAS in a first step. The three basic functionalities of an access server are called AAA that stands for authentication, authorization and accounting. The AAA server stores the log-ins and passwords of the subscribers of one or several ISPs. The connection between a user's modem to an AAA is carried out by means of the *Point-to-Point Protocol* (PPP).

The PPP protocol [3] enables a negotiation procedure to setup a logical connection between the terminal of an end user and an access server. Once this connection is established, PPP authorizes the transport of variable size layer 3 datagrams between the user's terminal and the remote database of the service provider. In other words, layer 3 datagrams transit via the ISO or the ATM switch on which is connected the NAS or the BAS, respectively, to reach the service provider through the Internet. PPP may transport datagrams coming from different layer 3 protocols (e.g., IP and IPX). As specified in IETF RFC-1661, PPP is a layer 2 (data link) protocol that includes three main functions:

- It provides a method for the encapsulation of datagrams coming from various layer 3 protocols.

- It defines a control procedure named LCP (Link Control Protocol) for establishing, configuring, testing, maintaining and releasing a layer 2 connection.

- It includes a family of control protocol enabling the establishing and configuring of various layer 3 protocols. These control protocols are called *Network Control Protocol* (NCP).

9.4.1 Encapsulation

Like the *High Level Data Link Control* (HLDC) Data Link Protocol, PPP is bit-oriented. HDLC is widely deployed as a layer 2 protocol in packet switching networks such as X.25 networks. Like HDLC, PPP can be used on point-to-point connections using different kinds of transmission medium (copper pairs, optical fibers, satellite link, and so forth). These connections can be either semipermanent or switched. The former are established on the long term by a carrier. The latter are set up in real time by an end user. Figure 9.7 illustrates the format of a PPP frame.

Like an HDLC frame, a PPP frame begins and ends by a flag byte made of six "1" preceded and followed by a "0." In the standardization documents the value of the bytes composing a PPP frame are written under the form of two blocks of 4 bits corresponding to an hexadecimal code. Whereas the one byte address field is set by the sender in HDLC, this field is unused with PPP and fixed to suite of ones. Similarly, the control field is fixed whereas it is programmable in HDLC. The value of the PPP control field corresponds in the case of HDLC to unnumbered frames; this means that the sliding window and the error recovery HDLC mechanisms are not used in PPP. The information field of a PPP frame is of variable size. It starts with a 2-byte protocol identifier (see Figure 9.7), followed by the payload itself. The protocol identifier specifies the type of used layer 3 protocol. The payload is upper-bounded to 1,500 bytes. We shall see in the next section that this value can be in fact negotiated at call setup. A PPP frame ends by a 2-byte FCS field. A simple version of the PPP frame may be negotiated at call setup in the case of low speed links. This version does not use the address and the control fields and the protocol identifier field is reduced to a single byte.

9.4.2 LCP Protocol

The behavior of the PPP protocol is specified by a finite state automata given in Figure 9.8. The automata consists of five states: dead, established, authenticate, terminate, and network. When the PPP connection is inactive, that is, no messages are sent, the automata is in the dead state. Messages can be exchanged in any of the other 4 states of the automata. These messages are specified by the *Link Control Protocol* (LCP) protocol that enables to negotiate several options about the format of the encapsulation of the payload. Thus, as mentioned in the previous section, the maximum size of the payload can be negotiated. LCP

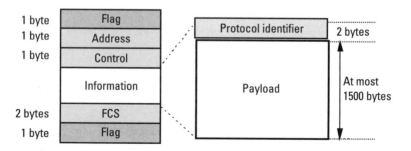

Figure 9.7 Format of a PPP frame.

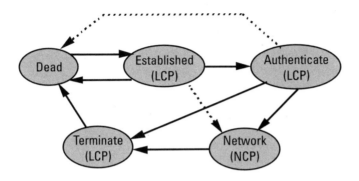

Figure 9.8 Finite state machine of the PPP automata.

messages are used to set up and release a PPP connection. In case of successful connection setup, the automata transits from the "Dead" state to the "Established" state. The quality of the transmission medium may be tested during this transition. In the case of ADSL access, it is in general necessary to identify the end user at the origin of the connection request. This explains the transition from the established state to the authentication state.

Authentication itself can be accomplished by means of two other protocols called *Password Authentication Protocol* (PAP) and *Challenge-Handshake Authentication Protocol* (CHAP), defined in the RFC-1334 and RFC-1994, respectively. The PAP and CHAP protocols correspond to layer 2 authentication mechanisms. A specialized protocol named IPsec is in charge of authentication at layer 3. The IPsec protocol has already been introduced in Section 1.4 dedicated to IP-VPN. It also provides means for authorization, encryption, and so forth. It can be used with IPv4 and IPv6. Without going into the details, the PAP protocol may be assimilated to a "log-in" procedure onto a remote computer. The end user that wants to open a PPP connection sends a user name followed by a password. In practice, PAP is rarely used because of its lack of confidentiality, the set of log-in and password being sent in clear into the network. CHAP enables a much better confidentiality than PAP because it relies on the exchange of a secret key known by the AAA server and by the end user exclusively, the value of this key being encrypted during its transmission.

If the end user authentication fails (the end user ID is not recognized by his or her network access provider), the automata returns automatically to the dead state. If this authentication is successful, the automata enters the network state. In case authentication is not necessary, the transit of the automata through the "Authenticate" state is useless (dotted line between the established and network states). Once in the network state, the type of routing protocol at layer 3 is negotiated. In most cases, this negotiation consists in assigning an IP address to the end user. The operations carried out in the network state refer to the NCP protocols described in the next section. The automata enters the terminate state as soon as the end user or the AAA server desires to close the connection. In that case, LCP messages are used to transit from the terminate state to the dead state. Let us remember that the LCP protocol enables to set up and release a layer 2 connection and to negotiate associated options like the maximum size of the packets.

9.4.3 NCP Protocols

The PPP protocol authorizes several simultaneous layer 3 sessions on the same layer 2 connection. For that purpose, PPP uses several NCP protocols according to the type of the considered layer 3 protocol. Thus, the NCP protocol called *IP Control Protocol* (IPCP) is used instead of IP. NCP protocols enable negotiation of options for the various considered layer 3 protocols. For instance, IPCP enables the temporary allocation of an IP address to each IP session in order not to monopolize this IP address when both sources at each end of the point-to-point connection are silent. NCP protocols enable compression of IP and TCP headers from 40 bytes to 4 bytes according to the RFC-1332.

9.5 NAS Servers and the RADIUS Protocol

9.5.1 NAS Servers

In the context of Internet access by means of voiceband analog modems, we have seen that an end user must first establish a point-to-point connection between his or her modem and an AAA server called a NAS. As it is illustrated in Figure 9.9, the NAS is connected to one of the switches of the PSTN on which is connected the subscriber line. The database of the content provider is connected to a node (an IP router in general) belonging to the network of an NSP. The POP of this NSP is itself connected to one of the switches of the PSTN. As mentioned above, the role of a NAS is to identify the end user, to check that this end user is authorized to access to the requested database, and to meter the volume of information exchanged during the connection between the end user and the database.

A NAS server is connected to a switch of the PSTN, for instance, by means of several parallel T1/E1 LLs. The end user begins to dial on his or her telephone set the phone number of his or her IAP (in most cases this number is dialed automatically by the modem of the end user). Via a phone connection through PSTN, the modem of the end user and one of the available modems at the NAS server begin to communicate by exchanging physical layer information. Such information may consist for example of the quality of the subscriber

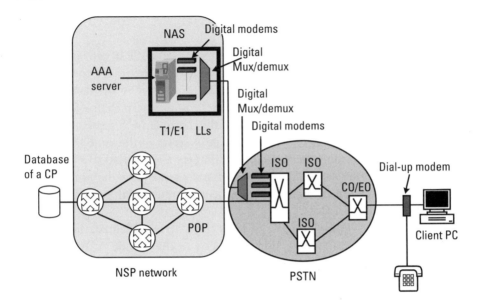

Figure 9.9 Access to a remote database by means of a voiceband modem.

line in terms of the SNR. On the basis of this information, both modems decide upon a common bit rate that will be used during the connection of the end user with the database of the CP. This bit rate corresponds to a certain amount of bits assigned to each QAM symbol. Once the dial-up modem of the end user and the digital modem of the NAS have completed the physical layer initialization, these two pieces of equipment proceed to the establishment of a point-to-point layer 2 connection between them. For that purpose, the dial-up modem sends a *set asynchronous balanced mode* (SABM) unnumbered HDLC frame to the NAS modem. The NAS modem answers back to the dial-up modem by a UA unnumbered acknowledgment HDLC frame. At this step, the PPP protocol may be activated in order to authenticate the end user at the NAS and to decide the use of an optional compression algorithm during the data transfer between the database and the dial-up modem. For that purpose, the LCP protocol is activated. One proceeds then to the negotiation of the layer 3 parameters by means of the NCP protocol. The IPCP assigns an IP address to the end user. If the calling party (the end user) is recognized as a subscriber of the ISP, the exchange of information between the end user and the requested database may then occur. In summary, IP packets generated by the end user are encapsulated into PPP frames. Once these frames arrive at the NAS, the IP packets are extracted from the received PPP frames before being forwarded through the NSP network to the CP database.

9.5.2   RADIUS Protocol

In practice, several NAS servers may be used by the same ISP in case of large dispersion of the subscribers on the territory. These NAS servers are connected to different IP routers of the same NSP. The access control to the database of a content provider relies in that case on a distinct equipment in charge of the AAA functions. A single AAA server is shared by the multiple NAS servers. Communications between the NAS servers and the AAA server of the same ISP are carried out by means of the *Remote Access Dial-In User Service* (RADIUS) Protocol once the end user has sent his or her log-in and password to the nearest NAS. Figure 9.10 presents an example of network configuration with two NAS servers and their associated AAAs.

The RADIUS Protocol (RFC 2138) is client-server–oriented, the multiple NAS servers being the clients sending requests to the unique AAA server. As mentioned above, the end user authentication is based on the usage of a pubic key algorithm under the control of the AAA server. If a connection request to a database is accepted by the AAA, the considered NAS enables the transfer of IP packets between the end user (e.g., terminal 2) and the database via several IP routers in the NSP network. The accounting function is activated at the AAA server on demand of NAS 2. The RADIUS Protocol makes it possible to charge the end user pro rata of the data volume exchanged between the terminal and the database of the CP. The NAS and AAA servers are not only used for the access to the databases of content providers but also for the access to intranets of enterprises. In that case, an enterprise owns and manages its own NAS and AAA servers connected on one of its LANs. In most cases, an enterprise owns a single NAS, the AAA server and the NAS corresponding to the same piece of equipment. The NAS of the enterprise is directly connected to the closest router of a NSP. We describe in the following section how the concept of a "tunnel" enables a member of an enterprise far from his or her office to connect economically to his or her intranet.

9.6   The L2TP Protocol

9.6.1   The Tunneling Principle

The tunneling concept has been introduced in order to facilitate the routing of IP packets in internet networks [4]. Initially, the basic idea of this concept was inspired by the service offered by SVCs in connection-oriented packet-switching networks. Thus, in X.25, frame relay, or ATM networks, all the packets assigned to the same SVC join the same destination. Similarly, all the IP packets belonging to the same tunnel join the same destination. The tunneling concept consists of encapsulating IP packets in a tunneling frame, the header of this frame

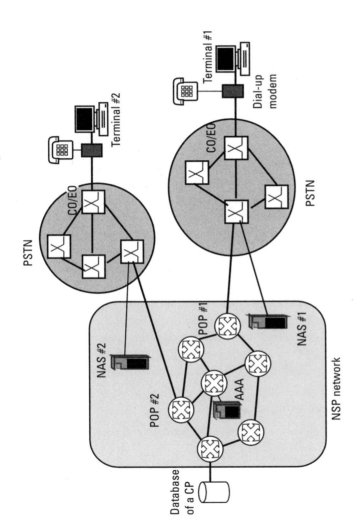

Figure 9.10 Example of network configuration with multiple NAS servers.

identifying the associated tunnel. Procedures enabling the setup and release of a tunnel have been defined. In practice, the management of IP tunnels is simpler and less expensive than the management of ATM SVCs. Three distinct protocols have been proposed by the IETF for tunneling:

- The *Point-to-Point Tunneling Protocol* (PPTP) protocol [5];
- The *Layer 2 Forwarding* (L2F) Protocol;
- The *Layer 2 Tunneling Protocol* (L2TP).

PPTP is used for tunnels exclusively through IP networks. L2F and L2TP can be used to create tunnels via any type of packet switching network (X.25, IP, frame relay, ATM). The L2TP protocol makes it possible to manage several tunnels between the same two endpoints. The reason of the multiplicity of these tunnels may be justified by QoS criteria. Thus, depending on the quality of the traversed physical links, two different tunnels may be used for real-time and nonreal-time data, respectively. In that case, the session between the two endpoints is referred by its identifier and by the identifier of the tunnel. A flow and congestion control mechanism is implemented by means of a sliding window mechanism and control messages proper to each tunnel. In practice, virtual channels activated in frame relay or ATM networks are frequently used as tunnels for IP packets. The main usage of L2TP consists in extending a PPP connection established between an end user and a NAS as far as a remote server (for instance, the private gateway of a corporate LAN) located higher in the network. We refer in the following only to L2TP, which is the most recent version of tunneling protocol; L2TP is inspired by PPTP and L2F themselves inspired by PPP. The L2TP protocol inherits the main functions of PPP:

- Authentication of the end user;
- Dynamic allocation of IP addresses to the end users;
- Optional data compression;
- Encryption of useful data;
- Multilayer 3 protocol support.

Certain constraints must be satisfied to apply tunneling:

- The two entities at each end of the tunnel must use the same layer 3 protocol (in most cases the IP protocol).
- These two entities must use the same tunneling protocol.

- In case of tunneling, authentication of a user who wants to connect remotely to the intranet of his or her enterprise is carried out by the nearest NAS and AAA servers of an ISP and not by the distant NAS and AAA servers of the enterprise; because in that case dynamic allocation of IP address to the user is not under the control of the enterprise, a fire-wall may filter some of the messages of this user at the border of his or her intranet.

- A tunnel must be established in a single IP autonomous domain, the routing of the packets being under the control of the source node (the source node determines the destination address); several tunnels may be concatenated within an autonomous domain.

Two types of tunneling are possible, compulsory and voluntary. Compulsory tunneling corresponds to a tunnel that is created a priori without the intervention of the end user. This tunnel is imposed on the end user. Voluntary tunneling is created at the initiative of the end user.

9.6.2   Example of Tunneling in the Case of Analog Access

Let us consider the case of a member of an enterprise out of his or her office who wants to connect remotely to his or her intranet. Instead of having to use a costly long-distance call to be connected to the NAS of his or her enterprise, this user may use a local call to be connected to the NAS of the nearest ISP. Figure 9.11 illustrates the network configuration associated to this example.

In a first step, the user (his or her dial-up modem) initializes a PPP session with the NAS of the nearest ISP. In order to know if the identity of the end user is recognized as one of the subscribers of the ISP, the NAS of the ISP consults in a second step its AAA server by means of the RADIUS protocol. If the user is not recognized, the AAA server may redirect the RADIUS request to another remote AAA depending on the same NSP (in such a case, the first AAA server behaves like a proxy). If the user's identity is recognized, the NAS of the ISP initializes in a third step a point-to-point tunnel by means of the L2TP protocol with the NAS of the enterprise. The advantage of this tunnel is to enable a remote consultation of the user's identity by the AAA server of the enterprise. Thus, the NAS of the enterprise identifies the user by its login and password in its own AAA server (step 4). One notices that the AAA and the NAS servers of the enterprise correspond to the same computer.[5] If the end user is recognized by the NAS of the enterprise, a PPP session between the NAS of the ISP and the

---

5. Let us recall that an AAA server and a NAS server correspond to distinct equipment only in the case of multiple NAS servers belonging to the same administrative entity; this is not the case in our example where the enterprise owns a single NAS.

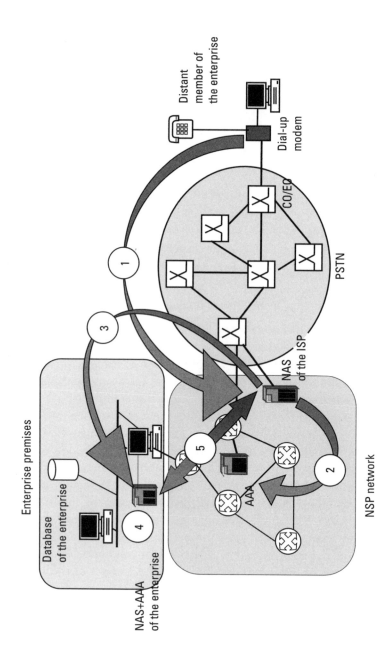

Figure 9.11 Example of remote access to an intranet.

NAS of the enterprise may be activated (step 5). The end user may then communicate bidirectionally with his or her intranet (i.e., with the database of his or her enterprise). The NAS of the enterprise operates like a home gateway. These five steps are illustrated on Figure 9.11. One notices that in our previous example, two PPP connections are put end to end, one between the dial-up modem of the user and the NAS of the ISP and another one from the NAS of the ISP and the NAS of the enterprise. We show in Figure 9.12 the various protocol layers taken into account once these two PPP connections are activated.

The end user sends IP packets encapsulated into PPP frames to the NAS of the ISP. The source and destination addresses of these IP packets correspond to the address that has been assigned by the NAS of the ISP to the end user, and to the address of the NAS of the enterprise, respectively. These PPP frames are de-encapsulated by the NAS of the ISP. The IP packets contained in these PPP frames are then re-encapsulated at the NAS of the ISP into new PPP frames before being sent to the NAS of the enterprise. These frames corresponding to the second PPP connection have the illusion that the NAS of the ISP is directly connected to the NAS of the enterprise thanks to the established tunnel. The tunneling operation consists in practice of encapsulating the second hop PPP frames into L2TP frames. The transport of these L2TP frames between the two NASs occurs under the control of the UDP protocol.[6] The UDP layer represented in Figure 9.12 at the NAS of the enterprise and at the NAS of the ISP is only used to control the flow of packets between these two equipment. At the

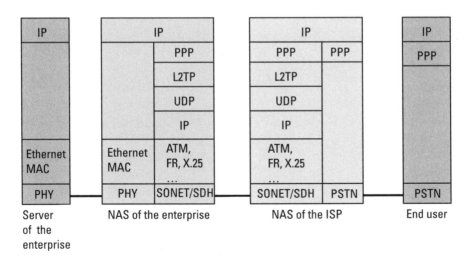

Figure 9.12 Protocol stack in the case of L2TP tunneling and voiceband access.

---

6. Let us recall that UDP is a datagram layer 4 protocol unlike TCP that is connection-oriented.

NAS of the ISP, upstream L2TP frames are encapsulated into IP packets with a source and a destination address that correspond to the IP address of the NAS of the ISP and to the IP address of the NAS of the enterprise, respectively. In many cases, the IP network of the NSPs is itself mapped onto an ATM network. Such an approach, called *overlay*, makes it possible to associate ATM switches with some of the IP routers of the NSP. The main motivation of IP overlay ATM networks is to lighten the complexity of address lookup by imposing IP packets to transit via ATM virtual connections. Thus, in our example, a semipermanent ATM connection may be set between the NAS of the ISP and the NAS of the enterprise. We have seen in Section 1.3.2.3 the benefits of MPLS that promotes in similar situations a peer-to-peer approach instead of an overlay approach. In most cases the NSP network uses at the physical layer the SONET/SDH technique (see Chapter 1).

Figure 9.13 depicts the format of IP packets transferred between the NAS of the ISP and the NAS of the enterprise via the L2TP tunnel. For simplicity, we have neglected in Figure 9.13 the headers proper to the UDP layer and to the layers 1 and 2 (SONET/SDH and ATM). Once they arrive at the NAS of the enterprise, the IP packets sent by the end user are extracted from the received datagram after several operations:

- A UDP segment is extracted from the received IP packet.

- An L2TP frame is extracted from the UDP segment.

- A PPP frame is extracted from the L2TP frame.

- The IP packet generated by the end user is extracted from the PPP frame.

- This IP packet is encapsulated into a MAC frame (for instance, an Ethernet frame) at the enterprise database.

In our description, we have made the assumption that the NAS of the enterprise and the NAS of the ISP are both connected on the network of the same NSP. We have underlined above that an L2TP tunnel could not go through different IP autonomous domains, that is through the networks of multiple NSPs. In summary, the advantage of a concatenation of tunnels is to enable two distant pieces of end equipment (in our example, the dial-up modem of a remote member of an enterprise and the database of this enterprise) to ignore during the data transfers the nature and the complexity of the various subnetworks that must be used between each of them.

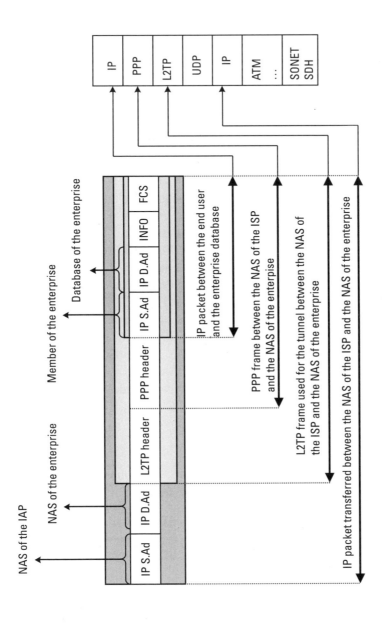

Figure 9.13 Formats of IP packets exchanged on the L2TP tunnel.

9.7 ATM-Based ADSL Protocol Architectures

9.7.1 Network Configuration

Section 9.2 discussed the main reasons that ADSL vendors have considered ATM as a transmission technique between an ATU-C and an ATU-R. Ideally, it should be convenient to set up and release in real time ATM SVCs between an ATU-R and its ATU-C or the NAS of an ISP. We have outlined in Chapter 1 the difficulty of managing in real time many SVCs in ATM public networks. This is the reason why most of the NSPs operate their ATM infrastructure as a cross-connected network with semipermanent virtual circuits (PVCs). Like in the case of an IP-based NSP network, a dedicated server called *broadband access server* (BAS) is used when the network of the NSP is ATM-based (see Figure 9.14). As mentioned at the beginning of Section 9.4, the role of BAS is quite similar to the role of a NAS.

A semipermanent virtual path (PVP) is established between each ATU-R and the ATU-C in the DSLAM (a unique VPI is used between a given ATU-R and the ATU-C). Similarly, a set of PVCs is established between the DSLAM and each of the BAS servers of the various ISPs. A DSLAM behaves like of VP cross-connect. It may change the value of the VPI of the ATM cells generated by a given ATU-R in order to forward them to a requested BAS. In theory, if we want any end user to be able to be connect with any ISP, the number of VPCs increases dramatically. If, for instance, M users are connected to a DSLAM and if N BAS servers are reachable from this DSLAM, the amount of PVCs to manage at the DSLAM should be equal to $(N+1) \cdot M$. Indeed, for each of these M users, a VPI from the ATU-R and the ATU-C plus N VPIs from the DSLAM

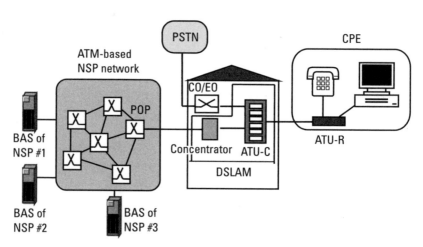

Figure 9.14 End-to-end network configuration in the case of ATM-based NSP networks.

and each of the BAS servers is necessary. Let us notice that VPI mapping carried out at the DSLAM does not modify the VCIs of the ATM cells.[7] The role of concentrator of a DSLAM is described in Chapter 8. We have seen that in many cases, this concentrating function is carried out by means of an ATM switch. We recalled in Section 1.3.1 the main characteristics of the ATCs. In most cases, only two ATCs are considered in the management plane of the DSLAM concentrator: CBR and UBR. In the case of the CBR ATC, a PVC may be assimilated to an LL between the ATU-C and the required BAS. An end user may wish to access successively to the networks of different NSPs. In that case, the DSLAM carries out a mapping between the desired NSP and the corresponding VPI. In other terms, the VPI proper to the connection between this end user and its DSLAM is translated by the DSLAM into a new value enabling to forward the user cells toward the right BAS.

9.7.2 BAS Server

Like a NAS used in IP networks, a BAS used in ATM networks is in charge of authentication of the end users. It checks the access authorization of the end user to the database of such or such content provider. Finally, it meters the volume of information exchanged between an end user and the database of a CP. Like in IP-based NSP networks, a single BAS is used in the case of private usage (intranet of an enterprise), multiple BAS servers being used in public networks. In the former case, the AAA function is carried out by the same computer as the BAS. In the latter case, the AAA server corresponds to a specific equipment in the network of the NSP. Like a NAS, a BAS provides dynamically an IP address to an end user after having established a PPP connection with its terminal equipment. Figure 9.15 illustrates the protocol architecture corresponding to the network configuration of Figure 9.14. The concentrator of the DSLAM groups the individual ATM connections coming from the end users to forward them toward the different BAS servers onto dedicated PVPs. Like in the case of analog access described in Figure 9.12, IP packets generated by the end user are encapsulated into PPP frames between the end user and the BAS.

9.7.3 L2TP Aggregation

The protocol approach we have described for ATM-based NSP networks in the previous sections is not scalable if the amount of ADSL connections increases considerably. For instance, a DSLAM may serve up to 1,000 subscriber lines (M = 1,000). If 10 ISPs are reachable from this DSLAM (N = 10), the amount of PVCs to manage is equal to 11,000 [i.e., $(N + 1)M$] at the DSLAM and to 10,000 (i.e., NM) within the ATM network. In order to solve this problem, the

---

7. This operation is a form of tunneling for the ATM VCIs.

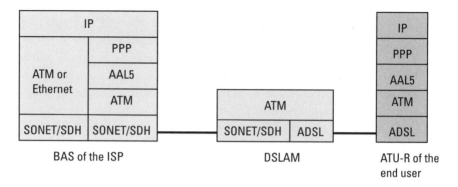

Figure 9.15   End-to-end protocol stack in the case of ADSL access.

usage of the L2TP tunneling protocol has been considered. The basic idea of this approach consists in linking each end user to a L2TP connections concentrator (a tunnel concentrator) called *L2TP Access Concentrator* (LAC). The LAC is able to determine all the individual ATM connections coming form the end users that require access to the same BAS. Instead of simply setting up an end-to-end path, where one VPC links an end user to the DSLAM and then another VPC links the DSLAM to the requested BAS, a LAC associates all these VCs onto the same virtual path. In other terms, several ATM cells arriving at the DSLAM are regenerated toward the same BAS with a common VPI. Figure 9.16 illustrates the principle of a LAC.

Each of the two end users *A* and *B* depending or not of the same DSLAM are connected to a LAC by means of their dedicated PVC.[8] One assumes that both users *A* and *B* request to be connected to databases depending on the IP network of the same NSP. In comparison to the architecture described in Figure 9.14, the routing of upstream ATM cells toward the various IP-based NSP networks does not rely on ATM addressing but on PPP addressing. We have represented in Figure 9.16 with a dotted line the L2TP tunnel linking the LAC and the BAS of NSP 1 and this BAS to the POP of NSP 1. The routing of the upstream data received at the POP of NSP 1 from user A and user B requires the use of a new server called LNS (L2TP Network Server). The PPP connections dynamically established by the end users end at the LNS. Figure 9.17 depicts the protocol stack associated with the network configuration given in Figure 9.16.

Because potentially thousands of ADSL connections can be activated simultaneously in the same ATM network, the L2TP aggregation technique

---

8. The fact that in our example, the LAC is located at the input of the ATM network is not characteristic, an LAC may be connected on any of the ATM switches of the network of the NSP.

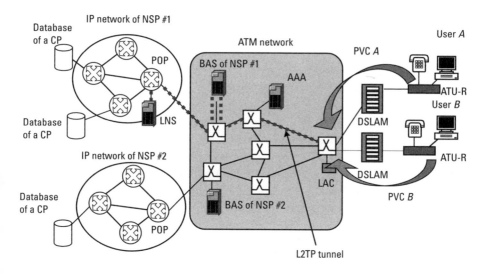

Figure 9.16 Example of L2TP aggregation in the case of ADSL access.

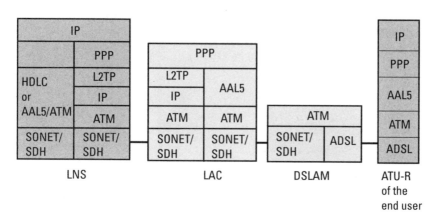

Figure 9.17 End-to-end protocol stack with L2TP concentration.

may be expected to be deployed in the next few years. On one hand, the carrier in charge of the ATM network (an NSP) may serve thousands of end users with LAC concentrators. On the other hand, the NSPs benefit from a simplified interface of their LNS with the ATM network because they have to rent a single PVC from the ATM carrier. In the longer term, we shall have to deploy multiple PVCs between each LAC and each LNS in order to satisfy services with differentiated QoS. Nevertheless, the LAC and LNS alternative should remain a scalable solution. The L2TP aggregation is also called *L2TP access aggregation* (LAA) architecture.

9.7.4 PTA Aggregation

LAC and LNS servers will be progressively introduced by the vendors and the operators. Before the deployment of this equipment in carriers' networks, another approach called *PPP terminated aggregation architecture* (PTA) is considered to be a short-term solution. The PTA approach consists of concentrating ATM PVCs from the end users, not at the level of a LAC but of a BAS. PPP connections transporting the IP packets from the end users use ATM PVCs between each ATU-R and the BAS. These IP packets are de-encapsulated at the BAS to be redirected toward the POP of the requested NSPs. Figure 9.18 illustrates an example of a network configuration using PTA aggregation.

Each end user activates a PPP connection over a dedicated PVC between its terminal and the BAS. The BAS concentrates these ATM cells onto a limited number of ATM PVCs in direction of the POPs of the various NSPs. Thus, the more distant the BAS from the DSLAMs, the less advantageous the PTA aggregation. Unlike the L2TP approach described in the previous section, the PTA approach does not allow the setup of PPP connections from the end users to the

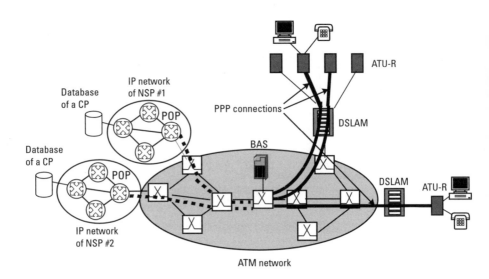

Figure 9.18 Example of a network configuration with PTA aggregation.

POPs of the NSPs. One of the drawbacks of the PTA approach is then that one cannot benefit from some of the PPP capabilities such as encryption and data compression. Indeed, let us recall from Section 1.4 that the provision of VPNs able to guarantee the confidentiality of the transported data is one of the key services required by enterprises. Nevertheless, the IPsec technique may be used in order to ensure confidentiality at the IP layer. Figure 9.19 illustrates the protocol stack associated to the PTA configuration of Figure 9.18.

9.8 TCP Windowing in ADSL Access Systems

The IP protocol is best effort in the sense that some packets may be lost or duplicated by mistake within the network. The TCP protocol is a universal standard for the end-to-end control of data communications in IP networks. Like HDLC, TCP uses a sliding window mechanism for flow control. Whereas HDLC windowing is based on a sequence numbering of the frames, TCP windowing is based on the amount of correct bytes dynamically received by the destination. Thus, a TCP window equal to k means that the sender may send up to k bytes without receiving an acknowledgment from the receiver. A TCP acknowledgment (ACK) corresponds to a small size TCP segment of 40 bytes. An ACK message may acknowledge the good reception of multiple IP packets. In this case, the optimal size of the TCP window is the one that enables the best network utilization. This optimization depends on the arrival and size statistics of the TCP segments. It also depends on the bit rate and the propagation delays proper to the physical links used along the TCP connection. More globally, the efficiency of TCP windowing depends on end-to-end delay, including queuing and processing delays in the intermediate nodes. This principle of TCP windowing is illustrated in Figure 9.20.

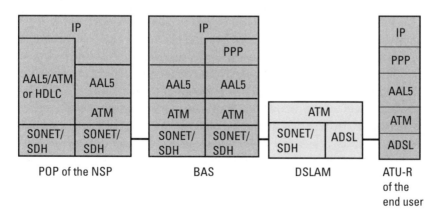

Figure 9.19 End-to-end protocol stack in the case of ADSL access with PTA aggregation.

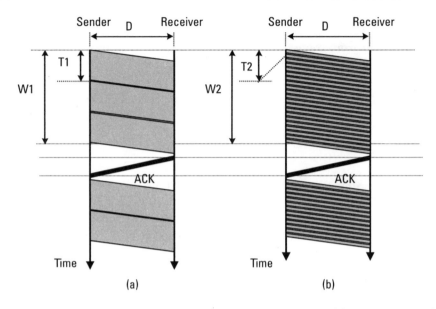

Figure 9.20 Impact of the window size on protocol efficiency under (a) low bit rate and (b) high bit rate.

Figure 9.20 is a very simplified representation of a windowing mechanism. One compares in Figure 9.20 the impact of the window size W on network utilization for the same physical network configuration. The two scenarios represented in Figure 9.20 consider the same distance D between source and destination and the same fixed size of the TCP segments. For both scenarios, there is no packet loss and the sender is assumed to be saturated (i.e., with always a pending TCP segment waiting for transmission). At the opposite site, the receiver has no pending packet. The transmission speed is the only difference between the two scenarios. In Figure 9.20(a), the transmission duration of a TCP segment is T_1, whereas in Figure 9.20(b), this duration is T_2. One assumes that in scenario B, the speed of electronics of the TCP cards at the sender is much greater than in scenario A. In other words, T_1 is much larger than T_2. In comparing each part of Figures 9.20, we see that the same network utilization is possible only if W_2 is much larger than W_1. Such a situation is frequent in very long haul networks like geostationary satellites links or in very high-speed networks. In practice, if the average bit rates of the sender and of the receiver are very asymmetrical and variable, it may be very difficult to find the optimum value of their respective flow control window. In such situations, the first end node that has consumed its transmission credit must remain inactive during a certain period of time. Such an asymmetry between the two end nodes exists in ADSL access systems. The TCP protocol is implemented at each end of the connection, for instance, in the end user terminal and in the remote database. Again, downstream bit rates are in general greater than upstream bit rates in ADSL.

At the beginning of the TCP connection, the size of the sender window W is set to a minimum value W_{min}. In the absence of any packet loss, the value of W is increased exponentially up to a maximum value W_{max}. This procedure is known as the "slow-start" mechanism [see Figure 9.21(a)]. In case of network congestion, some IP packets may be lost. Such a situation results in a strong reduction of W. The sender window is authorized to increase again but only after a certain delay and linearly [see Figure 9.21(b)]. This second TCP mechanism known as "congestion avoidance" induces in long-haul networks or in high-speed networks a waste in network capacity due to its intrinsic inertia. Indeed, due to the congestion avoidance mechanism, the return to W_{max} after a packet loss is very progressive (linear) and requires a silence period.

Experimentally, it has been observed that the slow-start and the congestion avoidance mechanisms induce a low network utilization if the capacity of the buffers of the sender and of the receiver is greater than the amount of bytes that could be theoretically sent end-to-end during a round-trip time.[9] This parameter is known in the literature as the delay-bandwidth product. Two new mechanisms named *fast retransmit* and *fast recovery* have been proposed for TCP flow control [6]. Investigations in real environment and by simulation have shown that special attention should be dedicated to the dimensioning of TCP flow control parameters in order not to lose in TCP the benefit of ADSL at layer 1. Thus, the maximum window size of 8 KB frequently implemented in analog modems is not well-suited to the ADSL environment. As an indication, a window W_{max} at least equal to 64 KB is necessary with downstream data rates on the order of 2 Mbps [7].

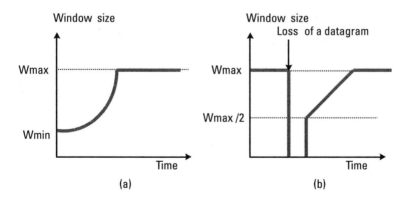

Figure 9.21 The (a) slow-start and (b) congestion avoidance mechanisms in TCP.

---

9. The round-trip time refers to the propagation delay on the way and return between the sender and the receiver.

9.9   Conclusion and Perspectives

We have described in this chapter from an end-to-end perspective the various protocol aspects of ADSL access systems. We have first underlined the main motivations that drove at the beginning of the 1990s modem vendors, service providers, and network providers to consider ATM as the best solution for point-to-point communication between the ATU-C and the ATU-R. We have described the various physical layer data formats used at the reference points of ADSL systems. We have then presented the network elements and the protocol stacks necessary in the global architecture for the authentication, the authorization, and the accounting of an end user connected to the remote database of a content provider. Our analysis has considered, as a first step, access to the Internet by means of dial-up modems before considering access by means of xDSL modems. We have outlined the importance of the right dimensioning of the TCP flow control mechanisms to avoid losing the benefit of ADSL in TCP. In Chapter 12, we shall see that since 2000, several vendors and carriers have promoted the replacement of the ATM technique in the access and even in the metro with Ethernet-based protocols.

References

[1]   ATM Forum Technical Specification, *Traffic Management 4.1*, ftp://ftp.atmforum.com/pub/approved-specs/af-tm-0121.000.pdf, 1999.

[2]   American Standard for Telecommunications, *Interface Between Network and Customer Installation: Asymmetrical Digital Subscriber Line (ADSL) Metallic Interface*, document T1E1.4/98-007R4, 1998.

[3]   IETF, *PPP: Point-to-Point Protocol*, RFC 1483, http://www.ietf.org.

[4]   IETF, *L2TP: Layer 2 Tunneling Protocol*, RFC 2661, http://www.ietf.org.

[5]   IETF, *PPTP: Point-to-Point Tunneling Protocol*, RFC 2637, http://www.ietf.org.

[6]   IETF, *TCP Slow Start Congestion Avoidance, Fast Retransmit and Fast Recovery Algorithms*, RFC 2001, http://www.ietf.org.

[7]   Garcia, J., *TCP/IP Performance over ADSL*, Internal ENST Report, ENST and Hewlett-Packard Laboratories, Bristol, England, 1999.

Selected Bibliography

IETF, *PPP over AAL5*, RFC 2364, http://www.ietf.org.

IETF, *TCP Slow Start Congestion Avoidance, Fast Retransmit and Fast Recovery Algorithms*, RFC 2001, http://www.ietf.org.

10

Voice and Video over DSL

10.1 Introduction

In the 1990s, carriers' networks were still essentially used for the transport of connection-oriented voice calls, PSTN being the predominant infrastructure in carriers' networks. Since 2000, the volume of traffic generated by LAN interconnection and Internet applications has progressively exceeded traditional voice traffic. A prospective market analysis reveals that the market of T1/E1 LLs should stagnate more or less in the next few years. On the other hand, two innovative services such as IP-VPNs (see Section 1.4) and video connections via packet-switching networks should emerge in telecommunications networks from 2000 to 2005. Traditional circuit-switched telephony services via PSTN has remained for decades a major source of revenue for ILECs. Until the 2000s, voice services have been estimated as 90% of the revenues of the operators, whereas, in volume, data traffic and voice traffic were sharing about 50% each of cable bandwidth. Traditional ADSL services aim to provide high-speed Internet access. A new generation of ADSL modems, known as *voice over DSL* (VoDSL), dating from 1999 enables the provision of multiple digitized voice channels using a fraction of the DMT subchannels dedicated to data traffic. Two market segments are targeted by VoDSL: residential users and SMEs [1]. Figure 10.1 illustrates traditional ADSL and VoDSL.

This chapter is divided into two sections. Section 10.2 describes the principle and characteristics of VoDSL. The standardization of VoDSL is today well advanced and several vendors provide the various equipment required for the provision of VoDSL services. Since 2000 VoDSL services have been offered from several CLECs. Currently, many experiments are carried out in various countries for the provision of video channels via ADSL access systems. Several

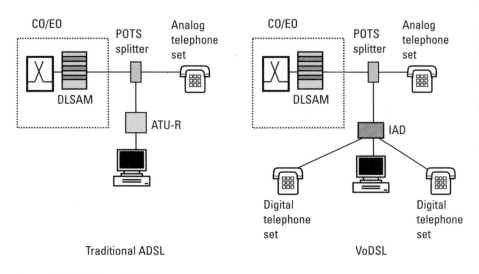

Figure 10.1 ADSL and VoDSL.

test beds have shown the feasibility of video channel distribution via packet-switching networks and ADSL access systems. Section 10.3 deals with video over DSL.

10.2 VoDSL

10.2.1 Introduction

Chapter 7 highlighted the problems inherent in the cohabitation of an analog telephone line with digital data channels. Several VoDSL equipment vendors proposed suppressing the analog voice channel in the ADSL system to replace it with digital telephony transported on additional DMT subchannels. With this evolution called *full digital line* (FDL), the problem of the POTS splitters becomes deciduous. VoDSL assumes, like ISDN, a digitization of the voice signal at the customer premises. If one neglects the impact of the voice coders' compression in terms of processing delays, digital voice signals are less sensitive to physical channel perturbations than analog voice signals, and erroneous bits may indeed be corrected at the receiver. An important argument in favor of the FDL approach is the possibility for the same carrier to provide via a single bill voice and data services. SMEs and call centers[1] are highly interested in cost

---

1. A call center consists of a group of operators in charge of answering in real time the information requests of the customers of an enterprise with the help of a phone line and a PC connected to a local database.

savings brought by VoDSL. For instance, instead of paying to an ILEC several tens of subscriber lines, a company in charge of a call center can now benefit from multiple voice circuits provided by a CLEC via a single subscriber line.[2] In the context of FDL, the problem of digital telephone set powering remains open. Indeed, at least one of these terminals should have the capability to remain operational in case of lack of electrical powering at the customer premises. Investigations are carried out for a remote powering of these telephone sets.

Traditional voice communications via class 5 COs are characterized by a standardized quality of audio signals and a high reliability of PSTN. VoDSL services must offer a comparable level of quality of the audio signal and the same network reliability. The current quality of PSTNs is mainly due to *signaling system number 7* (SS7), which provides rapid call setup and enables intelligent signaling. Similar to class 5 switches, VoDSL must also be able to provide annex services such as caller identification, call waiting, three-way calling, incoming call's indication, voice mail, and prepaid services. VoDSL needs the installation of two dedicated pieces of equipment in the network called *integrated access devices* (IADs) at the customer premises and *voice gateways* (VGWs) within the carrier's network. The IAD that replaces the ATU-R is able to manage multiple digitized voice connections in addition to traditional ADSL services. Thus, several analog or digital telephone sets and several PCs may be connected simultaneously onto the same IAD. In the case of SMEs, a *private branch exchange* (PABX) may be directly connected to the IAD. Typically, digital telephone sets are ISDN telephone sets that digitize directly voice signals. Analog telephone sets connected to the IAD are digitized within the IAD by codecs quite similar to the codecs used in EOs. The digital samples inherent to the various voice calls are packetized within the IAD. A signaling gateway called the VGW serves as a signaling interface for packetized voice traffic between a packet-switching network and traditional PSTN.

10.2.2 DSL Forum VoDSL Models

Two reference models for VoDSL called *broadband loop emulation service* (BLES) and *voice over multiservice broadband networks* (VoMBN), respectively, have been specified by the DSL Forum. The former, based on voice over ATM, is standardized, whereas the latter, based on voice over IP, is not.

10.2.2.1 BLES Model

The BLES model relies on the establishment of virtual telephone lines between an IAD and a class 5 switch. This class 5 switch may be the property either of an ILEC or of a CLEC. In the former case, the VGW is placed at the CO, close to the DSLAM and to the class 5 switch. In the latter case, the class 5 switch is

---

2. ILECs are also interested by the provision of VoDSL services.

reachable from the DSLAM via a packet-switching network (ATM, frame relay, or IP networks). Figures 10.2 and 10.3 depict VoDSL services offered by an ILEC and a CLEC, respectively.

An AAL2 VP/VC is used for each IAD between the customer premises and the PSTN switch for the transport of packetized voice. Several voice calls may be multiplexed in the same AAL2 VP/VC as it is described in Section 10.2.10. A BLES virtual line considers only basic signaling information such as the detection of the on-hook or off-hook position of a handset and the generation of a ringing signal. Such tasks as interpretation of telephone dialing, the generation of the acoustic tones inherent to this dialing, and call control procedures are out of the scope of the BLES model. Such services are under the responsibility of a class 5 PSTN switch. BLES is based on the ATM Forum AAL2 *loop emulation service* (LES) also known as *voice over ATM* (VoATM). VoATM is standardized by ITU-T in recommendations I.363 and I.366. VoIP consisting of transporting UDP/IP packets in an AAL5 ATM connection can also be used in VoDSL. Nevertheless, VoATM is preferred in most cases because of its better capability for QoS provisioning and bandwidth optimization inherent to silence suppression. According to the LES reference model, a *customer-premises interworking function* (CP-IWF) implemented within the IAD provides to the end-user analog interfaces (POTS) and ISDN basic rate interfaces. A *central office interworking function* (CO-IWF) implemented at the PSTN switch provides either a V5.1 or V5.2 or a GR-303/TR-008 signaling interface. The V5.1 and V5.2 interfaces described in Section 10.2.6 are used in Europe, Asia, and Latin America. The GR-303/TR-008 interface is used in North America.

10.2.2.2 VoMBN Model

In the future, telecommunications networks should facilitate a double convergence between fixed and mobile services and voice and data services. The concept of a *next generation network* (NGN) based on a packet-oriented architecture has been introduced for that purpose. An important aspect of NGN is the definition of signaling protocols enabling interoperability between packet-oriented data networks like IP networks and traditional PSTNs for voice services provisioning. The provision of telephony services within packet-oriented networks like IP networks is also considered in NGN. On one hand, ILECs want to reroute fax and dial-up modems traffic from PSTN toward IP networks. Let us recall that PSTNs are today saturated by such traffic (see Section 2.7). On the other hand, ISPs want to provide nationwide and international voice services on their networks in compatibility with PSTN SS7 signaling. Figure 10.4 illustrates the typical architecture of a VoMBN.

For the longer term, one introduces the concept of "soft switch" that aims to replace existing COs by two pieces of equipment: media gateways and *media gateway controllers* (MGCs). The promoters of soft switches estimate that this

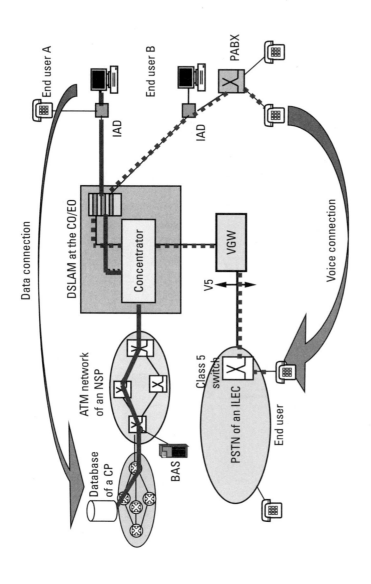

Figure 10.2 Network architecture of the BLES model with VoDSL provided by an ILEC.

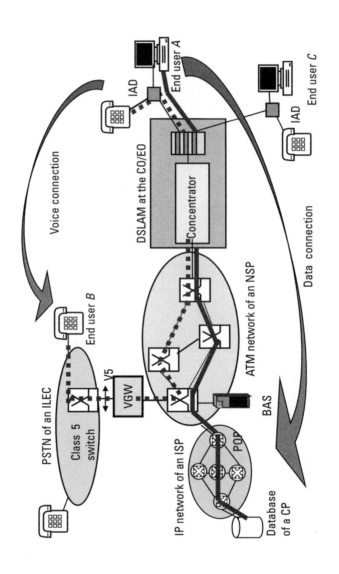

Figure 10.3 Network architecture of the BLES model with VoDSL provided by a CLEC.

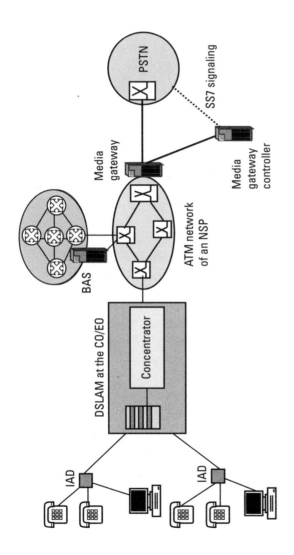

Figure 10.4 The VoMBN architecture.

technology should facilitate in the future computer telephony integration. A media gateway proceeds to digitization, compression, and packetization of voice signals arriving from PSTN before being forwarded to a packet-oriented network. An MGC manages voice call routing, call transfers, and annex services similar to those provided by SS7. The basic principle of a soft switch is to distribute signaling to MGCs. In comparison to a class 5 switch, a soft switch separates service control from call control and from connection control in order to facilitate service providers to rapidly and flexibly deploy new services. Different protocols can be used for VoMBN: H.248/Megaco and SIP/H.323. The H.248 and H.323 protocols are under development within ITU-T, whereas Megaco and SIP protocols are under development within IETF.

10.2.3 Voice Coders

We list in Table 10.1 examples of voice coders standardized by the ITU-T. The G.711 PCM codec universally used in PSTN does not use compression. Each codec requires a certain processing delay indicated in the fourth column of Table 10.1.

For information purposes, Table 10.2 gives an idea of the amount of equivalent circuits that can be obtained on standard-quality ADSL access system for various DSL line speed and compression techniques [2]. For instance, G.711

Table 10.1
Examples of Voice Coders

Type of Coder/ Decoder	Compression Rate	Characteristics	Delay
G.711	1/1	PCM at 64 Kbps	0.75 ms
G.726	2/1	ADPCM at 32 Kbps	1 ms
G.728	4/1	LD-CELP at 16 Kbps	5 ms
G.729A	8/1	CS-CELP at 8 Kbps	10 ms

Table 10.2
Performance of G.711 VoDSL Systems Without and with Compression

DSL Line Speed	Number of Circuits Without Compression	Number of Circuits with Compression
384 Kbps	6	40
768 Kbps	12	80
1.1 Mbps	17	110
1.5 Mbps	23	150

enables 6 or 12 channels at 64 Kbps on a 384-Kbps or a 768-Kbps DSL line. Let us recall that voice calls require symmetrical bit rates. Different coding schemes using distinct compression techniques can be used in the same AAL2 *virtual circuit connection* (VCC).

10.2.4 Principle of Packetized Voice

As mentioned above, VoDSL consists in using in the local loop ADSL access systems with some of the DMT subchannels dedicated to the transport of digitized voice channels.[3] The main objective of VoDSL is to facilitate the provision of cost-competitive long-distance telephone calls in comparison to traditional POTS service. For that purpose, one assumes that each voice call is transported in the major part of the traversed path through packet switching networks rather than PSTNs. The cost of VoDSL voice calls is cheaper than the cost of POTS connections because of statistical multiplexing inherent to packet switching. With packetized voice, a *speech activity detector* (SAD) is necessary in the telephone set in order to distinguish the measured amplitude voice signal from background noise. Thanks to SAD, one determines activity periods (or bursts) and inactivity periods (or silences) in the voice signal. Although voice sampling is applied during the whole duration of a call, only the samples belonging to activity periods are taken into account. These samples are grouped into packets either of fixed or variable size. Two modes of voice packets transfer are possible within a packet switching network. The first and easiest mode consists of using circuit emulation. Circuit emulation does not consider statistical multiplexing, all the samples of the voice signal being encapsulated into data packets. In other words, circuit emulation requires a bandwidth within packet-switched networks equal to the peak bit rate of the codec. The other mode refers in fact to what is known as packetized voice. It assumes that packetization occurs after silence detection. In ATM networks, the flow of ATM cells obtained after packetization is characterized by a *peak cell rate* (PCR) during bursts and by a *sustainable cell rate* (SCR) (i.e., a mean rate associated to the duration of the connection).

10.2.5 VoDSL Network Devices

10.2.5.1 Integrated Access Device

The physical layer of an IAD is identical to the physical layer of an ATU-R (i.e., digital information coming either from the PCs or the telephone sets is mapped onto DMT subchannels). An IAD must also manage the signaling channels associated with the multiple voice channels activated between the customer premises and the VGW. According to the ITU-T and ATM Forum specifications, ATM voice cells are transported from the IAD to the DSLAM by means

---

3. FDM-ADSL is preferable to EC-ADSL for this type of usage.

of the AAL2 layer. An IAD assigns different priorities to the various data and voice ATM connections. Four components are implemented within an IAD:

- *Hybrid:* Hybrids, also called 4-to-2 wire transformers in Europe, have already been introduced in Chapter 2 for the description of voiceband modems. A hybrid within an IAD enables to connect an analog telephone set in addition to several telephone sets and PCs. Between the hybrid and the modem itself, the IAD proceeds to voice quantization and packetization for upstream traffic.

- *Voice coder:* The voice coder makes it possible to digitize and compress voice signals arriving from analog telephone sets. It can also be used to compress the data directly provided by digital telephone sets. Two modes of operation are possible, either with or without SAD corresponding to packetized voice or circuit emulation, respectively. At reception, voice decoding is carried out, "codec" being the generic term for the coder/decoder.

- *Packetizer:* In the upstream, a packetizer groups the voice samples provided by the codec. The packetization delay is a key performance parameter in packetized voice. Too-large packets may induce prohibitive delays, too-small packets may cause a degradation of protocol efficiency in terms of overhead bits. In the downstream, a depacketizer may reduce the jitter introduced by network crossing.

- *Queuing device and scheduler:* These two elements are included in the modem itself. A queue is ideally assigned to each ATM connection. At least, two queues are necessary for real-time and nonreal-time traffic, respectively. Ideally, real-time and nonreal-time queues are assigned to each voice and to each data connection, respectively. A scheduler must determine the optimum order with which the pending ATM cells of the different queues are served by the modem. Two main types of scheduling policies are possible: work-conserving and nonwork-conserving. The former makes it possible to optimize the xDSL channel utilization and is adapted to TCP/IP flows. The latter is better suited to real-time traffic like voice channels [3]. These two policies are described in Section 12.2.2.4.

10.2.5.2 VGW

A VGW enables the bidirectional transit of voice connections between a packet-oriented network (e.g., IP, ATM, and frame relay) and a circuit-switching network (PSTN). It may be installed either within a DSLAM at the EO/CO or within the network of a NSP. The former configuration corresponds to the case where the carrier in charge of the VoDSL service is an ILEC. The latter configuration is adopted when this carrier is a CLEC. Once it has been digitized by an IAD, a voice call is transferred from the IAD to the CO/EO into ATM cells via the ADSL link. In order to explain the role of a VGW, let us consider the example of Figure 10.3 where end user A communicates with end user B.

In the upstream direction, a VGW receives the data packets issued by the calling party (end user A) that uses a derived phone. These packets must be converted by the VGW into a telephone circuit in the output PSTN. Depending on the adopted codecs[4] in the IAD and the transfer mode within the packet switching network,[5] the VGW has to rebuild the A-law or μ-law samples of the voice connection. These samples are then written into T1/E1 frames. In most cases, the data packets received by the VGW are ATM cells. They may also be IP packets or frame relay frames if an IP network or a frame relay network has a POP connected to the output PSTN. The VGW first tries to compensate for the jitter introduced on each data packet. It then extracts the bits transported on the received packets to rebuild a G.711 stream for PSTN. This operation assumes also a signaling adaptation between the input packet switching network and the output PSTN. If the IAD codec uses packetized voice, a background artificial noise is inserted by the VGW during silence periods. At last, the VGW must also perform echo control. We describe the origin of such echoes in a following section.

In the downstream direction, the called party uses a derived line. The VGW receives T1/E1 frames from and input PSTN and must convert these frames into the data packet format of the output packet switching network. The VGW must proceed to silence suppression on the downstream flow if the IAD codec operates with packetized voice. Voice packetization at the VGW is based on the A-law or μ-law samples that are inserted in the payload of data packets. Again, like for upstream direction, a signaling adaptation must be carried out by the VGW.

10.2.6 Signaling Interfaces

A VGW connects to a class 5 switch (class 5 switches have been described in Section 10.2.1). For the purposes of DSL, two types of signaling interfaces are possible between a VGW and a class 5 switch: the GR-303 interface and the V5.2 interface. The former is used in North America and is standardized by ANSI. The latter is used in Europe and is standardized by ETSI. Because the cost of a switch is directly proportional to the number of its line interface cards, both of these interfaces enable to proceed to voice traffic concentration. The V5 interface standard considers two variants V5.1 and V5.2. The V5.1 interface is used

---

4. Voice codecs are characterized by their output bit rate.
5. This transfer mode may assume or not silence suppression.

for the bidirectional transfer of E1 frames between an EO and a CO with a statistical assignment of time intervals to each of the phone calls. Let us recall that an E1 frame transport 30 voice calls or 15 narrowband ISDN connections. The V5.2 interface is used for the bidirectional transfer of 16 E1 frames between an EO and a CO. The V5.2 enables a dynamic assignment of the 30 time intervals of an E1 frame only to active phone calls. Thus, the V5.2 interface makes it possible to concentrate voice traffic coming from the end users. For that purpose, it uses a signaling protocol known as the *bearer channel connection* (BCC) protocol enabling the EO to work with the modems of a DSLAM. A GR-303 interface is quite similar to a V5.2 interface. It manages the bidirectional transfer of 28 T1 frames between an EO and a CO. Section 10.2.9 is dedicated to a more detailed description of signaling protocols inherent to voice traffic.

10.2.7 Typical Network Configurations

10.2.7.1 VoDSL and CVoDSL Services Provided by an ILEC

Figure 10.2 illustrates an example of network configuration when an ILEC provides VoDSL services. Upstream digitized voice channels are directly forwarded at the output of the DSLAM toward a VGW. Two types of AAL latyers may be used: AAL1 for CBR voice and AAL2 for packetized voice. The asynchronous ATM/AAL2 voice packets are converted into a TDM flow before being redirected toward the class 5 switch of the ILEC. The VGW converts ATM cells into PSTN frames. A class 5 switch interprets user signaling sent under the form of a dial tones. It proceeds to call routing and generates records for billing. It is able to manage enriched valued-added services like the transfer of a call toward a new number when the requested number is not available, the special treatment of 800 numbers charged on the called party, audio conferences, and automatic recall. If codecs without silence detection (i.e., without the use of SADs) are used, the fixed bit rate provided by these codecs is transported over an ATM connection on the xDSL link by means of AAL1. Two types of AAL1 are possible according to the ITU-T specifications: structured *circuit emulation service* (CES) and unstructured CES [4]. The former encapsulates directly E1/T1 frames containing voice samples of the various open calls into a single semipermanent ATM connection. The E1/T1 successive frames are rebuilt at the end of the PVC into the separate calls. Structured CES enables fractional E1/T1 services that is CES at $N \times 64$ Kbps. The value of N corresponding to the average number of active calls during a certain period of time (a few hours typically) may be adapted according to the statistical activity of the voice calls.[6]

---

6. Again, we refer to macroscopic voice calls activity and not to the microscopic activity of a given voice call in terms of bursts and silences.

A group of VoDSL vendors has introduced the concept of *channelized voice over DSL* (CVoDSL) making it possible to bundle several VoDSL services on a subscriber line without the need for packetization in the higher layers (ATM or IP). This approach, which is mainly supported by ILECs, consists of using DMT subchannels for the transport of voice calls at the DS0 rate (64 Kbps). Indeed, if derived lines are intended to directly join PSTN at the output of the DSLAM, the cost of packetization/depacketization is useless.

10.2.7.2 VoDSL Service Provided by a CLEC

Figure 10.3 illustrates an example of network configuration when VoDSL services are provided by a CLEC. The ILEC being the owner of the class 5 switch located in the CO/EO close to the DSLAM, the VGW is located between an ATM switch[7] of a NSP and a remote class 5 switch of an ILEC. End user *A* is simultaneously connected to user *B* for a voice call and to a Web site located on the network of an ISP. In most cases, the same service provider provides VoDSL services and high-speed access to the Internet. Transport of voice calls over an ATM network is referred to as VTOA (voice traffic over ATM). In our context, a PVC represented by a dotted line in Figure 10.3 is established between the DSLAM and the VGW. The AAL1 CES is the simplest transport protocol[8] for VTOA. In that case, all the codecs used at the IAD must be noncompressed and none of them uses a packetizer. An ATM AAL1 CES is more or less equivalent to a LL in the sense it does not exploit the gain inherent to statistical multiplexing. CES is justified for ILEC VoDSL service, however, not in the context of a CLEC's VoDSL service. Indeed, a gain on network utilization is requested in order to enable CLEC to be really competitive with ILEC. This gain comes from statistical multiplexing in the network of the NSP. This is the reason why CLEC VoDSL service must be based on AAL2 [5]. We describe in Section 10.2.10 the main characteristics of the new AAL2 layer specifically designed for packetized voice traffic. As described in Chapter 9, access to a service provider via an ATM network requires the use of a BAS server to carry out authentication, access authorization, and accounting of the end user. An ATM PVC carried in a L2TP tunnel can be set between a LAC and the BAS in order to limit the amount of ATM connections within the NSP network. Data packets are transported between the DSLAM and the POP of the NSP by means of a second ATM PVC using AAL5.

Transport of voice calls over a frame relay network is referred to as frame-based VoDSL. If the network of the NSP is IP-based, one refers to VoIP services. In the case of VoIP, the RTP or RTCP protocols introduced in Chapter 1 must be used at layer 4 in order to facilitate traffic resynchronization at the

---

7. The network of the NSP may in fact also be based on frame relay or IP technologies.
8. The AAL layer operates end-to-end and may be compared in that sense to a layer 4 protocol.

receiver [6]. The protocol architectures inherent in the three possible technologies for the NSP network (ATM, IP, and frame relay) are described in the following sections.

10.2.7.3 VoDSL Service Provided by Two CLECs

Another configuration is possible when both the calling party and the called party use a telephone set connected to an IAD. As illustrated in Figure 10.5, the two different NSPs that provide a VoDSL connection to these two parties via their respective packet switching network are equipped with their own VGW. We have represented in this figure with a dotted line the path linking two distant end users A and B connected to a derived line.

Let us notice that if the ATM networks of the two NSPs are directly interconnected without a PSTN, we are in the context of VoATM where VGWs become useless.

10.2.8 Impairments and Distortions

Two main forms of impairment degrade the quality of the perceived signal at the receiver: echoes and packet jitter. Both these impairments are presented in this section.

10.2.8.1 Various Forms of Echo

Two types of echoes may be considered in VoDSL network configurations: talker echo and listener echo. These echoes are observable only in the case of derived phone to POTS phone connections. For example, let us refer to the VoDSL user as the calling party and to the POTS user as the called party. Talker echo and listener echo are due to the imperfection of the hybrids used at the CO/EO of the called party and of the IAD.

Talker echo affects the audio quality perceived by the calling party, whatever the type of his or her telephone set (analog or digital). As illustrated in Figure 10.6, when the voice signal generated by the calling party arrives at the CO/EO of the called party, a fraction of the energy of this signal is in general reflected back into PSTN. This reflection occurs because of the imperfect adaptation between the hybrid of the CO/EO and the characteristic impedance of the called party's subscriber line. In practice, the VoDSL user hears an attenuated and delayed echo of his or her own voice. Talker echo is observed by the calling party whatever the type of its telephone set (analog or digital). Talker echo also exists for the network configuration of Figure 10.2. Meanwhile, the delay between the echo and the reference signal is in this case negligible.

The listener echo affects the audio quality perceived by the calling party only if this party uses an analog telephone set. A fraction of the energy of the voice signal generated by the called party is reflected twice as it is illustrated in

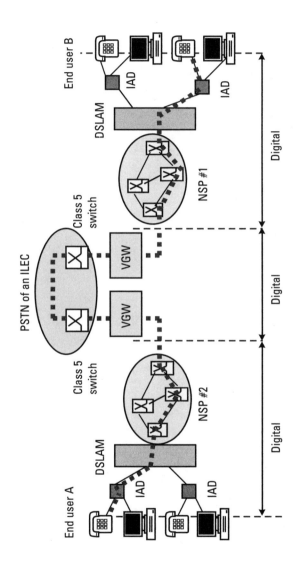

Figure 10.5 Network configuration VoDSL services provided by different CLECs.

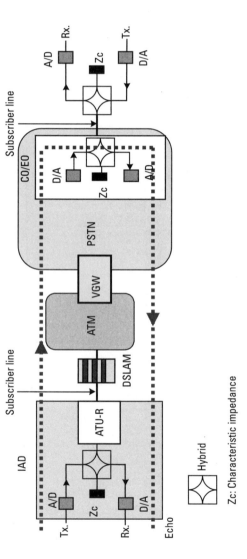

Figure 10.6 Illustration of talker echo.

Figure 10.7. A first reflection occurs at the hybrid of the IAD, the second reflection occurs at the hybrid of the CO/EO of the called party.

An echo controller must be used to reduce the impact of talker and listener echoes. An echo controller is made of two stages. A first stage estimates the echo and subtracts it from the received signal. A second stage is able to cancel the remaining echo entirely. Echo controller must be placed as close as possible to the echo sources. Thus, the echo controller implemented in the IAD compensates the echo due to the hybrid of the IAD. The echo controller implemented in the VGW compensates the echo due to the hybrid of the CO/EO of the called party. If the telephone sets used by the VoDSL end users are digital, there is no need for an echo controller in the IADs. If any of these telephone sets is analog, the hybrid implemented in the IAD of this telephone set may induce an echo at the other party. The ITU-T has standardized at the end of the 1990s cost-effective echo controllers that may be used in the VoDSL environment [7].

10.2.8.2 Sources of Distortion

In addition to talker and listener echoes, two main source of distortion may degrade the quality of a voice signal: packet jitter and packet loss. According to ITU-T recommendations G.114 and G.131, [8, 9], end-to-end delays from the talker to the listener up to 150 ms remain acceptable for an interactive dialogue, echo control is used if necessary. This 150-ms delay budget includes propagation delays and voice coding delays due to A/D conversion and to digital coding. Whereas echo control refers to an end-to-end delay upper-bound value, jitter control refers to strict constraints on packet interarrival times at VGWs and IADs. Several factors induce a fluctuation on packet interarrival times:

- Each packet is subject to a random queuing delay in each node of the NSP/ISP data network. This queuing delay depends on traffic statistics and internal architecture of the switches or routers used in the NSP/ISP networks.

- Some packets may be lost due to buffer overflow in the switches or the routers of the NSP/ISP networks. Packets may also be lost because of erroneous bits detected in an intermediate node. For instance, an ATM cell is discarded by an ATM switch in case of corrupted header.

- Time multiplexing of voice packets with data packets in the IADs induce a time distortion. We have mentioned in Section 10.2.5.1 that queuing strategies and service disciplines are necessary in the IADs in order to minimize the impact of this multiplexing on jitter. Two types of multiplexing are possible: continuous time or discrete time. Ideally, continuous time multiplexing techniques used in IP and frame relay networks seem less penalizing in terms of jitter than discrete time

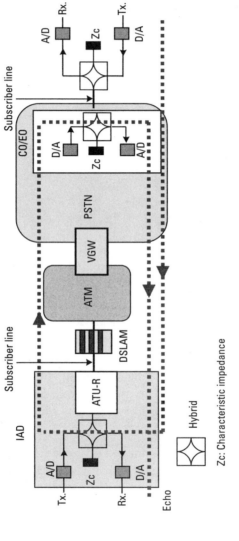

Figure 10.7 Illustration of listener echo.

multiplexing used in ATM. Indeed, the discrete nature of ATM induces in itself a time bias at the multiplexing stages.

- Voice packetization delay is also an important factor in voice signal distortion. In the case of upstream traffic, packetization occurs in the IAD. It is due to the delay needed for the reception of sufficient samples to fill an ATM cell or an IP packet. The higher the compression rate, the larger the voice packetization delay. Thus, it has been shown that voice packetization delay accounts for more than 15% of the entire delay budget with G.711 PCM encoding [10]. Packetization delay must remain under the minimum observable packet interarrival (we assume an ATM transfer mode on the ADSL link). As an indication, if one neglects AAL2 overheads, packetization delay is equal to 6 ms with a G.711 codec and to 12 ms with a G.726 codec.

A VGW or an IAD proceeds to encoding and decoding of the data received from a packet switching network to adapt, for instance, AAL2 ATM cells to E1/T1 frames. The duration of these operations strongly depends on the speed of the DSP of the VGW or IAD.

The compression/decompression delay differs between the various types of codec. Table 10.1 shows that this delay may reach up to 10 ms with G.729 codec.

EC used in VoDSL induces additional delays with about a 0.5 ms average value [11].

10.2.9 Signaling Protocols

Several signaling protocols may be adopted for different purposes in a VGW. We summarize here the main tasks dedicated to such signaling protocols:

- *End-user localization:* This functionality is particularly important for computers because such equipment does not have a permanent IP address. In most cases, NSPs/ISPs use *Dynamic Host Configuration Protocol* (DHCP) for dynamic assignment of IP addresses.

- *Session establishment:* Signaling protocols authorize the called party to accept or refuse a call establishment request received from a calling party. The called party may decide to redirect the call toward another user.

- *Session negotiation:* In the case of multimedia communications (voice and data access to the Internet), several parameters may be negotiated before the opening of the call, for instance, for the choice by both parties of a common compression and coding algorithm.

- *Dynamic management of a multicast group:* This functionality refers to group communications. It enables new users to join the group or some members to leave the group.

These various operations are supported in the ITU-T H.323 and the IETF SIP signaling protocols described next.

10.2.9.1 H.323

The H.323 Recommendation of the ITU-T represents a set of signaling protocols for multimedia communications (voice, video, data) over packet-switched networks that covers the functionalities mentioned above [12]. The first version of H.323 dating from 1996 (H.323v1) defines the components, procedures and protocols necessary to provide multimedia communications on LANs. The last version dating from November 2000 (H.323v4) extends these functionalities to wide area IP networks. H.323 covers various aspects such as packetization, call control signaling, flow synchronization, setup and release of individual flows, and so forth. It introduces the concept of the H.323 gateway, which enables interoperability between an H.323-compatible terminal equipment with other end equipment such as POTS telephone sets, ISDN terminals, and ATM terminals. A detailed description of H.323 characteristics is given in [13]. H323 has been implemented by several software and hardware vendors. Nevertheless, there are still many interoperability problems between pieces of remote end equipment because of the complexity of this signaling protocol.

10.2.9.2 SIP

An alternative to H.323 is the *Session Initiation Protocol* (SIP) defined by the IETF. SIP is a signaling protocol developed after H.323 that operates at the application layer. In comparison to H.323, the designers of SIP wanted to develop a simpler signaling protocol in order to facilitate interoperability between different vendors. The aim of SIP is to enable the transport of SS7 information through IP backbones. SIP uses the same syntax and semantics as *Hypertext Transfer Protocol* (HTTP) and operates on a client-server mode. SIP enables end-to-end services (e.g., call control, call setup, call release, and call transfer) via wide area IP networks. Like H.323, SIP is implemented by several software and hardware vendors. It is compatible with the *Real-Time Transport Protocol* (RTP), which enables the transport of voice packets over UDP in providing synchronization tools between the sender and the receiver. Interoperability between SIP-compatible equipment from different vendors is under testing. A more detailed description of SIP is given in [14].

10.2.9.3 Megaco, Sigtran

Currently, VGW vendors consider with a particular interest signaling protocols such as Megaco (media gateway control) or Sigtran (signaling transport). Megaco has been specified initially by the IETF. Its development has been transferred to the ITU-T under the name H.248. The H.248 specifications are today about to be finalized. On the other hand, Sigtran is still under study within IETF in the perspective of NGN). Both Megaco/H.248 and Sigtran protocols include information related to signal encoding/decoding techniques and information related to video, voice, fax, and data signaling between PSTN and packet-switching networks.

10.2.10 Protocol Architectures for VoDSL

We have underlined that VoDSL services could be offered via different types of packet switching networks. In most cases, VoDSL services are provided via an ATM backbone. ATM cells received at the DSLAM from the IADs are forwarded to the ATM backbone of a NSP/ISP (see Figure 10.3) on a PVP. Figure 10.8 depicts the protocol stack used from the IAD to the VGW. The ATC used through the ATM backbone for the transport of packetized voice cells is real-time VBR (rt-VBR) over AAL2. The CBR-ATC is adopted for circuit emulation (see Section 10.2.4) with AAL1 (essentially by ILECs). One assumes in Figure 10.8 that DS3/E3 physical links at 45 Mbps/32 Mbps, respectively, are adopted within the ATM backbone.

The AAL2 layer has been designed by ITU-T for low bit rate (DS0, E0, fractional T1/E1) delay sensitive services over ATM networks [15–17]. Packetized compressed voice is the typical application for AAL2. VoDSL and *Universal Mobile Telecommunication System* (UMTS) are two major fields of application for AAL2. As mentioned above, packetized voice is subject to a bandwidth/delay trade-off: small packets enable low packetization delays but are expensive in terms of overheads bits. For instance, a G.723.1 codec generates compressed

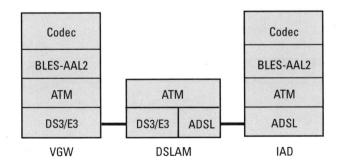

Figure 10.8 Protocol stack with ATM-based VoDSL.

voice bits at 6.3 Kbps. Such a bit rate imposes a 61-ms packetization delay for filling the 48-byte payload of an ATM cell. Reminding that voice interactivity imposes end-to-end delays under 150 ms, such a packetization delay is in general incompatible with the queuing delays in the intermediate nodes of the ATM network and imposes the use of echo controllers. A basic solution to this problem is to authorize partially filled payloads within ATM cells, bit stuffing being used for the remaining bits. Another alternative solution specified in the AAL2 standard consists of multiplexing the bits issued by different codecs into the same ATM cell.

Figure 10.9 illustrates the AAL2 compressed voice encapsulation process. Voice samples obtained at the output of a given codec are grouped into variable size blocks. As long as the upper bound of packetization delay is respected, several bursts of the same source may be grouped in a block. One associates with each block a *connection identifier* (CID) to form voice packets also called mini-cells. In our example, three derived lines of the same IAD are simultaneously active (three CID values). These derived lines may use codecs with different bit rates,[9] the size of a minicell being adapted to the speed of the associated codec. One notices that a voice packet may be transported into two successive ATM cells after segmentation.

Figure 10.10 details the overhead fields used in AAL2. Voice packets of different voice channels issued by the same IAD are multiplexed into a single AAL2 connection, thanks to the 8 bits CID field (in fact, up to 248 voice connections originated from the same IAD may be multiplexed onto the same ATM connection, the eights remaining CIDs being used for signaling purposes). The header of a block is 3 bytes long. The 6-bit *length indicator* (LI) field specifies the length of the block payload. The 5-bit *user-to-user indication* (UUI)

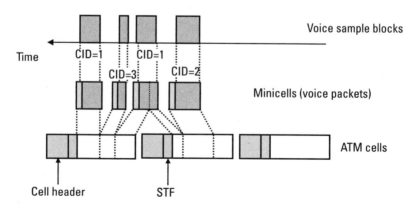

Figure 10.9 AAL2 packetization process.

---

9. This explains partially the variable size of the voice samples blocks.

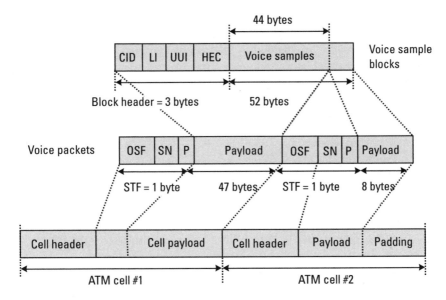

Figure 10.10 Protocol architecture for VoDSL services using VoATM.

is used for an end-to end signaling channel. A 5-bit *header error control* (HEC) protects the CID, LI, and UUI fields. The payload of a block is at most 64 bytes long.[10] A voice packet is obtained by adding a 1-byte *start field* (STF) header to each block. This header is made of a 6-bit *offset field* (OFS) that serves as a pointer indicating the position of the beginning of the following block (in our example, this pointer indicates the beginning of bit stuffing). The OSF field is followed by a 1-bit modulo 2 *sequence number* (SN) and a 1-bit parity bit. In the example of Figure 10.10, one considers a voice sample block of 52 bytes. Two voice packets are necessary to transport in two successive ATM cells this block.

10.3 Video over DSL

10.3.1 Introduction

VoD over subscriber lines was one of the first objectives of xDSL designers at the beginning of the 1990s (see Chapter 6). This ambitious target was rapidly abandoned due to technical limitations inherent to the state of the available technology in these times. These technical limitations concerned several domains:

---

10. We recall that a voice packet may be transported into two successive ATM cells.

- Video coding techniques that were requiring too high bit rates;
- The difficulty of managing client-server applications on a point-to-multipoint basis through public networks;
- The too limited capacity of copper wire access systems;
- The inefficiency of transport networks to manage multimedia communications.

During the 1990s, analog video channels distribution has been widely deployed to either geostationary satellite systems[11] or to coaxial cable networks. It is only after 2000 that the feasibility of video distribution over existing copper wires has been reconsidered. Today, three main advances make video distribution over subscriber copper lines technically and economically realistic:

- First, video coding techniques have considerably improved, enabling today TV channels distribution at less than 1.5 Mbps with the MPEG-4 standard and at 2–4 Mbps with the MPEG-2 standard [18].
- Advances in the field of client-server applications in the Internet enable development of cost-effective software tools for the provision of VoD services to multiple end users from a central video server.
- Thanks to the progress in the field of signal processing and high-speed integrated electronics, a new generation of xDSL modems has emerged and is today available on the shelves. ADSL modems authorize on average 500 Kbps in the upstream and 4–6 Mbps downstream. Such bit rates are widely sufficient for MPEG-4 video channel distribution. Meanwhile, the bit rates today commercially available on ADSL systems are much lower than these target values.
- Transport networks based on ATM or new generation IP techniques (DiffServ, MPLS) have the capacity to multiplex asynchronous flows and real-time traffic with QoS objectives thanks to more efficient control planes.
- Video content has also evolved during these last 20 years. In addition to commercial TV channels, video distribution may be used for the access to stored video databases. The purpose of these stored video sequences may be educational or commercial, the use of video sequences being coupled with Web access services.

We have seen in Chapter 3 that unbundling has motivated many CLECs to provide competitive telephony services (such as VoDSL or digital telephony

---

11. Such satellite systems are also known as direct broadcast over satellite

over cable networks). This competition in the fruitful segment of telephony services has in one way pushed the ILECs to provide innovative services to the end users. Video distribution over subscriber lines is an example of these innovative services. A key point in the commercial strategy of the ILECs interested in VoD services over copper wires is the possibility of offering the end users a selection of innovative services. Thus, a subscriber receives only one bill, which charges for all the services he or she has subscribed to. Interactive video is a more ambitious service than VoD. As an example, interactive video could consist of enabling an end user to stop the projection of a TV sequence, to rewind or fast-forward this sequence and to restart it in another language. Interactive video requires then the possibility for the consumer to interact with the content on the video server located within the public network. Interactive video remains a prospective topic [18] and is out of the scope of this book. VoD is characterized by a much simpler activity between the central video server and the end users.

10.3.2 Network Architectures

VoD assumes the transfer of stored images from a central server to the end users. There are two modes of transmission of stored video over the Internet, the "download mode" and the "streaming mode." With download mode, the consumer downloads the entire video file and then plays back the video file that has been stored locally. The time necessary for downloading the whole file is a major constraint of the download mode. No real-time constraints are imposed on the network. The streaming mode does not need a download of the whole file at once but step by step. Video streaming presents stringent constraints in terms of bandwidth, end-to-end delay and loss requirements to the network.

The download mode does not impose any stringent constraint in terms of QoS to the network. The higher the network capacity, the shorter the downloading. Meanwhile, short downloading is possible only if a certain residual BER is satisfied in order to prevent retransmissions. Figure 10.11 illustrates the main functions to be carried out for streaming VoD. Raw video and raw audio signals are first compressed and stored at the central server. Upon client requests, a streaming server gets the compressed audio/video information from the storage device. The application layer maps the bits of compressed information onto the protocol stack inherent to the underlying network (IP or ATM). This mapping takes into consideration the QoS constraints that will be required in the transport network. According to the current state of the technology, the adaptation of the streaming server to network constraints is preconfigured by the service provider. These constraints vary in general according to the technology and state of the network (congestion control and error control).

Within the network, cells, packets, or frames may be delayed or dropped. Continuous media distribution services enabling QoS supervision, multicast

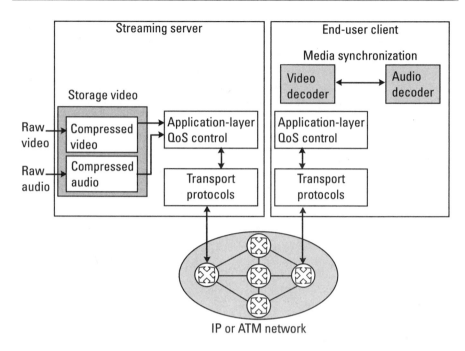

Figure 10.11 Video streaming architecture.

communication, and content replication are required within the network. At the receiver, the application layer tries to resynchronize the flow of received packets. Finally, the video flow and the audio flow are rebuilt from the bits extracted at the application layer. These two flows are themselves resynchronized together [19, 20].

10.3.2.1 IP Unicast Versus IP Multicast

The typical architecture on which VoD should be provided is illustrated in Figure 10.12. A central video server is connected to the IP network of an ISP. This IP network can be reached from the end users via the ATM network of a NSP. A PVP links each DSLAM to the nearest POP of the ISP. With web access, an end user must first connect to the BAS of the ISP for authentication, authorization, and accounting. If the IP network does not support multicast, it is necessary to duplicate at the central video server as many copies of a video sequence as the number of end users that require simultaneously this same sequence. These unicast flows tend to overload the IP network. In the ATM network, they must be transported in dedicated VCs before reaching the DSLAMs. Figure 10.12 illustrates such a duplication from the central video server and end users A and B. Both an ILEC and a CLEC may propose video streaming services. One may

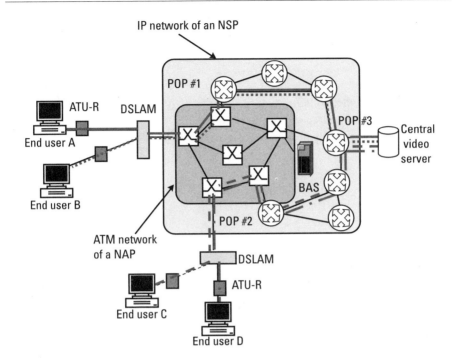

Figure 10.12 Example of VoD with IP unicast + ATM + ADSL network infrastructure.

conclude then that the IP unicast approach is not scalable. In our example, end users *A*, *B*, *C*, and *D* are supposed to request almost simultaneously the same video sequence from the central video server. With IP unicast, the two VCs used by *A* and *B* share the same VPC between their DSLAM and POP 1. Similarly, the two VCs used by *C* and *D* share another VPC between their DSLAM and POP 2.

IP multicast uses the concept of multicast group. An arbitrary group of users that require the same information at the same time are automatically associated in the same multicast group. The members of a group can be dispatched geographically. A tool called *Internet Group Management Protocol* (IGMP) is used to manage the dynamic existence of the groups. IGMP assigns a common IP multicast address to the members of same group. Some members may decide to leave a multicast group and new members may join an existing group. The scalability of IGMP is limited due to the complexity of IP multicast trees computation. The benefit of IP multicast concerns only the IP network. With IP multicast, the same amount of VCs are required within the ATM network (2 + 2 in our example). A multicast IP router has the capacity to duplicate IP packets. For instance, in the example of Figure 10.12, only two duplications of the information are carried out at the server to serve POPs 1 and 2. Each of these POPs itself duplicates the information as many times as the amount of end users that require the same sequence (2). One notices that IP multicast does not change traffic load on the two PVCs. Ideally, IP multicast should be managed at the DSLAM in order to lighten also the load on the ATM network. This means that a multicast IP router should be implemented at the DSLAM. In that case, duplications are really minimized within both the IP and ATM networks.

10.3.2.2 Video Staging

Video staging is an alternative approach to IP multicast to lighten traffic load within the ATM and IP networks. The basic idea consists of implementing in addition to the central video server several proxy video servers as close as possible to the DSLAMs. A predetermined amount of video information from the video sequence is stored a priori in the proxy servers. This information corresponds, for instance, to the most frequently requested bits in the video stream. Figure 10.13 illustrates and example of video-staging architecture. As in the case of video streaming services, video staging services may be provided by a CLEC or by an ILEC. For the content provider, video staging is more expensive in terms of equipment than a single central video server. Meanwhile, the cost effectiveness of this approach strongly depends on the type of information that must be stored in the proxies.

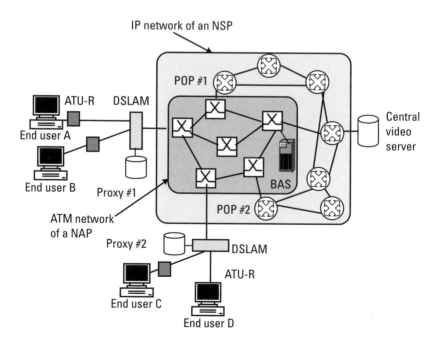

Figure 10.13 An example of video staging network architecture.

Figure 10.14 illustrates the protocol stack used in network configurations depicted in Figures 10.12 or 10.13. Let us consider a central video server and a unicast IP network. In a first step, a PPP session is set up between the ATU-R of the end user and the BAS to proceed to AAA. Once AAA has successfully been carried out, IP packets received at the BAS (more precisely at the ATM switch on which is connected the BAS) are forwarded to the POP of the IP network. Within the POP, IP packets are extracted from ATM cells and forwarded hop-by-hop to the central server. One notices that MPEG-4 needs a transport protocol able to provide temporal reference to the decoder. The RTP and UDP protocols that has been introduced in Chapter 1 are used for that purpose.

10.3.3 Video Compression Techniques

A video sequence is made of a set of fixed images called *video frames*. Typically, 30 frames per second are necessary to give the human eye the illusion of continuous motion. Digital video coding refers then to three dimensions: the two dimensions of the images and the time structure with which these images have been generated by the coder. Spatial and temporal redundancies may exist between pixels of the same image and between pixels of successive images. In other terms, the values of many pixels may then be predicted. In addition, the human eye has a limited capacity to find spatial details in an image. For all these reasons, the amount of bits necessary to encode each image of a sequence fluctuates. A picture is a set of lines made themselves of a certain amount of pixels. A pixel is characterized by its chrominance, luminance, and intensity. Today, two video compression techniques are available: MPEG and H.263. MPEG encoding uses DCT and entropy encoding[12] for taking into account spatial redundancies (known as *intraframe coding*) and temporal redundancies (known as *interframe coding*) [2].

10.3.3.1 MPEG-1 and MPEG-7

MPEG-1 was designed with the idea of storing moving pictures and audio on compact disks at a low bit rate. MPEG-7 offers a set of tools to build and manage an interactive video server.

10.3.3.2 MPEG-2

MPEG-2 has become very popular because it has been used as the compression mechanism for DVB and DVD. The MPEG-2 standard defines three types of pictures: intrapictures (I), predictional pictures (P) and bidirectional pictures (B). I-pictures are coded using only the information of the picture itself. P-pictures are coded with forward prediction that consists in using the previous I-picture or P-picture. B-pictures use bidirectional prediction. Instead of

---

12. Entropy encoding refers to information theory.

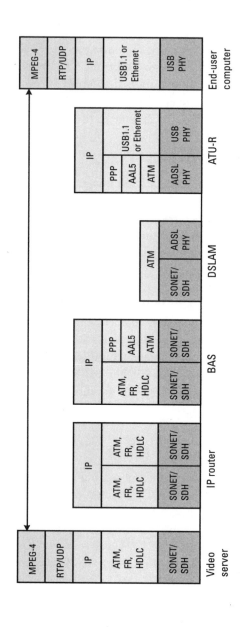

Figure 10.14 Typical protocol stack for MPEG-4 streaming.

estimating the value of an image (or of a block within an image) from a previous image (or block in the previous image), it is possible to also use the information contained in the predicted next image with motion prediction. Such a bidirectional prediction needs a delay at the encoder for waiting the coding of the next image. Thus, an MPEG-2 video sequence is made of successive regularly spaced I-frames, P-frames, and B-frames being inserted between two successive I-frames. MPEG-2 enables good quality video for data rates of a few megabits per second. Figure 10.15 illustrates the principle of *group of pictures* (GOP) in MPEG-2 video coding.

10.3.3.3 MPEG-4

MPEG-4 video compression technique has been developed for low bit rate networks [21]. MPEG-4 is object-based. MPEG-4 objects are part of a scene that can be accessed or manipulated independently. MPEG-4 that achieves higher compression ratios than MPEG-2 is very well suited to video streaming in IP best-effort networks. With MPEG-4, each scene is decomposed into *video objects* (VOs) that are coded individually. Each VO may use several scalability layers called *video objects layers* (VOLs), a base layer and two enhancement layers. Each VOL is itself made of a succession of snapshots called video object planes (VOPs) that are coded using *intracoded VOP* (I-VOP), forward *predicted VOP* (P-VOP) and *bidirectional VOP* (B-VOP). The structure of a group of VOPs is based on the same logic as the MPEG-2 GOPs (see Figure 10.15). The I-VOPs are spaced every 480 ms, B-VOPs and P-VOPs being interleaved as indicated in the figure and generated every 40 ms.

As an illustration, Figure 10.16(a) depicts successive MPEG-4 VOPs corresponding to a 60-minute sequence of the *Star Wars IV* movie.[13] The vertical axis indicates the lengths in bytes of MPEG-4 frames whereas the horizontal axis gives the index of the successive frames. One notices that if the maximum IP

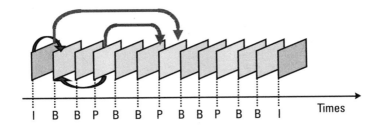

Figure 10.15 Principle of MPEG-2 video coding.

---

13. Let us remark that a VOP is a subset of a picture as it is defined in MPEG-2. In other words, the VOP rate is higher than the image rate.

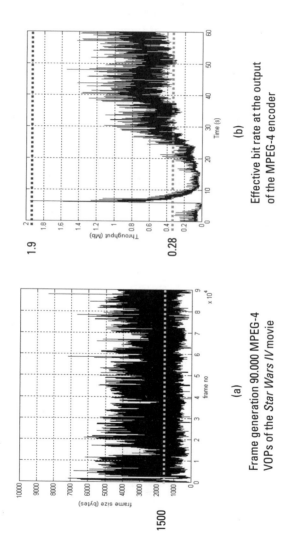

(a)

Frame generation 90.000 MPEG-4
VOPs of the *Star Wars IV* movie

(b)

Effective bit rate at the output
of the MPEG-4 encoder

Figure 10.16 An example of MPEG-4 video trace: (a) size of the generated frames and (b) effective bit rate.

packet payload is set to 1,500 bytes within the IP network of the ISP, a great majority of these MPEG-2 frames must be segmented. Figure 10.16(b) illustrates the effective bit rate at the output of the MPEG-4 coder for the same video sequence. One notices that the maximum bit rate at the output of the MPEG-4 encoder is about 1.9 Mbps, whereas the average bit rate is 0.28 Mbps.

10.3.3.4 H.263

MPEG-4 and H.263 are two video compression techniques that both exploit spatial redundancy by means of DCT and predictive coding to reduce temporal redundancy. We have seen in the previous section that MPEG-4 is object-based oriented (an image is decomposed into independent objects that are encoded separately). Each VOL is coded at different layers: a base layer and two enhancement layers. Like MPEG-1, H.263 is frame-oriented and not object oriented. H.263 may operate either with fixed bit rate or with variable bit rate. Without going into the technical details of H.263 encoding, comparison analysis have shown that variable bit rate H.263 enables in general higher compression ratios than MPEG-4 [22].

10.3.4 Current Industry Activities

In the case of a unicast IP network and of a central video server, the benefit of DiffServ IP networks has been outlined in Chapter 1. For that purpose, mapping rules between the DiffServ EF and AF forwarding classes with MPEG-4 VOL base layer and enhancement layers has been proposed. Within the ITEA-BRIC European research project, MPEG-4 streaming video via ADSL access has been demonstrated in a test bed. This test bed includes an *electronic program guide* (EPG) enabling consumers to choose their video sequence via a Web site. This EPG uses MPEG-7 tools [23].

10.4 Conclusion and Perspectives

VoDSL is technically mature and cost-effective for SMEs and residential users. It should benefit from a large deployment in the coming years. Several vendors (e.g., Alcatel, Cisco, RAD, 3Com, Zhone, and Netopia) are proposing IADs on the market. This success will probably induce the end of narrowband ISDN services. VoDSL market will contribute to the generalization of unbundling and competition between ILECs and CLECs. A performance study of VoDSL services reveals that for realistic network configurations, there is no perceptible difference in the audio quality between a voice call over PSTN and a VoDSL call using LES over AAL2 [24]. In the mean time, traditional POTS services should be progressively abandoned to the benefit of additional derived phone lines.

Thus, all the problems inherent to the design of POTS splitters should disappear due to the emergence of *full digital lines* (FDLs). Even if a lot of work has been done in the field of signaling protocols for multimedia networks, we have seen that the two leader approaches H.323 (ITU-T) and SIP (IETF) still need improvements in terms of interoperability. CVoDSL offers new commercial opportunites to ILECs by using ADSL as simple layer 1 pipes.

We have presented in this chapter the main constraints and characteristics of video distribution over packet switching networks and ADSL access systems. Two main approaches are possible, either based on a unique central video server or based on additional proxy servers close to the DSLAMs. For the former approach, IP multicast authorizes a certain level of scalability within the IP network. Ideally, a multicast IP router should be installed at the DSLAM. Nevertheless, the complexity inherent in the IGMP protocol remains an open problem. The video staging approach using proxy servers is an alternative to IP multicast. The feasibility of MPEG-4 VoD over network configurations quite similar to those considered in this chapter has been recently demonstrated within the ITEA-BRIC European project [23].

References

[1]   Verhoeyen, M., "Provision of Telephony Services over DSL," *Alcatel Telecom Review*, October 2000.

[2]   Tannenbaum, A., *Computer Networks*, 3rd ed., Upper Saddle River, NJ: Prentice Hall, 1996.

[3]   Varma, A., and D. Stiliadis, "Hardware Implementation of Fair Queuing Algorithms in ATM Networks," *IEEE Communications Magazine*, December 1997.

[4]   de Prycker, M., *Asynchronous Transfer Mode: The Solution for B-ISDN*, 2nd ed., Upper Saddle River, NJ: Prentice Hall, 1993.

[5]   *Voice and Multimedia over ATM-Loop Emulation Service Using AAL2*, ATM Forum document FB-VMOA-O145.000.

[6]   Huitema, C., *Routing in the Internet*, Upper Saddle River, NJ: Prentice Hall, 2000.

[7]   *Digital Network Echo Cancelers*, ITU-T Recommendation G.168, April 1997.

[8]   *One-Way Transmission Time*, ITU-T Recommendation G.114, February 1996.

[9]   *Control of Talker Echo*, ITU-T Recommendation G.131, August 1996.

[10]   Segev, R., and B. Wiseman, *Physical Layer Voice Transport (PLVT): A New VoDSL Method*, DSL Forum 2000-121, 2000.

[11]   Fardid, R., and G. Wetzel, *Access Networks Delay for VoDSL*, ADSL Forum 2000-054, 2000.

[12] Thon, G. A., "H.323: The Multimedia Communications Standard for Local Area Network," *IEEE Communications Magazine,* December 1996.

[13] *Packet-Based Multimedia Communications Systems,* ITU-T Recommendation H.323, February 1998.

[14] Schulzrinne, H., and J. Rosenberg, "The IETF Telephony Architecture and Protocols," *IEEE Network Magazine,* May–June 1999.

[15] *B-ISDN ATM Adaptation Layer Specification: AAL2,* ITU-T Recommendation I.363.2, September 1997.

[16] *Segmentation and Reassembly Service-Specific Convergence Sublayer for the AAL2,* ITU-T Recommendation I.366.1, June 1998.

[17] *AAL Type 2 Service-Specific Convergence Sublayer for Trunking,* ITU-T Recommendation I.366.2, February 1999.

[18] Merriman, P., "Architecture for Video Services over DSL," *Alcatel Telecom Review,* October 2000.

[19] Wang, X., and H. Schulzrinne, "Comparison of Adaptive Internet Multimedia Applications," *IEICE Transactions on Communications,* Vol. E82-B, No. 6, June 1999, pp. 806–818.

[20] Zhang, Q., et al., "Robust Scalable Video Streaming over Internet with Network-Adaptive Congestion Control and Unequal Loss Protection," *Packet Video Workshop,* Kyongju, Korea, April 2001.

[21] Soares, L. D., and F. Pereira, *MPEG-4: A Flexible Coding Standard for the Emerging Mobile Multimedia Applications,* Technical Report, MOMUSYS Project, 1999.

[22] *Video Coding for Low Bit Rate Communications,* ITU-T Recommendation H.263, December 1995.

[23] Koubaa, M., and M. Gagnaire, "A Performance Study of MPEG-4 Video Streaming in IP Networks," *IEEE LAN-MAN Conference,* Stockholm, Sweden, August 12–14, 2002.

[24] Buchli, M. J. C., et al., "Voice over DSL Quality Study," *Alcatel Telecom Review,* July 2001.

Selected Bibliography

The E-Model: A Computational Model for Use in Transmission Planning, ITU-T Recommendation G.107, December 1998.

11

Wireless Local Loop

11.1 Introduction

The objective of this chapter is to give an up-to-date overview of wireless access systems also called WITL systems.[1] These systems may be classified in two categories: narrowband and broadband. In this introduction, let us first analyze the benefits and drawbacks of wireless access systems compared to xDSL technologies. For a CLEC, wireless access is probably the most cost-effective and efficient way to enter the market of high-speed access to the Internet. We list below the main benefits of WITL for such carriers:

- Unlike xDSL technologies, wireless technologies are not subject to unbundling (see Chapter 3). Because ILECs are also newcomers to wireless access, the competition between operators is in this case fairer and favors technical innovation.

- In comparison to wired access systems, it is not necessary to dig up the streets to install WITL systems. This is why WITL systems have been widely deployed in Eastern European countries during the 1990s to replace obsolete fixed telephone networks.

- In case of commercial failure, a WITL system can be easily taken down to be reused in another location.

- WITL systems may be engineered at an incremental cost proportional to the market demand. In general, WITL systems operate on several frequencies which may be used either in FDD or in *time division duplexing* (TDD). FDD consists of separating the upstream channels from the

---

1. The acronym RITL, which stands for *radio in the loop,* is also used.

downstream channels in the frequency domain whereas TDD does this separation in the time domain. WITL equipment today available on the shelves are modular. A module corresponds to a transmitter/receiver operating on a given frequency (case of TDD) or on two frequencies (case of FDD). Once the capacity of a module is about to be saturated, the operator may install a new module. Such a property is a guarantee of fast return on investment for a carrier.

- Unlike xDSL systems which are based on point-to-point links, a wireless access system is point-to-multipoint oriented. In the downstream direction, a base station close to the CO or close to the nearest IP router naturally broadcast data to a set of end users. The area covered by a base station is called a cell. The size of a cell depends on the admissible SNR and on the end users' density. The point-to-multipoint nature of WITL is particularly interesting in the case of video channel distribution. In the upstream direction, a base station enables to reduce the amount of interface cards at the CO or at the IP router by concentrating upstream traffic.

- WITL systems may be seen as the ideal complement of a radio-mobile network. In most cases, handover[2] and paging[3] procedures are much more simple in WITL networks than in mobile networks. A non-negligible fraction of radio-mobile subscribers use their handset as a fixed phone once at home. One may guess that bistandard phones like GSM-DECT or GSM-LMDS could reduce the operational cost of mobile networks; with such a bistandard phone, a user may easily switch from the mobile mode to the wireless access system mode and thus lighten the load of the mobile network.

Depending on the frequency range of the adopted system, certain constraints may have to be considered. We list below some of the limitations of WITL systems:

- Radio resources are expensive because they are rare. In several countries, frequency bands used by the army have been released in order to enable WITL.

- Broadband WITL systems operate in general at frequencies higher than 10 GHz. For such frequencies, the quality of a radio signal fluctuates. Disruptive effects such as shadowing and parasite reflections degrade,

---

2. The term *handover* refers to the necessity of switching during a call from one base station to another because of the user's mobility.

3. The term *paging* refers to the procedure of localization of an end user within a cell.

for instance, SNR and, consequently, reduce the achievable throughput per hertz. In addition, foggy or rainy weather also degrades SNR due to a loss of energy of the electromagnetic field. Attenuation of the electromagnetic field increases with frequency.

- Frequencies above 20 GHz imply a direct view between transmitters and receivers. Direct visibility requires high points to install the antennas. In metropolitan areas, such high points, which are already in demand for radio-mobile base stations, may be rare and expensive.

In Section 11.2, we give a quick summary about radio propagation. We then describe in Section 11.3 the main narrowband systems exploited today commercially: *digital enhanced cordless telephone* (DECT), *public access communication system* (PACS), and *personal handy-phone system* (PHS) technologies. Section 11.4 is dedicated to the presentation of broadband WITL systems. Three different approaches are considered: *local multipoint distribution service* (LMDS), "two-layer" LMDS, and IEEE 802.16. LMDS is the only broadband WITL system available today on the market. The "two-layer" LMDS is a recent evolution of LMDS aiming at surmounting difficulties inherent to radio propagation at high frequencies. The IEEE 802.16 standard is still under definition for high-speed wireless access systems. Section 11.5 briefly describes some of the techniques used for wireless network planning.

11.2 A Reminder on Radio Propagation

11.2.1 Attenuation

Propagation of an electromagnetic wave in free space is subjected to attenuation. By simplification, let us consider omnidirectional antennas. If P_r and P_e stand for the received power and the transmitted power, respectively, the ratio between P_r and P_e is given by:

$$\log\left(\frac{P_r}{P_e}\right) = 10 \cdot \log\left(\frac{\lambda}{4\pi \cdot d}\right)^2 + G_R + G_T \qquad (11.1)$$

where d is the distance between the sender and the receiver, λ is the wavelength of the electromagnetic field, G_R and G_T being the gain expressed in decibel of the receiver antenna and of the transmitter antenna, respectively. This equation can be expressed as a function of the frequency f of the wave, knowing that $C = \lambda \cdot f$, where C stands for the propagation speed of light:

$$\log\left(\frac{P_r}{P_e}\right) = 20 \cdot \log\left(\frac{c}{4\pi \cdot d \cdot f}\right) + G_R + G_T \qquad (11.2)$$

The higher the frequency and the higher the distance, the higher the attenuation. In general, several beams issued by the same transmitter arrive at a receiver. The resulting wave arriving at the receiver antenna is the superimposition of the line-of-sight signal and of parasite signals. These parasite signals, due to multipath propagation, are strongly time-correlated with the reference beam. Three physical disruptive effects may be observed in this context: reflection, diffraction, and scattering. Reflection of a radio signal occurs when the size of the obstacle is large compared to the wavelength of the signal. Depending of the frequency range of a radio system, reflection may occur on buildings or on the ground. Depending on the geometry of the environment, diffraction of the different beams may also occur. Scattering occurs when the size of the obstacle is smaller than the wavelength of the signal [1].

11.2.2 Multipath Fading

Figure 11.1 shows an example of multipath propagation between a transmitting antenna and a receiving antenna. Due to the different paths used by the various beams (line-of-sight and reflected beams), the received signal is the summation of various signals lightly shifted in time[4] and amplitude.

Depending on the amplitude and phase distortions observed on the various paths, the resulting signal at the receiver may be subject to destructive interference. This phenomenon known as multipath fading is fortunately statistically rare due to the time fluctuations of the transmitted signal. Time dispersion between the different paths may cause ISI at the receiver.[5] To give an idea of the

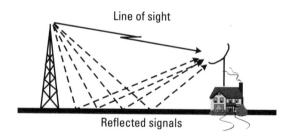

Figure 11.1 An example of multipath propagation.

---

4. Time shifting corresponds to a phase difference in the case of sinusoidal waves.

5. The definition of ISI is given in Section 6.4.1.

amplitude of the ISI phenomenon, let us consider, for instance, a length difference between two paths of 1 km.[6] This distance shift induces a time shift of 3.33 μs at the receiver. In case of a 300-Kbps data rate,[7] the duration of a bit being equal to 3.33 μs, destructive interference may be observed via the overlap between the signals associated with two different symbols. Equalization filters are used to reduce the impact of multipath fading. For an admissible BER, multipath interference imposes on each network configuration a maximum data rate. Multipath fading analysis is getting more complex in the case of mobile communications. The faster the mobiles, the lower the power of the received line-of-sight signal compared to the power of the received resulting signal. In case of mobility, a random phase frequency shift called *Doppler spread* is observed at the receiver. Note that the Doppler phenomenon may also be observed in the case of a fixed transmitter and of a fixed receiver but under a fast fluctuating environment. Such rapid fluctuations may be observed at high frequencies. In the context of WITL systems, we will see through the various technologies described in this chapter, that in most cases, only slow mobility (comparable to pedestrian mobility) is possible. Nevertheless, the impact of the Doppler's phenomenon cannot be neglected. Typical Doppler spread values (i.e., frequency shifts) are 10–250 Hz in suburban areas, 10–20 Hz in urban areas, and 10–100 Hz in the case of indoor networks (wireless LANs) [1].

11.2.3 Propagation Models

Broadband radio communication requires the availability of a large radio spectrum. At high frequencies, the power of a radio wave is attenuated by the presence of oxygen atoms (this is, for instance, the case with rainy or foggy weather). All the systems that will be described in this chapter are based on the concept of radio cell. Like in mobile networks, WITL requires network planning from the carrier, that is frequency assignment over the different cells of the network in order to minimize interference. Several types of analytical models are used for network planning:

- The simplest model, known as the Gaussian channel model, neglects multipath fading and provides an upper bound on the performance of the system. We have already mentioned this model also called AWGN in previous chapters. In the AWGN model, the power of the various noise sources is approximated by a white noise (i.e., with a constant power spectral density) with a Gaussian probability density function. This model remains acceptable in the case of multipath propagation

6. Such a distance difference between paths may be observed in WITL systems.

7. Such a data rate is typically offered to a user in the case of WITL systems.

system under the assumption that both the transmitter antenna, the receiver antenna and the environment are geographically static.

- The Rayleigh channel model is adapted to radio systems with fast fluctuating environment, for example with radio-mobile communications. For such environments, the weight of the line-of-sight path is comparable to the weight of multipaths in terms of SNR at the receiver.

Between the Rayleigh model and the Gaussian model, intermediate models can be taken into account thanks to the Rice model. In Rician models, the power of the envelope of the signal resulting from the superimposition of the line-of-sight path with the nondirect paths is assumed to be described by a probability density function such as:

$$p(x, y) = \frac{1}{2\pi\sigma^2} e^{\left(-\frac{1}{2\sigma^2}\left((x-m_x)^2 + (y-m_y)^2\right)\right)} \qquad (11.3)$$

where x and y stand for the coordinates of a point in the plane orthogonal to the line of sight centered at the receiver. The parameter m stands for the average amplitude of the direct path and σ^2 stands for the variance of the amplitude of the multipath beam. A Rice channel is characterized by its Rice factor K given by:

$$K = \frac{m^2}{\sigma^2} \qquad (11.4)$$

A Rice factor close to zero means that the weight of the direct path is negligible compared to the weight of the multipaths. This case corresponds then to a Rayleigh channel. At the opposite, when K is getting large, the channel is tending towards a Gaussian model. Operators may use numerical simulations based on the Rician model to investigate, for instance, the impact of foggy or rainy weather on SNR. It is possible to evaluate analytically the BER at the receiver knowing the modulation model and the propagation model. One may observe that for a given BER, the larger K, the lower the transmission power. In other words, for a given bit rate, the same BER, is easier to obtain by means of a Gaussian channel than with a Rice channel. Practically, network planners evaluate by means of experimental measurements the impact of multipath fading and attenuation on SNR in order to determine the size and the capacity of the radio cells. Multipath fading and attenuation depends on weather patterns and local topography.

11.3 Narrowband WITL Systems

It is generally admitted that narrowband WITL systems operate in the range of a few megahertz (i.e., under 10 GHz) whereas broadband WITL systems operate in the range of a few tens of gigahertz. Wireless access systems are based on digital radio transmission. The first WITL systems like DECT are based on TDD access. TDD consists in using a single frequency for the upstream and for the downstream channels between a user and his or her base station. Time is decomposed into periodic frames both at the base station and at the end user. A frame is itself divided into two subfields corresponding, respectively, to a certain amount of upstream slots and to a certain amount of downstream slots. The continuity of the frame structure in the time domain assumes that propagation delays are negligible. This assumption is widely respected knowing that most WITL cells have a radius under 5 km. The duration of a TDD frame must be low enough in order to be imperceptible to a human ear in the case of voice communications. Indeed codecs located at the customer premises must have the opportunity to send their samples with a certain periodicity to enable efficient decoding at the receivers. This constraint is satisfied by DECT, PACS, and PHS systems. The three systems being based on very similar approaches, we will focus our description only on the first of them [2]. The three systems use *adaptive delta pulse code modulation* (ADPCM) voice coding at 32 Kbps.[8] All of them are interoperable, not only with PSTN, but also with ISDN. Because they require much less bandwidth for the signaling channels, wireless access systems offer a higher capacity to the user than cellular systems for the same operational bandwidth.

11.3.1 DECT Systems

The DECT technology has been applied successfully to wireless access for more than 10 years. A DECT base station is called a *radio fixed part* (RFP). A DECT user may be connected, either directly to a base station by means of a *portable part* (PP) corresponding to his or her personal DECT handset, or by means of a *wireless relay station* (WRS). In the latter case, the WRS corresponds to a fixed antenna located on the roof of the customer's building. An indoor cabling system makes it possible to connect phones or personal computers of the end user to the WRS. The standardization of the DECT wireless access system began in 1992 at ETSI [3]. Several RFPs located in the same zone are connected to a switch called a *fixed part* (FP). An FP serves as interworking unit between DECT systems and PSTN or ISDN networks and is closed to COs. Figure 11.2 depicts an example of DECT systems used for wireless access system access.

---

8. As a comparison, GSM uses a lower quality voice coding technique at 13 Kbps.

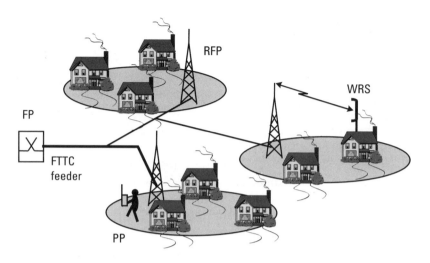

Figure 11.2 DECT network configuration.

A DECT system operates between 1,880 MHz and 1,900 MHz representing a 20-MHz bandwidth. This bandwidth is divided into 10 channels separated by 1.728 MHz from each other. Access to a channel is based on a TDMA
technique. Like GSM cellular networks, DECT wireless access system systems
use *Gaussian minimum shift keying* (GMSK) modulation for the air interface.
We have described in the previous chapters the principles of QAM modulation.
Without going into the details, let us say that GMSK modulation is an evolution of QPSK modulation, which is a QAM with a 4-point constellation. The
transmission power of an RFP is 250 mW. According to the offered capacity of
the system, this power allows cells with at most a radius of 5 km. A particular
phase and a particular amplitude modulates a frequency carrier every 2 time bits.
For a bit rate of W bps, QPSK needs $W/2$ Hz. The more abrupt the phase and
the amplitude transitions of the modulated signal, the larger the bandwidth. In
order to reduce the required bandwidth for a given data rate, GMSK consists in
smoothing the amplitude and phase transitions between the points of the constellation [4]. DECT technology has been standardized by ETSI as a wireless
access system in 1993. Several vendors have provided DECT equipment on the
market for about 10 years. Typically, a DECT wireless access system has the
capacity to offer the equivalent of a basic ISDN access to each user.

11.3.1.1 Frame Format

Figure 11.3 illustrates a TDD DECT frame format. Each 10-ms frame is
divided into two sections of 5 ms each and dedicated to the downlink and to the
uplink, respectively. In other words, on each of the 10 radio carriers, time is

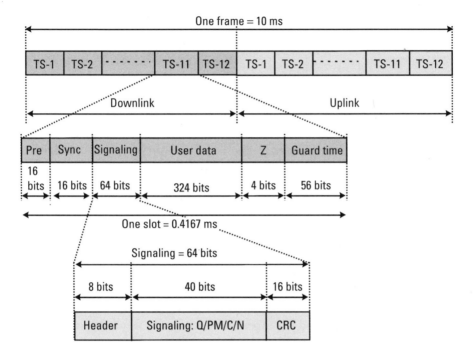

Figure 11.3 Format of a DECT TDD frame.

organized in 10-ms contiguous frames. A TDD frame is made of 12 downstream slots and of 12 upstream slots. Each slot called a TS representing a duration of 0.4167 ms is made of the following fields:

- *Pre (16 bits):* This field enables rough synchronization at the receiver.
- *Sync (16 bits):* This field enables bit synchronization at the receiver.
- *Signaling (64 bits):* This field carries the various signaling DECT fields necessary to the MAC layer.
- *Payload (320 bits):* This field is used for the transport of data or digitized voice. The quantum of allocated bandwidth is 32 Kbps.[9]
- *Z (4 bits):* This field is used for power control. Thanks to this bit, a terminal adapts dynamically its transmission power according to the measured SNR.
- *Guard time (56 bits):* This field enables to separate successive slots in the time domain.

---

9. In fact, the lowest capacity a user may reserve is half of a TS—that is, 16 Kbps.

Figure 11.3 also details the format of a signaling field. The header (8 bits) precise how are eventually multiplexed several signaling channels on the signaling field. The five signaling channels of DECT are called Q, P, M, C, and N. The Q channel is used to broadcast information in the downstream direction from the base station (RFP) to the PPs or to the WRSs. The P channel is used for paging in the downstream direction in order to contact an end user in the case of a downstream call. The M channel is used in both directions for the opening a communication channel (also called a *bearer*) between a user and a RFP. The N channel allows PPs and WRSs to exchange with the RFP their respective identities. The C channel is used for general control purposes. The header of the signaling field specifies how the C, P, Q, M, and N channels are multiplexed in the 40-bit payload.

11.3.1.2 The Dynamic Channel Allocation (DCA) Access Protocol

Dynamic access to a DECT WITL system is based on the DCA protocol. A DECT channel is defined by a given time slot TS(i) and a given frequency $f(j)$ chosen among 120 possible values. For commercial purposes, a DECT system may operate in a limited and less expensive mode for its first utilization. In that case only 12 bearers instead of 120 are in fact possible within a cell. For that purpose, a single receiver is active at any instant at each end of the radio link. In other words, at each instant, a single frequency may be detected at both ends of the radio system. The total available capacity of a DECT cell in this limited version is 12×32 Kbps. By convention, a bearer capacity of at most 352 Kbps

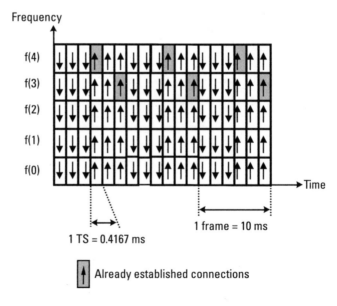

Figure 11.4 Example of DECT bearers.

$(11 \times 32$ Kbps) may be allocated to the same user. In the absence of any connection, a base station (RFP) scans every 100 ms the 12 upstream TSs on each of the 10 carriers. This procedure, called *bearer setup attempt,* enables the RFP to detect a connection request from a user. A multiframe structure 160-ms-long and grouping 16 TDD frames is also used in order to assign a given capacity to the signaling channels. Figure 11.4 illustrates an example of network utilization in the time and frequency domains once a new user wants to proceed to a bearer setup attempt. On this simplified representation of the system, the already used time slots are colored in gray. We see that for a given connection, a bearer consists in one or several DECT channels. A bearer may be either symmetrical or asymmetrical, the amount of time slots allocated in each direction being respectively either identical or different.

As it is illustrated in Figure 11.5, the scanning procedure of the RFP automatically adapts to the dynamic presence of active channels. The RFP that has the knowledge of active channels interrupts its scanning procedure on each active time slot in order to listen to the associated upstream traffic. For instance, during frame 1, the RFP listens to upstream traffic on frequency $f(0)$ except during TS(i) and TS(j) where it listens to frequencies $f(4)$ and $f(3)$, respectively. Similarly, during frame 2, the RFP listens to upstream traffic on frequency $f(1)$ excepted during TS(i) and TS(j) where it listens to frequencies $f(4)$ and $f(3)$, respectively.

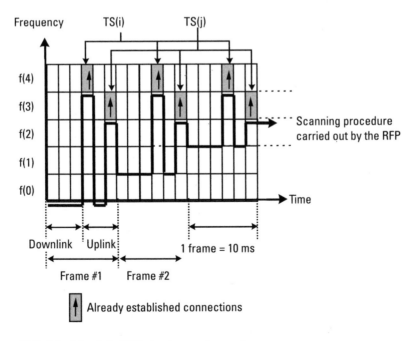

Figure 11.5 Principle of the RFP's background scanning process.

A bearer setup attempt carried out by a user (a PP or a WRS) consists of three steps:

1. First, the user (his or her PP or WRS) chooses the best base station with which to communicate. This choice is based on the determination of the best received signal.

2. Once the base station is selected, the user has to choose in which DECT channel (that is, on which frequency and in which time slot) its bearer setup attempt message will be sent explicitly to the RFP.

3. When the user sends his or her setup attempt message, other users located in the same cell may take simultaneously the same decision. In that case, a collision occurs in the considered DECT channel. A contention resolution algorithm similar to the binary exponential back-off is applied.

During step 2, the RFP informs (via the Q broadcast signaling channel) the inactive users of the cell about its scanning process. This information consists of the frequency and the time slot scanned by the RFP at each instant. Once the user (a PP or a WRS) is frame and bit synchronized with this process, he or she initiates his or her own scanning process. As it is illustrated in Figure 11.6, inactive users' scanning process is dephased by a frame duration in comparison to the RFP scanning process. In other words, when the RFP is listening to a free upstream channel in TS(i) at $f(j)$, the user is also listening to upstream channel $f(j+1)$ during the same time slot TS(i).

We have represented in Figure 11.6 two types of DECT channels. Some are clearly busy (gray), whereas others are more dubious (light gray). The gray channels are the ones the RFP has declared to the users as busy. The light gray channels are free in the considered cell even if a certain signal intensity can be measured by the users. Such channels typically correspond to the case where a parasite signal is received [for instance, on frequency $f(3)$ in Figure 11.6] from an adjacent cell.

After a cycle of 10 TDD frames, the user (his or her terminal) records and sorts the state of the DECT channels in a table called *radio signal strength intensity* (RSSI). By convention, the free channels (in black in Figure 11.6) are located at the head of this table, then the dubious channels (light gray) are classified according to their respective power level, from the lowest to the highest. By convention, channels with a measured power lower than −93 dBm[10] are assumed to be free. The user decides to choose the best of these channels to send his or her bearer setup attempt to the RFP, knowing that busy channels

---

10. A dBm represents the ratio in decibel of a power in watts divided by a reference power of 1 mW.

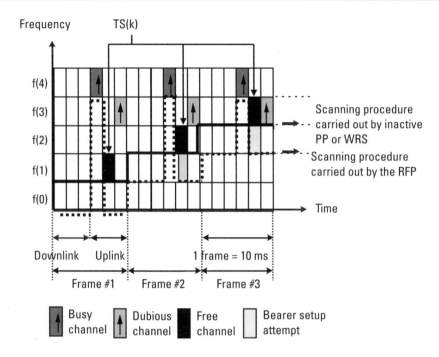

Figure 11.6 Principle of inactive users' scanning process.

(gray) are not available. We see in Figure 11.6 the utility of the time shift between the RFP's scanning procedure and the user's scanning procedure. After having confirmed over three successive cycles that the strength intensity on the chosen channel has not fluctuated (free channel [TS(k), $f(1)$] in black), the user may finally send his or her bearer setup attempt during the fourth cycle and be sure that the RFP will listen to his or her message. Let us assume that the end of the third scanning cycle of the user occurs at the end of frame 1 when it is listening to upstream traffic on $f(1)$. Then the user sends his or her bearer setup attempt during TS(k) on the same frequency $f(1)$ during frame 2. We see in Figure 11.6 that during TS(k) in frame 2, the RFP is effectively listening to $f(1)$. No collision occurs if this user is the only one to become active in the cell at this precise instant.

Figure 11.7 illustrates the messages exchanged between a PP or WRS and the RFP for bearer setup. The very first message of this dialogue is sent by the user on the chosen free channel. First, the RFP identifies the user as one of its subscribers. The FMID and PMID correspond to the user identification code. The RFP allocates, if it is possible, the amount of time slots corresponding to the desired capacity in the upstream and in the downstream directions. Once

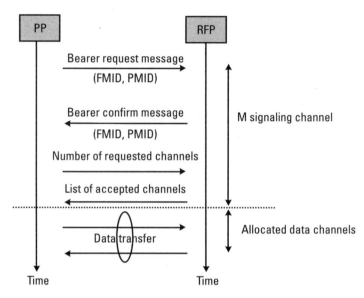

FMID = Fixed Part MAC Identifier (RFP identity)

PMID = Portable Part (or WRS) MAC Identifier (PP or WRS identity)

Figure 11.7 Bearer negotiation.

the user knows its dedicated channels, duplex transmission between the PP and its RFP may begin.[11]

For both data traffic and voice traffic, bandwidth allocation is based on the peak cell rate of the connection. In other terms, the DECT MAC protocol is circuit-switched–oriented with static bandwidth allocation per connection. An optional *automatic repeat request* (ARQ) algorithm may be used by a user to secure his or her data transfer.

11.3.1.3 Handover Procedures

Different forms of handover may occur in DECT systems in case of degradation of the observed SNR. Three types of handover are considered in DECT systems:

- *Bearer handover:* This form of handover concerns both types of terminal equipment, a PP or a WRS. Bearer handover consists of changing of DECT channels during a communication, the RFP and the FP

---

11. Strictly speaking, a DECT bearer is never full duplex because of the TDD frame structure. Nevertheless, the frame duration is low enough to give the application (a phone call, for instance) the appearance of full-duplex service.

remaining the same. Let us recall that a DECT channel corresponds to a couple [TS(i), $f(j)$]. In practice, bearer handover occurs in the case of interference with another DECT user located in a neighboring cell and operating on the same frequency.

- *Connection handover:* This form of handover concerns only the mobile handset (PP). Connection handover consists of changing of DECT channels and of base station (RFP) during a connection, the FP remaining the same. In practice, connection handover corresponds to an end user equipped with a PP that walks around his or her house, this user being located at the border of two adjacent cells. The base stations of these two cells depend on the same FP.

- *External handover:* This form of handover concerns only the PP. External handover consists in changing of DECT channels, of RFP and of FP during a communication. In practice, external handover corresponds to the same situation as in connection handover, the base stations of the two adjacent cells depending on different FPs.

Bearer and connection handover are qualified as "seamless handovers" because in DECT systems, all the PPs and the RFPs depending on the same FP are frame- and bit-synchronized. Changing of the DECT channel or the RFP is a relatively easy task that can be carried out in very short delays (less than 50 ms). In other terms, a user does not detect any interruption of his or her phone call in case of seamless handover. On the other hand, a few bits may become erroneous in case of data transmission during a bearer handover or during a connection handover. External handover is more complex than bearer handover and connection handover because it needs a complete resynchronization at the frame level and at the bit level. This resynchronization may take about 100 ms and is detectable at the ear by the user during a phone call.

11.3.2 PHSs

The PHS technology is due to the initiative of the telecommunication Ministry of Japan in 1989. The air interface of PHS systems was standardized by *Association for Radio Industry and Business* (ARIB)[12] in Japan in 1993 [5, 6]. PHS WITL systems are widely deployed in Japan with about 8 million subscribers in 1999. A PHS cell is qualified as a microcell because its radius of about 150m is much smaller than the 5-km radius of a DECT cell. There is a strong similarity between DECT and PHS. PHS systems operate in the band [1,906 MHz, 1,918

---

12. The ARIB is in charge of the standardization of the *third generation* (3G) of mobile networks. This 3G is based on wideband CDMA for UMTS.

MHz] as a WITL system. Like DECT systems, PHS systems are based on a combination of TDD and TDMA access. On each carrier of the operational band, time is divided into TDD frames. Whereas a DECT base station transmits with a 250-mW power, a PHS base station operates at 500 mW. Similar to a DECT terminal, a PHS terminal may consists of a mobile handset or fixed equipment. In the case of a mobile handset, a reduced mobility up to 10 km/hr is possible [7].

Two types of handover procedures are used in PHS systems. The simplest handover is activated in case of SNR degradation and is characterized by a simple changing of PHS channel, the base station remaining unchanged. A PHS channel corresponds to a given carrier frequency and a certain amount of TSs used in the TDD frames on this frequency. The other type of handover is activated when a user moves along the border between two adjacent cells. In this case, handover is more complex because it also implies a changing of base station. Because PHS systems are based on microcells, both handover procedures are seamless. They can be compared in terms of complexity, respectively, to the "bearer handover" and to the "connection handover" of DECT systems. The spectrum capacity of a PHS WITL system is 12 MHz instead of 20 MHz for DECT systems. The [1,906 MHz, 1,918 MHz] band is divided into several carriers separated by a 300-kHz gap from each other. Up to 40 carriers are available for public access. PHS systems use DQPSK for the air interface. Such a technique is simpler and less expensive to implement than the GMSK used in DECT systems, particularly for the demodulator. Let us simply recall that DQPSK modulation consists in a four-point QAM constellation, each point of the constellation corresponding to a 2-bit symbol. Whereas the absolute phase of the received signal is measured for each symbol in a QAM modulation, DQPSK considers the phase difference between the last two successive received symbols. This technique makes it possible to reduce the impact of phase distortion induced by the air interface.

A PHS TDD frame is made of four downstream time slots and of four upstream time slots. Among these four full-duplex channels, one is dedicated to signaling purposes (TS1 and TS5). The duration of the whole TDD frame is 5 ms, in comparison to the 10 ms of a DECT TDD frame. Whereas up to 120 full duplex DECT channels are possible in the same cell, 40×4 (i.e., 160) full-duplex PHS channels are possible per microcell. Among these 160 channels, only 40×3, or 120, are accessible for the user's information (voice or data). The quantum of bits that may be allocated to a user in a frame is 160 bits per time slot. This quantum corresponds to an ADPCM voice call at 32 Kbps. The setup of a PHS bearer is comparable to a bearer setup attempt and uses the *dynamic channel allocation* (DCA) protocol. With a PHS TS being 240 bits long, the transmission rate on each FDMA carrier is 384 Kbps.

11.3.3 PACS

The PACS technology is inspired by the PHS technology and by the *wireless access communication system* (WACS) technology proposed by Bellcore in the United States for the PCS (Personal Communication System) 1900 standard. PACS has been standardized in 1996 by ANSI. Two variants of PACS systems have been developed. A version based on TDD frames operates as a wireless PBX. Another version based on FDD frames has been adopted for public WITL systems [8]. Like in DECT, a PACS end user may use his or her mobile handset to be connected to a base station called a *radio part* (RP). He or she may also use an indoor wiring and a fixed directive antenna located on the roof of his or her building to communicate with the RP. The radius of a PACS' cell is comparable to the radius of a DECT cell, that is around 5 km. Like PHS systems, PACS systems use DQPSK modulation for the air interface. Like DECT and PHS systems, PACS systems are based on a combination of TDD and TDMA access. We list below specific characteristics of PACS systems:

- PHS and DECT systems enable only limited mobility of the end users (lower than 10 km/hr), whereas PACS systems authorize mobility up to 100 km/hr. This property imposes on PACS systems the capability of managing in real time handover procedures.

- Spatial diversity is the second original aspect of PACS systems. This technique aims at reducing the natural disparity in terms of SNR between the downlink channels and the uplink channels of a radio access system. This disparity is due to the higher transmission power of a base station than the transmission power of a user's terminal. Selective attenuation may occur on specific frequencies of a radio interface. It has been shown that antenna diversity, also called spatial diversity, makes it possible to reduce this selective attenuation by using two reception antennas instead of a single one at the base station [9]. The two antennas must be located at the same height, oriented in the same direction and be distant from one another by 10–20 wavelengths.[13] A typical value of the gain obtained on SNR by means of spatial diversity is 5 dB.

- Unlike DECT and PHS systems, which use TDD, PACS systems use separate frequency sub-bands for upstream traffic [1,850 MHz, 1,910 MHz] and for downstream traffic [1,930 MHz, 1,990 MHz]. Such sub-bands are in the same range of frequencies as DECT and PHS systems. Each of these two 60-MHz sub-bands is divided into 200 channels of 300 kHz each. On each of these carriers, time is divided into 2.5-ms TDMA frames made of eight time slots each.

---

13. This wavelength refers to the frequency carrier used for transmission.

In terms of network planning, PACS channels consists of a couple of frequency carriers (one for the uplink, one for the downlink) and of time slots' indexes of these frequencies assigned by means of the *Quasi-Static Autonomous Frequency Assignment algorithm* (QSAFA). Without describing the QSAFA algorithm, let us simply say that it is less efficient than the DCA algorithm adopted by DECT and PHS systems in terms of bandwidth utilization.

11.4 Broadband WITL Systems

Three variants of broadband access technologies are considered in this section: LMDS, "two-layer" LMDS, and the IEEE 802.16 standard. LMDS networks are today deployed in several countries [10]. The "two-layer" version of LMDS is still under study but is currently implemented in experimental equipment [11]. The IEEE 802.16 is still under definition.

11.4.1 Local Multipoint Distribution System

11.4.1.1 LMDS Network Configuration

Since the 1990s LMDS technology has been widely adopted by manufacturers and operators. Like narrowband WITL systems, LMDS systems are based on a cell configuration. Because of its capacity, an LMDS system is suited either to residential access or to small enterprises. Figure 11.8 illustrates the configuration of an LMDS cell. A base station is called an air interface unit (AIU), whereas users' antennas are called *network interface units* (NIUs). Several AIUs are

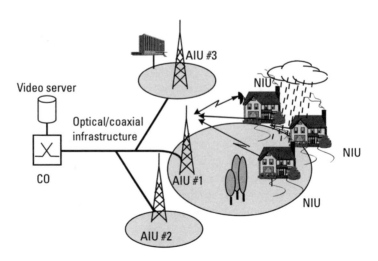

Figure 11.8 Configuration of a LMDS access system.

connected to the nearest CO by means of an optical fiber tree (or of a coaxial tree) in charge of concentrating the upstream traffic. The size of an LMDS cell is comparable to the size of a DECT cell, that is between 1 and 7 km.

LMDS operates in millimeter-wave bands at much higher frequencies than narrowband systems. The bands allocated to LMDS systems differ from one country to another. Thus, the [28 GHz, 29 GHz] band has been reserved in the United States, whereas several sub-bands are allocated in Europe between 25 GHz and 43 GHz. At such frequencies, a direct view between a base station and user's terminals is almost mandatory. This direct view is feasible only if high points are available. We mentioned in the introduction of this chapter that high points are already exploited in many cases for 2G mobile communication systems (like GSM). The frequency spectrum dedicated to an LMDS cell is 1 GHz. Like PACS systems, LMDS systems use two different bandwidths for upstream and downstream communications, respectively. Two variants of LMDS have been standardized: LMDS-DAVIC and LMDS-ETSI. The LMDS-DAVIC version 1.5 dating from 1999 can be considered a de facto standard because it has been adopted by a large number of manufacturers. DAVIC is an American association of manufacturers, telcos, and service providers implied in video distribution. LMDS-ETSI is under definition within the ETSI. Both standards specify the air interface and the MAC layer of LMDS systems. Whereas LMDS-DAVIC and LMDS-ETSI specifications of the physical layer are very similar, the two standards differ on several aspects at the level of the MAC layer. In the following, we choose to describe exclusively LMDS-DAVIC, which is the most deployed version.

11.4.1.2 LMDS Air Interface

LMDS systems operate with FDD to separate upstream channels from downstream channels. The main benefit of FDD over TDD is that it makes it possible to choose a modulation technique proper to each direction of transmission. Knowing that downstream traffic typically is more voluminous than upstream traffic, knowing also that transmission power is greater at the AIU than at the NIUs, it seems logical to use different modulation techniques for the uplinks and for the downlinks. The main drawback of FDD over TDD is that is requires a wider frequency spectrum. Figure 11.9 illustrates LMDS bandwidth allocation at an *intermediate frequency* (IF). We can see that a 150-MHz subband is used for a certain amount of 2-MHz upstream channels. A 850-MHz subband is used for a certain amount of 40-MHz downstream channels. A DQPSK modulation is used for upstream channels whereas several types of modulation are proposed for downstream channels: QPSK, 16-QAM, 32-QAM, or even 256-QAM. The data rate offered by each upstream or each downstream channel strongly depends on the adopted modulation technique.

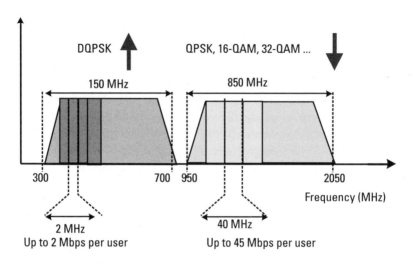

Figure 11.9 Frequency allocation for LMDS WITL.

Figure 11.10 illustrates the basic elements of an LMDS-DAVIC communication system. In the downstream direction, data coming from the CO are assumed to be ATM-formatted. This means that in both LMDS standards, the network node located at the head-end of the LMDS network is ideally an ATM switch. As discussed in Chapter 4, a typical block diagram of a modem may

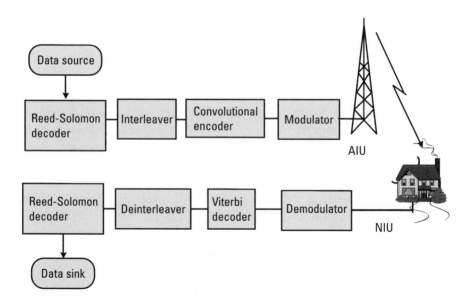

Figure 11.10 Block diagram of a LMDS system.

include a channel encoder, an interleaver, and a modulator on the transmitter side. In the case of a fluctuating propagation environment like radio communications, a concatenated encoder is in general necessary in order to strengthen the performance of the whole system in terms of SNR. We see from Figure 11.10 that an LMDS modem is very comparable in terms of complexity to an ADSL modem, apart the modulation technique which is single-carrier oriented in LMDS, whereas it is multicarrier-oriented in DMT-ADSL. We do not describe again the principle of each element of such a communication system; that has been done in detail in Chapter 6.

In the downstream direction, user's data are multiplexed with broadcast MPEG-2[14] television channels. In order to facilitate such a multiplexing of video streams with ATM cells, LMDS-DAVIC has adopted the DVB[15] physical layer. DVB is an international standard (ISO-IEC 13818) for digital television broadcasting. Different types of media are considered in DVB: satellite (DVB-S), cable (DVB-C) and microwave radio terrestrial systems (DVB-T). The DVB standard specifies a set of transmission tools enabling to multiplex by means of a generic frame format digital TV channels, digital audio channels and any form of data (IP packets, ATM cells, and so forth). DVB is itself based on the MPEG-2 standard. The DVB generic frame format corresponds to an MPEG-2 transport stream (also referred as MPEG-2 TS). The size of an MPEG2-TS is 188 bytes. DVB also specifies a concatenated encoder applied to each MPEG2-TS. The outer code of this concatenated encoder corresponds to a Reed-Solomon encoder RS(204, 188), the inner code being a convolutional encoder with a rate equal to 0.5 and a constraint length equal to 7 (see Chapter 6).

Within an AIU, successive 188 bytes MPEG2-TS arrive from the optical infrastructure at the input of a Reed-Solomon encoder. The RS(204, 188) Reed-Solomon encoder transforms each of these DVB frames into successive blocks of 204 bytes, as illustrated in Figure 11.11. Such an encoder is able to detect and correct up to $(204 - 188)/2 = 8$ bytes per block (see Section 6.7.2). In the context of LMDS, an MPEG2-TS frame (the term MPEG2-TS packet is also used) is made of 187 bytes of data corresponding either to MPEG-2 pictures or to ATM cells, with a 1-byte synchronization (SYN) overhead. One assumes that voice and data traffic are transported between the AIU and the NIUs on ATM virtual connections. In order to facilitate clock recognition at the receiver, an interleaving mechanism is applied to the SYN bytes of every eight successive MPEG2-TS. This mechanism is called *pseudobinary random sequence*

---

14. The MPEG-2 standard specifies a rule of compression of digital video channels. This standard has also been adopted for cable TV systems and satellite systems.

15. DVB is a consortium of about 300 companies in the fields of broadcasting, manufacturing, network operation, and regulation bodies aiming to promote international standards for digital video broadcasting.

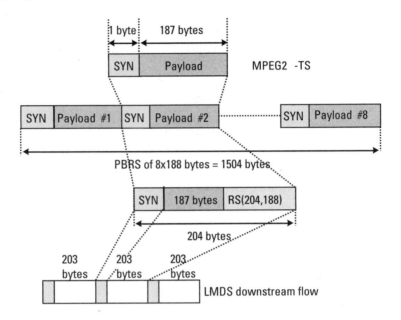

Figure 11.11 Data format in the downlink.

(PBRS). The first stage of the LMDS transmission block diagram corresponds to the Reed-Solomon encoding.

Figure 11.12 illustrates the impact of Reed-Solomon encoding on BER. In this example, which is not based on the LMDS-DAVIC Reed-Solomon encoder, we have plotted for a particular bit rate the fluctuations of BER versus SNR expressed as E_b/N_0 (E_b stands for the signal's power and N_0 stands for the power of AWGN). For that purpose, one considers three kinds of coding applied to a simple BPSK modulation with hard decoding.[16] The first coder (case 1) actually uses no coding at all, while the two other coders are RS(255, 199,28) and RS(255,239,8).[17] Simulation results show that for a given BER, the more efficient the coder, the lower the required transmission power. We see, for instance, that for a 10^{-4} BER, the highest $E_b/N_0 = 10.5$ dB is required when no RS encoding is applied. The same BER is obtained with an RS(255, 199,28) encoder for $E_b/N_0 = 6.7$ dB and with an RS(255,239,8) encoder for $E_b/N_0 = 8.7$ dB. In other words, a gain of almost 4 dB is obtained when the most powerful encoder is applied compared to the case of no RS encoding. This difference has an impact on the power consumption of the AIU and of the NIUs and on cell radius.

---

16. Hard decoding and soft decoding were introduced in Chapter 6.

17. The third parameter mentioned in the RS coders specifications corresponds to their error correction capacity (see Section 6.7.2).

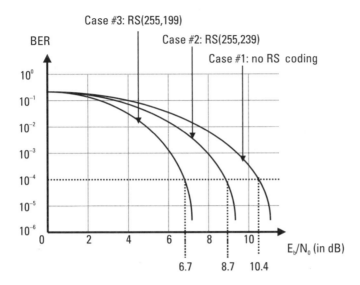

Figure 11.12 Impact of RS encoding on BER.

Let us come back to the RS(204,188) proper to LMDS-DAVIC. At the output of the RS encoder, successive 204-byte packets are passed to an interleaver. The main objective of an interleaver is to alleviate channel memory effects in the case of long burst of errors. Figure 11.13 illustrates the characteristics of the LMDS-DAVIC interleaver. All 17-byte blocks arriving from the RS encoder are enqueued line by line is a memory. As it appears in Figure 11.13, the memory of the LMDS-DAVIC interleaver corresponds to a matrix with 12 lines and 12 columns. The intersection between a line and a column corresponds to a 17-byte register. The memory capacity of line number "0" is null. The memory capacity of line number "1" is of one register. The memory capacity of line number "2" is of two registers and so on. In a first step, the serial data arriving at the input of the interleaver are buffered line by line, from line "1" to

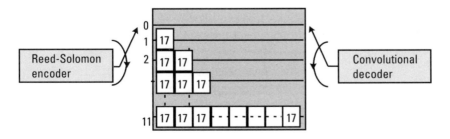

Figure 11.13 LMDS-DAVIC interleaver.

line "11." Interleaving is obtained by dequeueing the matrix column after column from column number "0" to column number "11." The successive 17-byte blocks obtained by this procedure are sent to the input of the next stage of the block diagram.

The third stage of the LMDS transmission block diagram corresponds to a convolutional encoder. The objective and principle of a convolution encoder is described in detail in Chapter 6. Figure 11.14 describes the principle of the LMDS-DAVIC convolutional encoder. Let us recall that the code rate R is the ratio of the number N_i of bits that form an input unit of the encoder over the length N_o of the output unit obtained after encoding. The code rate of this coder is then $R = 1/2$. The constraint length L of a convolutional encoder is given by the size of its shift register. We see that in the case of LMDS-DAVIC, $L = 7$ bits.

The two generator polynomials (see Section 6.9.3) associated to the two outputs S_1 and S_2 characterizing this connection diagram are:

$$G_1(D) = 1 + D^1 + D^2 + D^3 + D^6 \tag{11.5}$$

$$G_2(D) = D^1 + D^2 + D^3 + D^5 + D^6 \tag{11.6}$$

These two polynomials may also be expressed in base 8 as $(171)_8$ and $(133)_8$. At the output of the convolutional encoder, downstream data are sent to the modulator. We have mentioned previously that a manufacturer may implement either QPSK, 4-QAM, 16-QAM, or 256-QAM in the LMDS modulator. The main objective of adding a convolutional encoder to the RS encoder is to compensate for the bad quality of the transmission channel. As mentioned in Chapter 6, the association of Reed-Solomon encoding with convolutional encoding is called concatenated encoding, the former and the latter of these two encoders being called *outer coder* and *inner coder*, respectively. In order to illustrate the necessity of convolutional encoding in case of very noisy channels, we have plotted in Figure 11.15 two simulation results presenting a 16-QAM constellation as it is detected by a receiver. Each point represents a received symbol of 4 bits. The left-hand side and the right-hand side figures correspond to a E_b/N_0 of

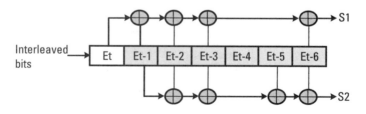

Figure 11.14 LMDS-DAVIC convolutional encoder connection diagram.

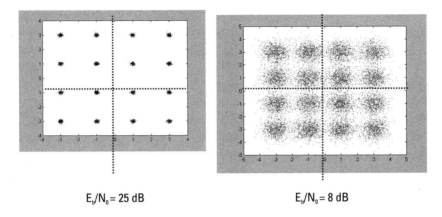

$E_b/N_0 = 25$ dB $E_b/N_0 = 8$ dB

Figure 11.15 Impact of noise on a 16-QAM constellation at the receiver.

25 dB and a E_b/N_0 of 8 dB, respectively. When $E_b/N_0 = 25$ dB, we are able to clearly separate the points of the constellation. By measuring the amplitude and phase of the received symbol, we can obtain the value of the associated symbol without any ambiguity. When $E_b/N_0 = 8$ dB, we see that the clouds of points overlap. Several points are in fact almost equidistant from two different reference points of the constellation. In other terms, the decoding of these signals is ambiguous. The Viterbi algorithm (see Section 6.9.2.2) implemented in the convolutional decoder makes it possible to reduce this ambiguity by choosing the most probable symbol's value corresponding to the received signal. Such a technique is known as the maximum-likelihood algorithm. We have described in details the principle of maximum-likelihood decoding in Sections 6.9 and 6.10.

At the receiver terminal, downstream data goes through four successive stages: the demodulator, the Viterbi decoder, a deinterleaver, and the Reed-Solomon decoder. Figure 11.16 depicts the impact of the adopted modulation technique on the performance of the system. We have plotted in Figure 11.16 the fluctuations of BER versus SNR E_b/N_0 for four different modulation techniques: 4-QAM, 16-QAM, 64-QAM and 256-QAM. To obtain a BER of 10^{-3}, we see that the required values for E_b/N_0 are 10 dB, 13 dB, 17 dB and 22 dB for 4-QAM, 16-QAM, 64-QAM and 256-QAM, respectively. The more efficient in terms of bits per second per hertz the modulation technique, the higher the required transmission power.[18]

From the NIU to the AIU, the transmitter and the receiver building blocks are roughly the same as those mentioned in Figure 11.10, but with different parameters. According to each environment, a carrier can refrain from using

18. Unlike simulation results presented in Figure 11.12 for which a given bit rate is considered, the four curves plotted in Figure 11.16 do not correspond to the same bit rate.

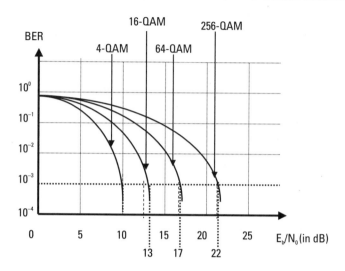

Figure 11.16 Impact of modulation on BER.

concatenated encoding for the uplink because the uplink bit rate is lower than the downlink bit rate. Figure 11.17 illustrates data formatting of upstream traffic. LMDS-DAVIC uplink does not consider DVD framing; a user's terminal is assumed to transmit ATM cells only. We see that each 53-byte ATM cell is encoded by means of an RS(63,53) encoder. The successive 63-byte blocks obtained at the output of the Reed-Solomon encoder are inserted in a MAC frame, including a 4-byte preamble and a 1-byte guard time. The preamble is introduced for each ATM cell in order to facilitate clock recovery at the AIU. Several vendors only consider RS encoding but not convolutional encoding or interleaving for upstream traffic because they estimate SNR constraints less stringent in that case than in the case of downstream traffic.

11.4.1.3 LMDS Access Protocol

In the downlink, DVB frames (MPEG2-TS) are used to multiplex digital voice, data and video channels. Only data and voice traffic are multiplexed in the uplink. This is the reason why the MPEG2-TS packet format is not used in the uplink. As mentioned above, downstream MPEG2-TS are used to transport either MPEG-2 video-coded information or ATM cells. Figure 11.18 depicts how ATM cells are encapsulated into MPEG2-TS. MPEG2-TS being indexed modulo 2, in each suite of two successive MPEG2-TS, up to 7 ATM cells can be transported. A certain amount of overhead bits have to be added in the payload of each frame for a complete mapping. Let us remember that the 188 bytes of a frame obtained after this encapsulation are forwarded to the Reed-Solomon encoder RS(204,188) implemented at the AIU.

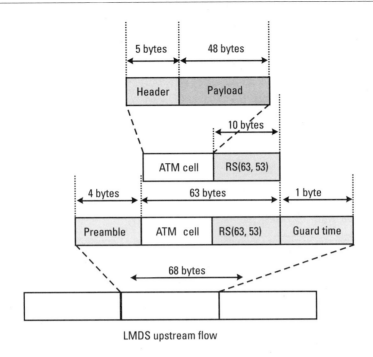

Figure 11.17 Data formatting for upstream traffic.

LMDS-DAVIC specifies on each upstream channel and on each downstream channel frame formats for which the quantum of bandwidth corresponds to an ATM cell. In order to distinguish these frames from the physical layer MPEG2-TS frames, we shall call them in the following "MAC-frames." Upstream and downstream MAC-frames have by convention the same duration of 6 ms in order to facilitate correct synchronization between the AIU and the NIUs. A downstream MAC-frame may transport up to 728 ATM cells (encapsulated in MPEG2-TS frames at the physical layer). An upstream MAC-frame may transport up to 24 ATM cells [without MPEG2-TS encapsulation (see Figure 11.17)]. In the LMDS-DAVIC standard, the field containing an ATM cell, either in an upstream MAC-frame or in an upstream MAC-frame, is called a *slot* (see Figure 11.19).

The MAC protocol proposed in LMDS-DAVIC is based on a request-permit mechanism. For any pending ATM cell, an NIU sends a request to the AIU in order to get a permit of transmission. Three types of slots are specified in the LMDS-DAVIC standard:

- *Contention slots C:* These slots transport MAC messages or user's data. A contention slot may be used by any NIU with pending ATM cells.

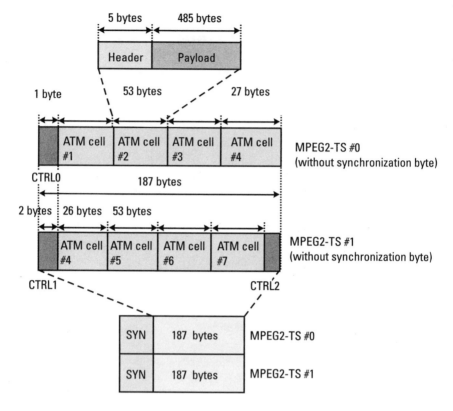

Figure 11.18 Downstream ATM cells encapsulation in MPEG2-TS frames.

Access to a contention slot is then random, collisions being possible between concurrent NIUs. Once it sends its bandwidth request in a [C] slot, an NIU activates a timer. In case of successful transmission, the AIU sends a permit of transmission to this NIU. This permit is supposed to arrive at the NIU in a delay Δ greater than a *round-trip time* (RTT) over the LMDS system. In case of contention, the AIU is unable to read correctly the bandwidth request. As soon as the NIU's timer becomes greater than Δ, a contention resolution algorithm is then activated. LMDS-DAVIC lets each manufacturer implement the contention resolution algorithm of its choice. Two mechanisms are proposed in the standard: the binary exponential back-off and the ternary tree [12].

- *Reserved slots NIU:* Reserved slots transport OAM cells, or user's data. A reserved slot is explicitly dedicated to a given NIU.

- *Polling slots P:* These slots transport OAM cells, or user's data. Polling slots are used exclusively to enable a new active NIU to insert into the

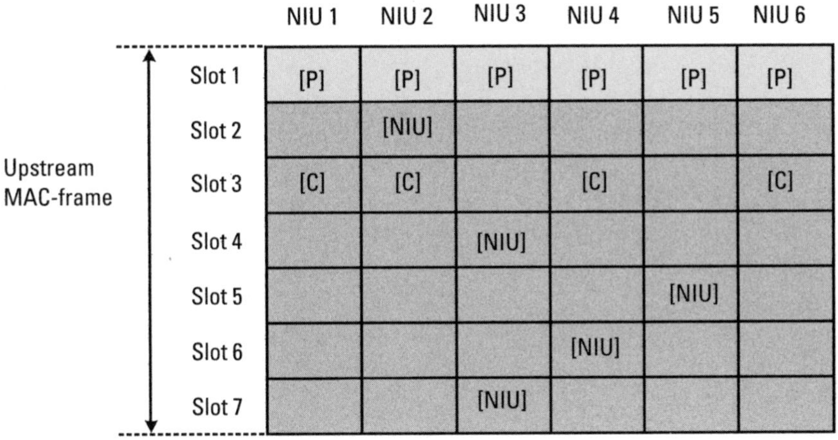

	NIU 1	NIU 2	NIU 3	NIU 4	NIU 5	NIU 6
Slot 1	[P]	[P]	[P]	[P]	[P]	[P]
Slot 2		[NIU]				
Slot 3	[C]	[C]		[C]		[C]
Slot 4			[NIU]			
Slot 5					[NIU]	
Slot 6				[NIU]		
Slot 7			[NIU]			

Upstream MAC-frame

Figure 11.19 An example of upstream MAC-frame.

traffic within a LMDS cell. This means that polling slots are not used to poll regularly active end equipment like in other access systems. The duration of upstream frames is designed in a way that each inactive NIU is polled at most every 2 seconds.

The distribution of contention slots ([C]) and of reserved slots ([NIU]) in each upstream frame is specified by the AIU to NIUs via a downstream broadcast channel Q. An example of upstream MAC-frame configuration is given in Figure 11.19. Polling slots are systematically located during the first time slot of the frame and are dedicated to a given inactive NIU. If N is the number of NIUs in a cell, at most N successive MAC frames are necessary to poll all inactive NIUs. The value of N decreases with the number of active NIUs. In the example of Figure 11.19, NIUs 2, 3, 4, and 5 are active. A polling slot located in the first position of the frame is dedicated to one of the inactive NIUs, for instance, NIU 6. A single reserved slot ([NIU] slot) is assigned to a given active NIU at any time slot. Slot 2 is dedicated to NIU 2, slot 4 to NIU 3, slot 5 to NIU 5, slot 6 to NIU 4 and slot 7 to NIU 3. We can see that bandwidth allocated to NIU 3 is twice the bandwidth allocated to NIUs 2, 4, and 5. In our example, slot 3 is a contention slot and can be accessed by active nodes[19] 2 and 4, and also by inactive nodes 1 and 6. In the case of NIU 2 and NIU 4, this contention slot enables these two stations to set up a second ATM connection.

---

19. By "active" node, we mean a node with an open ATM connection and with pending ATM cells for this connection.

In the LMDS-DAVIC standard, a solution is proposed for dynamic band-width allocation. In the case of a VBR ATM connection, the AIU generates reserved slots on a burst-by-burst basis. Indeed, a VBR data traffic (e.g., associated with a TCP/IP flow) is characterized by active and inactive periods, an active period corresponding to a burst. On the basis of the ATM specifications, the bit rate of a VBR source is approximated by a CBR traffic during each burst. One may then estimate that CBR traffic is a particular case of VBR traffic. At connection setup, one assumes that the AIU gets the knowledge of the parameters of the ATM connections. According to ITU-T specifications, these parameters are the following:

- *Peak cell rate (PCR):* This is the maximum data rate at which a connection may generate traffic during a burst.

- *Sustainable cell rate (SCR):* This is an intermediate value between the PCR and the *minimum cell rate* (MCR) of the connection.

- *Burstiness of the connection (B):* This parameter corresponds to the granularity of the connection. In general, it is given by the ratio of the sum of the mean burst duration $E[T_{on}]$ and of the mean silence duration $E[T_{off}]$, over the mean burst duration.

Equation (11.7) gives the expression of burstiness B:

$$B = \frac{\left\{ E[T_{on}] + E[T_{off}] \right\}}{E[T_{off}]} \tag{11.7}$$

We see in Figure 11.20 an example of ideal allocation of reserved slots to a VBR connection. At the beginning of each new burst, a bandwidth request is sent to the AIU. Once it has received this bandwidth request, the AIU begins to specify reserved [NIU] slots for the next upstream frames with a periodicity of $T_{pcr} = 1/PCR$. It is important to mention that propagation delay (the round-trip time) is negligible compared to T_{pcr}. By convention, as soon as the AIU detects an unused reserved slot for the considered connection, it stops its periodic [NIU] slots reservation. A CBR ATM connection can be assimilated to a VBR ATM connection made of a single burst during the whole connection duration.

We have in fact assumed in the example of Figure 11.20 an ideal band-width allocation. Thus, the transmissions of bandwidth requests in contention slots are assumed to be successful. Several factors may in practice have a disruptive effect on (NIU) slot distribution:

- Collisions for the transmission of bandwidth requests;

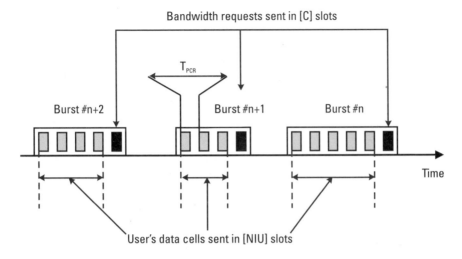

Figure 11.20 Burst-by-burst reservation for VBR connections.

- Multiplexed ATM connections within the same NIU;
- Large number of active connections in the system.

Several aspects of the LMDS-DAVIC standard remain unspecified or implementation-specific for manufacturers or operators:

- The strategy with which bandwidth requests are generated according to the type of the considered ATM connection: one request per ATM cell, one request per burst, one request per connection.
- The strategy with which bandwidth requests coming from different NIUs and associated with different types of ATM connections are satisfied at the AIU. Let us recall that the main types of ATM connections are (see Section 1.3.1): rtVBR, nrtVBR, CBR, UBR, GFR.
- Whether the permits are sent in real time by the AIU or merged.
- The rules with which it is determined in each upstream frame the amount N_c of contention slots; several solutions are proposed and left to the discretion of the manufacturer. For instance, a solution could consist of choosing N_c proportional to the current amount of active NIUs.

A few studies have investigated these open aspects of the standard [13, 14]. For instance, it has been shown that due to the negligible propagation delays in LMDS systems, it may occur that the service of a burst by the AIU is ending whereas a new burst belonging to the same connection is already generated at the NIU. In this case, a *burst merging mechanism* has been proposed in order to reduce the amount of access with potential contention. The burst merging mechanism has a positive impact on rtVBR connection conformance.

11.4.2 Two-Layer LMDS

A new version of LMDS known as "two-layer" LMDS is under investigation and development. The main motivation for this new version is to bypass the difficulties inherent in radio propagation in the millimeter frequency range (i.e., multipath fading, absorption in case of rainy weather, and strong attenuation). These difficulties prevent mobility of the end users. Figure 11.21 illustrates a two-layer LMDS network configuration. Many of the technical characteristics given in this section are not finalized and open to discussion. Two-layer LMDS systems are actually investigated within an ACTS European research program called CABSINET [15].

Two types of cells are considered: macrocells and microcells. Macrocells operate at high frequencies between 28 GHz and 40 GHz and are characterized by a radius of about 3 km. A macrocell covers several microcells operating at lower frequencies between 5 GHz and 17 GHz. The radius of a microcell is between 50m and 500m. The main benefit of two-layer LMDS configuration is that it does not require a direct view between the user's terminal and the base

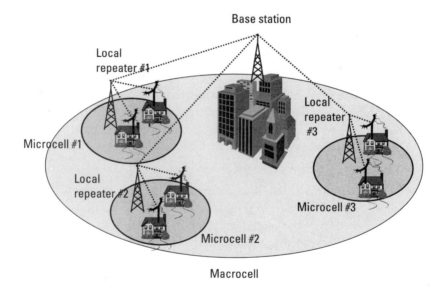

Figure 11.21 Configuration of a two-layer LMDS system. (*From:* [16]. © 2001 IEEE. Reprinted with permission.)

station. Unlike LMDS, two-layer LMDS enables a certain mobility of the end users. Line-of-sight remains mandatory between a base station (the base station is in this case the antenna located at the center of the macrocell) and *local repeaters* (LRs) located at the center of the microcells. In the CABSINET research program, a base station is not an omnidirectional antenna but a sectorized antenna[20] corresponding to a set of four directive antennas dividing a macrocell into four sectors.

Within a macrocell, a base station may be connected either to a local repeater or to a fixed user terminal. In the CABSINET program, the [40.5 GHz, 42.5 GHz] band is allocated to macrocells whereas the [5.725 GHz, 5.875 GHz] band is allocated to microcells. Within a macrocell, 50 downlink channels of about 40 MHz each are used. Several advanced modulation techniques are under investigation for these downstream channels in coordination with the DVB Consortium. Turbo-code 16-QAM modulation or *code OFDM* (COFDM) are some of these advanced modulation techniques combined with digital video distribution framing. The same kinds of modulation techniques are investigated and compared for upstream channels within macrocells but without digital video framing. Within microcells, DS-CDMA modulation is applied to both downstream and upstream directions [17]. Figure 11.22 illustrates the transmitter and receiver building blocks of a two-layer LMDS system.

As shown in Figure 11.22, a user's terminal can be fixed or nomadic. In the former case, the user's terminal is directly connected to the base station by means of a radio channel in the 40-GHz range. The equipment installed at the end user corresponds to a 10-cm diameter parabolic antenna pointed toward the base station and to a *DVB multiplex stream* (DVB-MS) set-top box. This set-top box is a modified version of a set-top box used for satellite access. It makes it possible to demodulate and to deframe the received signal in order to get one or several phone connections, a TV channel from a remote server, and data from the Internet. One notices that in the uplink, a fixed user's terminal is either a telephone set or a computer. A nomadic user is supposed to move at a low speed (a few meters per second). This user is connected indirectly to the base station via a local repeater by means of a radio channel in the 6-GHz range. In terms of offered services, we have seen that existing LMDS systems enable the transport of IP packets via a classical IP over ATM architecture. The main drawback of ATM is its cell tax of 5 bytes for every 48 bytes of user data. While this cell tax is not a real problem on cable networks and particularly on optical access networks because of their large potential bandwidth, this is not the case in radio environment. Investigations are currently carried out within the CABSINET program to facilitate the transport of IPv4 or IPv6 packets without using the ATM layer. Even while the modulation technique and the framing rules to be adopted for

20. Sectorized antennas are also used in cellular networks like GSM to increase cells' capacity.

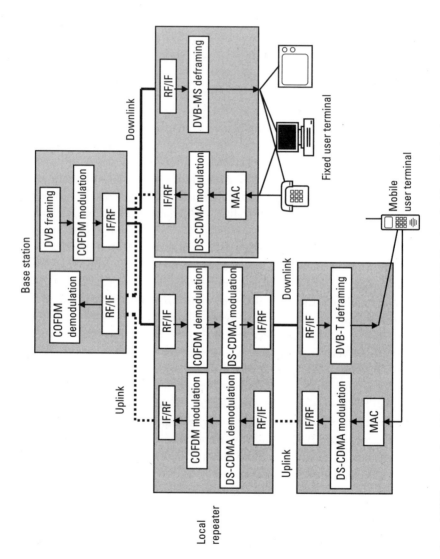

Figure 11.22 Two-layer LMDS protocol architecture. (*From:* [16]. © 2001 IEEE. Reprinted with permission.)

the two-layer LMDS are still under discussion, such systems aim to offer roughly the same data rates than LMDS.

11.4.3 IEEE 802.16

The IEEE standardization body published in June 2000 an "invitation to contribute for the specification of a new standard for fixed point-to-multipoint broadband wireless access systems." Like all the systems described in the previous sections of this chapter, an IEEE 802.16 system consists of a set of base stations covering microcells (cells with a radius under 500m) in which are located the end users. This new standard designated by IEEE as 802.16 covers both the physical layer and the MAC layer. The frequency range of the air interface is very open to discussion, frequencies between 10 GHz and 66 GHz being considered. Both real-time (voice and video) and nonreal-time applications should be transported on the system. The physical layer is specified in a very flexible way in order to be adaptable to the coming advanced technologies for radio communication systems.[21] Like LMDS systems, IEEE 802.16 systems will rely on a request-permit mechanism at the MAC layer. The first draft versions of the standard have been proposed in May 2001 [18]. These versions are subject to change. We have focused in the following only on the general concepts that should likely remain in the final version of the standard.

11.4.3.1 General Architecture

Figure 11.23 depicts the protocol stack of an IEEE 802.16 system. This LAN-oriented architecture is divided into several sublayers. In the upper part of the MAC layer, a *convergence sublayer* (CS) enables via a "parameterized protocol independent" *service access point* (SAP) to multiplex various types of traffic on the system. These types of traffic are assumed to be packet-oriented like IP, ATM, and Ethernet. The various data flows sharing simultaneously the system capacity are identified by a local CID. Among all the operations to be carried out at the MAC layer (e.g., framing, error detection, contention resolution, and bandwidth allocation), some of them are specific to given applications; others are common to several applications. The operations corresponding to the former case are grouped in a service-specific convergence sublayer, the others being associated with the common part MAC sublayer. The *physical layer* (PHY) is also made of two sublayers, a *physical medium–dependent* (PHY-PMD) sublayer and a *convergence sublayer* (PHY-CS). The PHY-PMD specifies the types of antennas to be used at the base station and at the customer premises, and the power budget constraints to be respected. Closely related to the type of antennas, the modulation techniques are also specified in the PHY-PMD sublayer.

---

21. New line codes and new modulation techniques should be developed in the coming years thanks to the advances in high-performance digital processors.

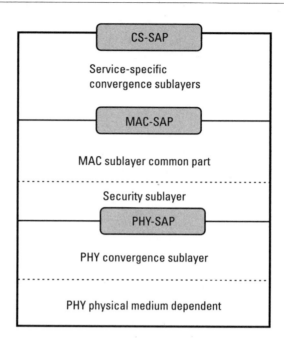

Figure 11.23 Basic protocol stack of an IEEE 802.16 system.

The PHY-CS deals with bit and packet synchronization and parallel/serial conversion. Optionally, it may also include bit error detection and bit error recovery.

11.4.3.2 Physical Layer

The modulation techniques proposed for the IEEE 802.16 standard are quite similar to those adopted for LMDS, that is QPSK, 16-QAM, and 64-QAM. Interleaving functions and auto-correcting codes are also within the scope of the standard. Whereas LMDS physical systems are based of a given modulation's technique, IEEE 802.16 systems make it possible to choose dynamically this modulation's technique. This choice is based on the required quality of service and on the SNR observed on the air interface. We have mentioned in the section dedicated to the LMDS air interface the benefits and drawbacks of TDD and of FDD. Whereas LMDS systems are based on FDD in which the uplink and the downlink use separated radio channels, IEEE 802.16 systems may operate either in FDD or in TDD in order to merge the benefits of the two techniques. When FDD is used, two types of user's terminal may be considered on the system, either full-duplex or half-duplex. As it is depicted in Figure 11.24, a half-duplex station cannot transmit when it receives information from the base station. Time is divided into successive fixed size frames in order to coordinate

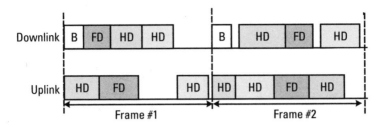

Figure 11.24 Full-duplex and half-duplex terminals with FDD IEEE 802.16 systems. HD: Half-duplex communications, FD: Full duplex communications, B: Broadcast channel.

the transmission of successive variable data packets. A packet cannot overlap over two (or more) successive frames [2, 19].

If TDD is used, the physical layer operates in a mode comparable to what we have described for DECT systems. On a given frequency, time is divided in successive fixed-size frames. Each frame is itself divided into two parts dedicated to downstream and upstream traffic, respectively. A frame is made of a succession of time slots. Unlike DECT systems, TDD IEEE 802.16 systems use a movable boundary between the downlink channels and the uplink channels in order to facilitate a greater flexibility in bandwidth allocation. The position of the boundary in each frame is set dynamically under the control of the MAC layer. Figure 11.25 illustrates the format of a TDD IEEE 802.16 frame.

11.4.3.3 MAC Layer

The IEEE 802.16 working group has specified several parameters for the format of the MAC-PDUs to be transported over the air interface. A MAC-PDU is made of three main fields: a 6-byte header, a variable size payload, and an optional *cyclic redundancy checksum* (CRC). The header of each MAC-PDU should contain at least the following information:

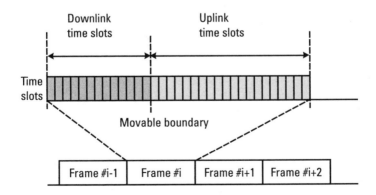

Figure 11.25 Format of a TDD IEEE 802.16 frame.

- A *connection identifier* (CID) of 16 bits: CID may be used either for voice, video, and data flows. The CID enables up to 64K simultaneous flows within a cell.

- A payload length coded on 11 bits: This field indicates the length in bytes of the payload of the MAC-PDU. A MAC-PDU payload is then between 0 and 2,047 bytes long.

- A *header type* (HT) bit in order to specify if the MAC-PDU corresponds to a bandwidth request or to user data.

- Two fields related to encryption of the payload: A bit *encryption* (EC) specifies if encryption is activated or not; a set of 2 bits called an *encryption key sequence* (EKS) is used to specify the applied encryption/decryption technique.

- An 8-bit header protection field: Some proposals submitted to the 802.16 working group consider a 1-byte-long *header checksum* (HCS).

- A 6-bit type field specifies the nature of the MAC-PDU payload: user data, bandwidth request, or management data (data used for managing multiple access).

- Three bits remain reserved for future use.

Figure 11.26 illustrates the format of a MAC-PDU. We see that the variable size payload may be followed in the MAC frame by an optional *cyclic redundancy checksum* (CRC). The MAC-PDU payload may include itself a subheader specifying three types of operation that can be applied at the MAC layer:

- Fragmentation subheader;
- Grant management subheader;
- Packing subheader.

In the case of a succession of small-size MAC-PDUs, several MAC-PDUs originated by or destined to the same node[22] may be concatenated in order to facilitate synchronization at the receiver. The succession of several concatenated MAC-PDUs is called a *burst*. Within a burst, a receiver is able to recognize the semantic of each MAC-PDU thanks to the CIDs. Figure 11.27 depicts an example of MAC-PDUs concatenation. On this example, we see that the upstream burst number $(n - 1)$ is made of two users' PDUs belonging to the same flow (CID = 1) and of a bandwidth request PDU (CID = 2).

A user's MAC-SDU may also be divided into several MAC-PDUs if the fragmentation process is activated. Fragmentation may be decided either on the

---

22. This node may be either the base station or a user's terminal.

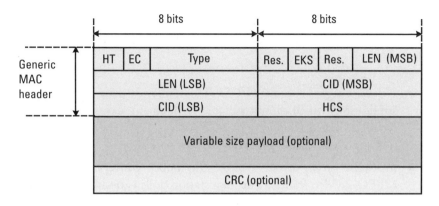

Figure 11.26 Format of an IEEE 802.16 MAC-PDU.

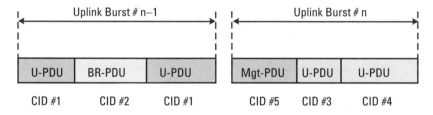

Figure 11.27 An example of MAC-PDU concatenation.

downlink channel by the base station or on the uplink channel by the terminal in order to consider fluid limit traffic. In that case, the fragmentation subheader makes it possible to rebuild the original MAC-SDU within the receiver. As mentioned above, multiple access in an IEEE 802.16 cell is based on a request-permit mechanism. Two modes of bandwidth allocation are used:

- *Grant per connection (GPC):* In this case, permits are allocated to a given terminal, and within this terminal, to a given connection.

- *Grant per terminal (GPT):* In this case permits are allocated to a given terminal but are anonymous for the connections. Service disciplines are necessary in order to assign correctly the permits to the right connections within a terminal. The draft standard considers the possibility to associate time stamps to the packets waiting for transmission within the terminal in order to facilitate permits' assignment.

Both the GPT and the GPC modes must be supported at the base station. At network initialization, a user terminal informs the base station of the mode adopted by the customer in his or her subscription. Different scheduling policies for the request-permit scheduling are used to satisfy different types of applications:

- *Unsolicited grant service (UGS):* This policy is designed for real-time applications which cannot suffer very fluctuant access delays like CBR circuit-switched telephony or PCM VoIP. For these kinds of applications, the base station generates permits periodically.

- *Real-time polling service (rtPS):* This policy concerns rtVBR such as MPEG-2 video frames. The base station polls periodically users' terminals in order to be informed of their bandwidth requests. The polling period duration is assumed to be compatible with the fluctuations of the bit rate of the source. On the basis of the results of polling, the base station generates permits for the transmission of variable size packets.

- *UGS with activity detection (UGS-AD):* This service can be compared to UGS but for which the source is inactive over very short periods (a few tens of ms). During the activity periods of the source, the base station generates permits automatically (that is, without explicit multiple requests) and periodically. As soon as the source gets inactive, an unused reserved slot is detected by the base station, which in this case switches to the rtPS mode. When it becomes active again, the base station that has the knowledge of the individual connections, reenters the UGS-AD mode. Packetized voice (see Chapter 10) over IP is an example of application for which this service has been designed.

- *nrtPS:* This service concerns applications that generate variable-size packets without real-time constraint. The base station polls periodically the end station with a longer period than in the case of rtPS. File transfers are an example of application corresponding to this type of service.

- *Best-effort service (BES):* Unlike the other service policies, best effort service is not based on a request-permit mechanism but on contention slots. In the upstream frames, a certain amount of contention slots are accessible for the transmission of best effort packets. Web access or electronic mail applications are some example of applications concerned by the BES service. In case of collision, a truncated binary exponential back-off algorithm is used for contention resolution.

11.5 Elements of WITL Network Planning

We have seen in the previous sections that narrowband and broadband WITL systems are based on cells or microcells. Within a cell or a microcell, either FDD or TDD techniques are adopted to enable bidirectional communication between users' terminals and base stations. Whatever the adopted WITL technology, carriers have to proceed to frequency planning in order to cover areas much larger than the radius of a cell. The problem of frequency planning consists of determining the optimal distance between two base stations operating on the same frequency in order to maintain interference between them below a certain threshold. The calculation of this distance strongly depends on the model adopted for describing the geometry of a cell [20]: circular, hexagonal, or rectangular. The hexagonal model and the rectangular model that approximate the circular model are the most popular.

11.5.1 The Hexagonal Model

Figure 11.28 illustrates the principle of the hexagonal model. This model is widely used for the planning of radio-mobile networks like GSM. Let us remark that the perimeter of a circle of radius R may be divided into six regular sections. The Euclidean distance between the two end points of each sections is equal to R. A frequency (in the case of TDD) or a set of frequencies (case of FDD) are assigned to each cell. Let us associate a color to each frequency, or to each set of frequencies. Network planning consists in allocating different colors to any couple of adjacent cells. In general, one estimates that the same frequency (color) may be reused in two different cells if these cells are in the worst case separated by a cell operating on a different frequency (with a different color). Thus, we can see from Figure 11.28 that only three colors (three frequencies) are necessary in order to build the whole network. We can see that the minimum distance

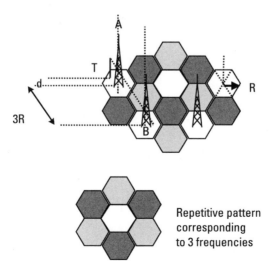

Repetitive pattern
corresponding
to 3 frequencies

Figure 11.28 Hexagonal pattern.

between the center of two cells using the same color equals three times the radius R of a cell. We have shown in the lower part of Figure 11.28 the set of the three colors assigned to a regular pattern made of seven cells. This pattern is repeated in order to cover, for instance, a whole city.

In radio communication, SNR is also called $C/(I+N)$ where C stands for the power of the reference signal, I stands for the summation of the power of the various interference, and N stands for noise. Noise can be assimilated to a white noise that can be easily reduced at the receiver by means of a passband filter. We can then ignore the value of N compared to the value of I. It is then possible to evaluate analytically the expression of C/I corresponding to the worst case interference. Let us consider a user's terminal located at a distance d from its base station. Using (11.1), the received power P_{r1} by terminal T from its base station A transmitting with power P_e is:

$$P_{r1} = P_e g_R g_T \left(\frac{\lambda}{4\pi \cdot d} \right)^2 \qquad (11.8)$$

where g_R and g_T stand for the gain of the receiver antenna and of the transmitter antenna, respectively. Let us recall that in (11.1), G_R and G_T were standing for these same gains but expressed in decibels. This user's terminal also receives the signal sent by base station B on the same frequency and distant from $(d + 3R)$. The power P_{r2} received by terminal T from base station B transmitting with power P_e is:

$$P_{r2} = P_e g_R g_T \left(\frac{\lambda}{4\pi \cdot (d + 3R)} \right)^2 \qquad (11.9)$$

We have then for the C/I ratio:

$$\frac{C}{I} = \frac{P_{r1}}{P_{r2}} = \frac{(4\pi \cdot (d + 3R))^2}{(4\pi d)^2} \qquad (11.10)$$

Expressed in decibels, C/I becomes:

$$\left(\frac{C}{I} \right)_{dB} = 10 \cdot \log \left(\frac{P_{r1}}{P_{r2}} \right) = 10 \log \left(\frac{(d + 3R)^2}{d^2} \right)$$
$$= 20 \log \left(1 + \frac{3R}{d} \right) \qquad (11.11)$$

The *C/I* ratio is minimum when *d* is maximum (that is, when the user's terminal is located on the cell boundary at distance *R* from its base station). We have in this worst case:

$$\left(\frac{C}{I}\right)_{dB} = 20\log\left(1+\frac{3R}{R}\right) = 20\log(4) = 12 \text{ dB} \qquad (11.12)$$

The hexagonal model may be used for DECT, PACS, PHS, and LMDS.

11.5.2 The Square Model

We have seen in Section 11.4.2 dedicated to the two-layer LMDS that a base station located at the center of a macrocell could be based on sectorized antennas. We have seen that such antennas divide a cell into four 90° adjacent sectors. Whereas the hexagonal model is well suited to omnidirectional antennas, a square model is better suited to sectorized antennas. Figure 11.29 depicts a set of adjacent sectorized antennas. Whereas three different frequencies are necessary in the case of the hexagonal model, we need four frequencies to build the whole

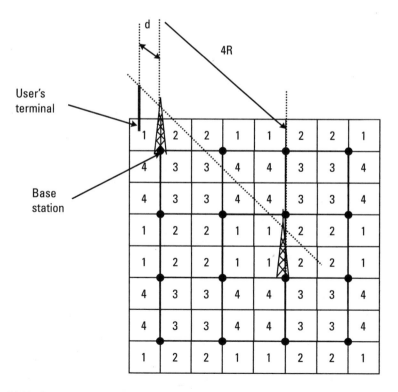

Figure 11.29 Square pattern with four frequencies.

network in the case of sectorized antennas. Frequencies 1, 2, 3, and 4 are used in each of the four sectors of a square cell.

Let us evaluate the worst case C/I under this kind of cell's model. Again, let us consider a user's terminal located at distance d from its base station A. The received power P_{r1} by this terminal from its base station is given by (11.8). This user's terminal also receives the signal sent by the base station B on the same frequency but distant from $(d + 4R)$. Again, using (11.1), the power P_{r2} received by this terminal from base station B transmitting with power P_e is:

$$P_{r2} = P_e g_R g_T \left(\frac{\lambda}{4\pi \cdot (d + 4R)} \right)^2 \tag{11.13}$$

The C/I ratio expressed in decibels is then :

$$\left(\frac{C}{I} \right)_{dB} = 10 \cdot \log \left(\frac{P_{r1}}{P_{r2}} \right) = 10 \log \left(\frac{(d + 4R)^2}{d^2} \right) \tag{11.14}$$
$$= 20 \log \left(1 + \frac{4R}{d} \right)$$

The C/I ratio is minimum when d equals R:

$$\left(\frac{C}{I} \right)_{dB} = 20 \log \left(1 + \frac{4R}{R} \right) = 20 \log(5) = 14 \text{ dB} \tag{11.15}$$

11.6 Conclusion and Perspectives

The objective of this chapter was to give an overview of existing solutions proposed for WITL access systems (also known as RITL systems). We have chosen to classify WITL systems into two categories: narrowband access systems and broadband access systems. The three main narrowband WITL systems, DECT, PHS, and PACS, have been presented. Today, there is no universal standard of narrowband WITL. We have seen that DECT systems are mature and widely deployed in Europe and in Africa. PHS and PACS are fundamentally very similar to DECT; this is why we have focused our presentation on this latter system. PHS and PACS are essentially used in Japan and in North America, respectively. PHS benefits from a real commercial success in Japan. Unlike DECT and PHS, PACS gives real mobility to the user.

In terms of broadband radio access systems, three solutions have been successively presented: LMDS, two-layer LMDS, and IEEE 802.16. We have focused our description on LMDS systems because they are today the most mature and the most deployed broadband technology. While the air interface of LMDS is well specified in the LMDS-DAVIC standard, this is not the case for the MAC layer. Vendors actually implement their own protocols at the MAC layer, the common point between the equipment being the data formatting based on MPEG-2 frames and ATM cells. The physical layer of IEEE 802.16 systems aims at a great flexibility, with factors such as the type of modulation and encoding chosen according to the environment and according to the desired service. The same kind of flexibility is expected at the MAC layer, with ATM, IP, and Ethernet being the typical clients of the IEEE 802.16 systems. The IEEE 802.16 systems are still under evaluation by means of experimental test beds.

In the coming years, the commercial competition between xDSL and WITL should become more active in the market of broadband access. Due to important advances in the field of radio communications, even at high frequencies (up to 40 GHz), broadband WITL systems enable a very efficient transmission. The main constraint of broadband wireless access stems from the very limited size of the cells. Picocells combined with macrocells, like they are defined in two-layer LMDS systems, are a promising approach. In the next few years, broadband WITL systems should concern SMEs rather than residential users. To be really competitive in terms of cost and performance facing other technologies, WITL systems require the development of new efficient network planning tools.

References

[1] Bing, B., *Broadband Wireless Access,* Norwell, MA: Kluwer Academic Publishers, 2000.

[2] Koffman, I., and V. Roman, "Broadband Wireless Access Solutions Based on OFDM Access in IEEE 802.16," *IEEE Communications Magazine,* April 2002.

[3] Padgett, J. E., C. G. Günther, and T. Hattori, "Overview of Wireless Personal Communications," *IEEE Communications Magazine,* January 1995.

[4] Brasche, G., and B. Walke, "Concepts Services and Protocols of the New GSM Phase 2+, General Packet Radio Service," *IEEE Communications Magazine,* Vol. 35, No. 8, August 1997.

[5] NTT DoCoMo, http://www.nttdocomo.co.jp.

[6] MOU of the PHS system, http://www.phsmou.or.jp.

[7] Dayem, R. A., *PCS and Digital Cellular Technologies: Assessing Your Options,* Upper Saddle River, NJ: Prentice Hall, 1997.

 [8]   Noerpel, A. R., Y. B. Lin, and H. Sherry, "PACS: Personal Access Communications System—A Tutorial," *IEEE Communications Magazine,* June 1996.

 [9]   Lee, W. C. Y., *Mobile Cellular Telecommunications Systems,* New York: McGraw-Hill, 1989.

[10]   Cornaglia, B., et al., "LMDS Systems: A Possible Solution for Wireless ATM Access Networks," *Proc. of the IEEE International Conference on ATM (ICATM'98),* Colmar, France, 1998.

[11]   Mähönen, P., et al., "40-GHz LMDS Systems Architecture Development," *Proc. of the ICT'98 Conference,* Vol. 1, 1998, p. 422.

[12]   Bonnarigo, D., M. de Marco, and R. Leonardi, "A Comparison of Back-Off and Ternary Tree Collision Resolution Algorithms in HFC Access Networks," *Proc. of IEEE Globecom'98,* Sydney, Australia, 1998, pp. 45–50.

[13]   Kuri, J., and M. Gagnaire, "ATM Traffic Management in a LMDS Wireless Access Network," *IEEE Communications Magazine,* September 2001.

[14]   Gagnaire, M., "ATM Traffic Management and Multiple Access in a Wireless LMDS Access Network," *Third Annual Symposium on Advanced Technologies,* NIST, Boulder, CO, September 2000.

[15]   ETSI standard ETS 300 744, *Digital Broadcasting Systems for Television, Sound and Data Services: Framing Structure, Channel Coding and Modulation for Terrestrial Broadcasting,* May 1996.

[16]   Mähönen, P., et al., "Wireless Internet over LMDS: Architecture and Experimental Implementation," *IEEE Communications Magazine,* May 2001, pp. 126–132.

[17]   Pahlavan, K., and A. H. Levesque, *Wireless Information Networks,* Reading, MA: Addison-Wesley, 1995.

[18]   IEEE Draft, *Local and Metropolitan Area Networks—Part 16: Standard Air Interface for Fixed Broadband Wireless Access Systems,* IEEE P.802.16/D3, 2001.

[19]   Eklund, C., et al., "IEEE Standard 802.16: A Technical Overview of the Wireless MAN Air Interface for Broadband Wireless Access," *IEEE Communications Magazine,* June 2002.

[20]   Himket, Sari, "Broadband Radio Access to Homes and Businesses: MMDS and LMDS," *Computer Networks Journal,* Vol. 31, 1999, pp. 379–393.

Selected Bibliography

Momtahan, O., and H. Hashemi, "A Comparative Evaluation of DECT, PACS, and PHS Standards for Wireless Local Loop Applications," *IEEE Communications Magazine,* May 2001.

12

Optical and EFM Access Networks

12.1 Introduction

Chapters 1 through 11 have described various access network technologies. After presenting the unbundling characteristics, Chapter 3 listed these various technologies. Subsequently, wired technologies based either on existing copper wires, coaxial cables, or the medium-voltage grid of an electricity supplier were described. This chapter deals with two innovative approaches for the last mile referring either to the transmission medium or to the protocol stack, respectively. On one hand, many investigations have been carried out during these last 10 years, especially within the research laboratories of major ILECs, in order to determine the potential uses for optical communications in access networks. Although several test beds have demonstrated the technical maturity of optoelectronic technologies, optical access systems have been considered during past years to be too expensive to be really competitive with other existing solutions like ADSL. On the other hand, the supremacy of ATM in the access or even further in public networks architecture has been more and more discussed by a few CLECs and equipment vendors during the last 10 years. These CLECs and equipment vendors estimate that Ethernet-based systems now have the capacity to provide QoS guarantees to the end users with a simplified and economic protocol architecture in comparison to ATM.

This chapter is organized in two major sections. Section 12.2 describes the main characteristics of optical access networks as they have been standardized within ITU-T. Such networks known as PONs and BPONs were designed for the provision of end-to-end ATM services. Section 12.3 introduces the EFM concept. In a first approach, the EFM concept is independent from optical access. We shall see that some versions of EFM are in fact based on optical access networks.

12.2 Optical Access Systems

12.2.1 The FSAN Initiative

At the beginning of the 1990s, the main characteristics of optical fibers compared to other transmission mediums were widely recognized. Optical fibers are characterized by a very wide bandwidth in the 1,500-nm range with a very low attenuation around 0.18 dB/km. According to the state of the technology at that time, this bandwidth enables transmission bit rates up to the gigabits per second range on point-to-point fiber links of a few tens of kilometers. Such numbers have driven most carriers to estimate that optical technologies were not suited to access networks because of their inherent cost of installation and their too advanced performances. Indeed, an access network as it was considered in the mid-1990s only needs a few megabits per second over a range limited to 5 km. In spite of this unfavorable context, a group of seven ILECs (France Telecom, Deutsche Telekom, British Telecom, NTT, KPN, Telefonica, and Telecom Italia) decided in 1995 to associate their efforts in order to demonstrate the feasibility of optical access networks. This association supported by the ITU-T took the name FSAN. Other members joined the FSAN initiative (including Bell South, GTE, Swiss PTT and Telstra) at the end of the 1990s.

In the mid-1990s (that is, before unbundling), the ILEC members of the ITU-T were considering ATM as the only protocol able to guarantee QoS criteria to the end users.[1] This is the reason why, since the beginning of their investigations, the FSAN members wanted to design optical access systems in order to provide end-to-end ATM connections. In recent years, new carriers such as Bell Canada, Verizon, and Qwest (United States) and Singtel (Singapore) decided to join the FSAN initiative. The FSAN Consortium cooperates with different standardization bodies such as ITU-T, ETSI, or the ATM Forum.

12.2.1.1 Which Topology for Optical Access Systems?

One of the very first problems considered within the FSAN was the choice of the best topology for the optical infrastructure. Two main configurations were compared and tested: point-to-point and point-to-multipoint. Point-to-point topology presents the advantage of simplicity, with each end user benefiting from the whole capacity of his or her dedicated transmission pipe. Nevertheless, several drawbacks are inherent to point-to-point topology. First, it is necessary

---

1. The main factors justifying the adoption of ATM by ILECs in the mid-1980s are detailed in Section 9.2.

to install for each end user an optical modem at each end of the dedicated optical fiber link. Based on the available technology of the mid-1990s, optical modems were much more expensive than ADSL modems. A second disadvantage of point-to-point topology is the lack of traffic concentration between the end users and the access node—that is, the nearest ATM switch.[2] A third drawback is the high cost of civil engineering to install the optical fibers. In comparison, point-to-multipoint optical topologies may be provided easily and at a moderate cost thanks to passive optical couplers. They are well suited to broadcast services like video channel distribution. A passive optical coupler makes it possible to merge an injection fiber with a transport fiber by simply heating at a very high temperature[3] the silica core of these two fibers. For all these reasons, the FSAN initiative has adopted the point-to-multipoint architecture.

The concept of PON described in the next section consists in an optical tree using a cascade of such couplers. According to the FSAN and ITU-T taxonomy, the optical modem used at the head of the optical tree is called OLT, whereas the optical modems used at the CPEs are called ONUs. End users connected to a point-to-multipoint optical access system cannot use the whole capacity of the optical fiber. For economies of scale, the same laser diode and the same photo-detector must be used in the optical modems of the end users.[4] In other words, end users who want access simultaneously to the fiber infrastructure must share the capacity of a common upstream optical channel by means of a MAC protocol.[5] This apparent drawback may be seen as an advantage if one designs judiciously this protocol MAC in order to proceed to upstream traffic concentration. We detail the benefits of PON systems in Section 12.2.2.

12.2.1.2 Which Range for Optical Access Systems?

Several studies carried out at the end of the 1980s analyzed the service-to-cost ratio of an optical access system for various coverage assumptions. Figure 12.1 illustrates the basic possible scenarios in this matter.

A first approach known as *fiber to the home* (FTTH) has the merit that it is simple, with each end user being equipped with his or her own ONU. In practice, FTTH means that either the optical fiber arrives in the private residence of the customer or it arrives in the customer's apartment in a building. The FTTH scenario was considered at the beginning of the 1990s as too expensive in terms

---

2. Chapter 2 explained that the cost of a switch is proportional to the amount of its line interface cards.

3. About 3,000°C are necessary to melt silica.

4. However today, this argument has lost much of its veracity with the emergence of WDM technologies.

5. The reader recognizes here a characteristic of HFC cable networks that was presented in Chapter 3.

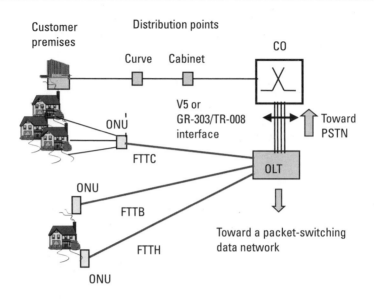

Figure 12.1 FTTx network configurations.

of equipment cost. A more economical approach known as *fiber to the building* (FTTB) consists of installing a single ONU per building. This assumes that within the building, a private cabling system makes it possible to link the end users to the ONU. Typically, ADSL or VDSL modems may be used on the existing private copper wire cabling system for that purpose. At last, the cheapest approach known as *fiber to the curb* (FTTC) assumes that the ONUs are installed at distribution points in the local loop. The termination of the access system from each ONU to multiple customer premises reuses in this case the existing copper wires of the local loop.

Synchronous services (digital voice), client-server services (Web browsing) as well as asynchronous data services (TCP/IP file transfers) are assumed to be transported via FTTx access systems in ATM cells. At the output of the OLT, voice traffic is either forwarded to a CO of the PSTN or to an ATM switch by means of AAL1 or AAL2 ATM connections. Between the OLT and the CO, two types of signaling interfaces are possible: the V.5 interface in Europe or the GR-303/TR-008 interface in North America. Concerning voice traffic redirected to an ATM switch at the output of the OLT, we introduced in the VoDSL discussion the AAL1 and AAL2 layers. AAL1 is used for *circuit emulation service* (CES) when voice activity detection is not available at the terminal equipment. AAL2 assumes *speech activity detection* (SAD) with packetized voice.

Several technical and economical investigations concluded at the beginning of the 1990s that thanks to the advances in xDSL technologies, the FTTC scenario was the most realistic [1, 2]. The basic idea of FTTC is to reuse the exiting copper pairs arriving from the end users at the distribution points in which are installed the ONUs. Each of these subscriber line terminations is equipped with a VDSL modem. A VDSL modem is installed within the home or apartment of the end user (like an ADSL modem), the other VDSL modem being placed at the distribution point. Thus, the downstream optical signal received in an ONU from the OLT is demodulated. The obtained data flow is demultiplexed by the ONU to be forwarded to the multiple VDSL modems installed in the same distribution point. In other words, the ATM cells contained in the downstream flow are dispatched according to their destination address to the right VDSL modem (or end user).

12.2.2 APON Access Systems

12.2.2.1 APON Configuration

We have underlined in the previous section that all types of traffic were transported via ATM connections between the OLT and the ONUs in PON access systems. This is the reason why the acronym APON has finally been adopted by the FSAN and ITU-T. Figure 12.2 illustrates the configuration of an APON access system. Two separate wavelengths are adopted for upstream and downstream transmissions, respectively. Wavelength $\lambda_1 = 1,550$ nm is used for downstream traffic at 155 Mbps or 622 Mbps. Wavelength $\lambda_2 = 1,310$ nm is used for upstream traffic at 155 Mbps. These downstream and upstream capacities are

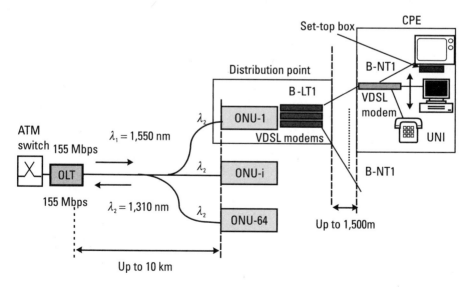

Figure 12.2 APON configuration.

dynamically shared between the various active ATM connections thanks to a MAC protocol.

As illustrated in Figure 12.2, up to 64 ONUs may be connected to an APON. Considering a 3-dB attenuation for each splitting point, the maximum range between an ONU and the OLT is equal to 10 km. Another version of APON with only 32 ONUs is also considered enabling a range of 20 km. The target downstream and upstream bit rates offered to each user of the APON system are 10 Mbps and a few megabits per second, respectively. Such figures are achievable on the VDSL links for a distance of about 1,500m, which corresponds to the average distance between the last distribution point in the local loop and the CPEs. According to the ITU-T taxonomy, VDSL modems installed at the distribution point and at the CPEs are called *broadband line termination of type 1* (B-LT1) and *broadband network termination of type 1* (B-NT1),[6] respectively. Up to 10 B-NT1 modems may be connected to the same ONU.

We have already listed in Section 12.2.1.1 a few benefits of point-to-multipoint versus point-to-point optical topologies. The fact that a PON does not require any remote powered electrical equipment like repeaters/regenerators in the access network is also a great advantage for a carrier. Indeed, this characteristic makes it possible to minimize the maintenance cost of the PON system.

12.2.2.2 APON Initialization

Before its activation for the provision of commercial services, a PON must be initialized by its carrier. This network initialization is based on three procedures: distance ranging, clock ranging, and power ranging.

Distance ranging consists in the evaluation of the RTT between each ONU and the OLT. This distance can be deduced from SNR at the receiver of an optical pulse sent with a given power by the OLT. On the basis of the highest value of these measured RTTs, each ONU k sets an artificial delay Δ_k corresponding to a fixed bufferization delay applied to both upstream and downstream traffic. In this way, propagation delays between the OLT and each ONU are artificially equalized. We outline in Section 12.2.2.3 the importance of distance ranging for the APON MAC protocol. Power ranging consists in equalizing the SNR for each propagation path between the OLT and the ONUs.

Clock ranging is the procedure enabling the OLT to demodulate correctly successive upstream optical bursts originated by different ONUs. Clock distortion is essentially due to propagation disparities between the ONUs and the OLT. In other words, distance ranging does not suppress clock distortion due to physical propagation of an optical or electrical signal. Clock distortion is much less noticeable at the ONUs for downstream traffic because in that case, the

---

6. Similar notations were introduced in Chapter 2 for narrowband ISDN modems.

physical distance between the sender (the OLT) and the receiver (a given ONU) is fixed. In order to determine burst by burst the ideal clock with which a burst must be demodulated at the OLT, a synchronization preamble must be added to each upstream ATM cell. In traditional LANs like Ethernet or token rings, a similar preamble associated to each data frame coupled with a PLL at the receivers is used for clock ranging. The PLL technique could have been adopted for APON packets but the FSAN has estimated that the number of requested bits for the preamble (several bytes) should have been prohibitive in reference to the size of ATM cells. Like in the great majority of optical systems, the OLT and the ONUs use an NRZ optical modulation. A "1" bit consists in applying a voltage level $+V$ enabling to activate a laser diode. Thus, the laser diode of an ONU transmits an optical burst on λ_1 during the bit duration. A "0" bit consists in applying a voltage level $-V$ for which the laser diode is blocked. Due to the high bit rate of the system (at least 155 Mbps) and to ONU-OLT distance disparities, *clock phase alignment* (CPA) at the OLT is very difficult to achieve. According to the state of the technology, the accuracy of the fine clock ranging is only 7 ns, that is, between a 0.1- and 0.2-bit duration. A new CPA technique based on an oversampling mechanism has been developed specifically for APON systems. This technique enables to proceed to CPA by means of only 3 bits [3].

Optical power ranging at the receiver is necessary if one wants to prevent any saturation at the photodetectors. Indeed, let us assume a fixed detection threshold at the OLT. This threshold must be set in accordance with the worst case, that is to the largest OLT-ONU distance. Knowing that the laser diodes implemented in the ONUs transmit with the same power, the OLT photodetector could be saturated when it receives an optical burst from the nearest ONU. The APON MAC protocol that we shall describe in Section 12.2.2.3 makes it necessary for the OLT to know in advance systematically the address of the ONU at the origin of the next upstream ATM cell. Thanks to this knowledge, the OLT is able to dynamically adapt its photodetection threshold according to the MAC address of the originating ONU. The fact an upstream preamble and a MAC address must be associated to each upstream cell implies the addition of an overhead to each upstream ATM cell.

12.2.2.3 APON MAC Protocol

The MAC protocol of an APON access system aims at three objectives:

- To prevent any collision between concurrent upstream transmissions;

- To share dynamically, fairly, and efficiently the available upstream capacity of the PON system between the various ONUs;

- To maintain the best possible conformance of the ATM connections initiated by the end users.

The APON MAC protocol is based on a request/permit mechanism. The basic idea of this protocol consists of sending a bandwidth request to the OLT for each ATM cell pending for transmission within an ONU. Thanks to its double knowledge of the traffic load and of the characteristics of the active ATM connections, the OLT may send a transmission permit to the ONU at the origin of a request. The aim of this permit is to inform the ONU at which instant it will be authorized to send its ATM cell. The OLT, toward which converges upstream traffic from the various ONUs, has the possibility to determine dynamically (cell after cell) the available bandwidth on the system. Two approaches are possible for enabling the OLT to have the knowledge of ATM traffic contracts initiated by the B-NT1s:

- Similar to an ATM switch, the OLT intercepts and interprets the Q.2931 signaling ATM cells; thus, the OLT knows in detail and in real time the QoS requirements of the various active ATM connections on the basis of the characteristics of their ATC:[7] CBR, rtVBR, nrtVBR, GFR, ABR, UBR.

- The OLT uses the *broadband bearer control protocol* (BBCP) protocol defined in the ETSI standardized VB.5.2 interface in Europe or the GR-303/TR-008 interface in North America. These interfaces specify the signaling messages that must be used between an OLT and an ATM switch.

In an ONU, a per-ATC queuing strategy is adopted. From an ATM perspective, this means that a specific FIFO queue is used for the pending ATM cells belonging to the same ATC. Figure 12.3 illustrates the configuration of an ONU. For instance, all the ATM cells of the various active CBR connections issued by the 10 B-NT1 modems connected to this ONU are enqueued in the CBR buffer. When a permit arrives in an ONU, a service discipline must be activated in order to determine which ATC queue must be served in priority.

In order to manage the request/permit mechanism, the FSAN has defined a frame structure for both downstream and upstream traffic. This frame format standardized by the ITU-T in the G.983 recommendation is depicted in Figure 12.4. Downstream traffic is made of a continuous flow of 53 bytes ATM cells. Permits are transmitted by bursts from the OLT to the ONUs in dedicated periodical *physical layer OAM* (PLOAM) cells. The reason why 26 permits are transported in each PLOAM cell is explained further. User cells transported from the OLT to the ONUs between PLOAM cells are not subject to the request/permit mechanism. At the same time user cells are transported from the ONUs to the OLT. These cells have received their permit thanks to previous PLOAM cells.

---

7. The concept of *ATM transfer capability* (ATC) was introduced in Section 1.3.1.

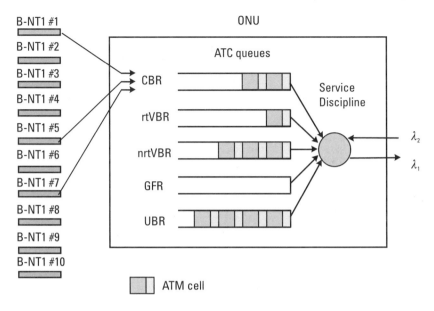

Figure 12.3 Per-ATC queuing strategy at the ONU.

One notices that upstream ATM cells are encapsulated in APON packets in order to take into account the 3 CPA bits mentioned in Section 12.2.2.2. An APON packet is made of a 3-byte header and of a 53-byte payload that is used to carry ATM cells.[8]

An APON packet is then 56 bytes long. The concept of APON packet is useless for downstream traffic because, as mentioned above, clock distortion between a sender and a receiver is in that case known in advance by the ONUs. The 3-byte header of an APON packet is used first to carry out burst by burst CPA. It is also used for the power- and distance-ranging procedures. A certain amount of bits within the APON packet headers is also used for a 2-bit guard time between successive upstream ATM cells because of the inevitable clock distortion inherent to propagation.

The FSAN G.983 framing presented in Figure 12.4 shows that for a 155 Mbps symmetrical traffic capacity, a downstream frame is made of 56 ATM cells of 53 bytes each, whereas an upstream frame is made of 53 APON packets of 56 bytes each. If one assumes that a request/permit is necessary for each upstream cell, the OLT must send over long periods of time as many permits as the amount of ATM cells that can be physically transmitted by the ONUs onto the medium. This is the reason why the two downstream PLOAM cells may transport 26 permits each. In the upstream frame, a specific packet known as the

---

8. A detailed description of the APON packet header is given in the standard.

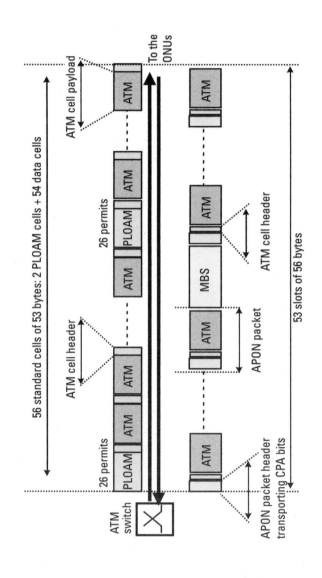

Figure 12.4   G.983 ITU-T frame format.

multiburst slot (MBS) is used for requests' transmission. The detailed format of a 56-byte-long MBS slot is given in Figure 12.5.

An MBS slot is made of eight subfields of 56 bits each and corresponding to eight ONUs. The basic idea of the MBS slot is to poll cyclically the 64 ONUs, with eight successive ONUs being polled during every G.983 cycle (or frame). As an indication, a cycle lasts 0.15 ms. In other terms, the 64 ONUs are polled in eight cycles, that is, in 1.2 ms. The RTT limited to propagation delays is about 0.1 ms for a OLT-to-ONU distance of 10 km. In practice, additional delays must be added for MAC protocol processing at the OLT and the ONUs. In the example of Figure 12.5, subfield *i* enables ONU *i* to send its pending requests to the OLT. One notices that only two types of requests are taken into account in the MBS format for ABR/GFR-rq and for rtVBR-rq, respectively. We have seen in Chapter 9 that GFR-ATC replaces today ABR-ATC for the transport of TCP/IP connections over ATM networks. The ABR/GFR-rq field is set to the value at the polling instant of the total number of pending ABR/GFR cells at the ONU. It is important to note that the OLT knows the index of the ONU at the origin of an ABR/GFR-rq field. On the other hand, the OLT does not know the precise reference of the ATM connections requiring these requests. Indeed, it has been considered within the FSAN that the price to pay in terms of wasted bandwidth for associating with each request and with each permit the VPI/VCI fields was unacceptable. By consequence, the permits are sent by the OLT to the right ONU, but within the ONU, the assignment of a permit to the right ATM connection is a difficult problem. In the next section, we describe a few possible approaches for ATM traffic management over APON systems. Some of these approaches are able to solve the problem of permit assignment.

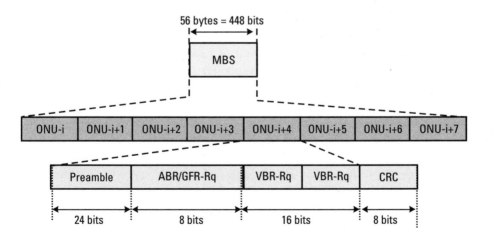

Figure 12.5 The multiburst slot format.

12.2.2.4 ATM Traffic Management over APON

Let us consider in a first step the case of CBR connections (fixed bit rate voice or video using AAL1). We have underlined above that the OLT has the knowledge of ATM traffic contracts. The basic rule mentioned in the previous sections consisting of associating a request/permit to each upstream cell can be lightened. In the case of voice over AAL1, it is useless that an ONU sends a request for each cell of a CBR connection. As soon as the call setup has been accepted by the ATM switch at the head of the optical tree, the OLT simply generates permits for this connection at a rate equal to T_{pcr} corresponding to the inverse of the PCR of the connection (PCR $= 1/T_{pcr}$). Because the permits cannot be transmitted immediately by the OLT, a queuing strategy and a service discipline must also be applied to the permits within the OLT.

Concerning rtVBR traffic (packetized voice, adaptive coding video), the OLT knows the PCR of the connections during their activity periods. Similarly, it seems reasonable to use a request only to indicate the beginning of a new burst to the OLT. Once it receives this request, the OLT reactivates a periodic generator with rate T_{pcr}. As soon as a permit is not used by the considered connection, the OLT stops to generate permits.

UBR traffic is best effort and does not benefit from any bandwidth reservation. This is the reason why UBR connections are the last ones to be served at the ONU. If a permit arrives and none of the other ATC queues uses this permit, then this permit may be used to send a UBR cell.

Today, most APON modem vendors only implement CBR and UBR ATCs in their MAC protocol. The way to manage VBR and GFR traffic is not specified by the ITU-T and is left to the initiative of the vendor. Research studies have shown that the use of a simple priority mechanism between the ATC queues for permit assignment within an ONU was not efficient. An intuitive approach drives, for instance, the manufacturer to affect priorities by decreasing orders from CBR, rtVBR, nrtVBR, GFR, and UBR (see Figure 12.3). More generally, it has been observed by means of computer simulations that service disciplines enabling an ATM cell transmission as soon as a permit is available may have for certain traffic scenarios a negative impact on the conformance of real-time connections.[9] Such service disciplines are called *work-conserving* because they use systematically a permit in case of at least a pending cell. Nonwork-conserving disciplines are more complex to implement but are very efficient for preserving the conformance of ATM connections through the optical tree. For that purpose, one associates to each pending cell a time stamp indicating the ideal time at which this cell should consume a permit in reference to the instant of transmission of the previous cell belonging to the same connection. The time stamp of a real-time cell is calculated at the arrival of this cell

---

9. CBR and rtVBR connections are two typical examples of real-time ATM connections.

within the ONU from a B-NT1. This time stamp decreases to zero with time. A pending cell becomes eligible for transmission—that is, this cell is authorized to use a permit, only if its time stamp is null or if it is equal to the smallest of the time stamps of all the pending cells within the ONU. It has been shown that an approach based on nonwork-conserving disciplines for real-time traffic is very efficient for preserving the conformity of ATM connections within an APON [4–6].

12.2.3 From APON to BPON

12.2.3.1 First Market Opportunities for PON

We have mentioned in the previous section the absence of real interest in APON systems in the community of telecommunications carriers and service providers during the 1990s. Since 2000, CLECs and ILECs began to reconsider their position about market opportunities for PON access systems. Both technical and economical factors explain this evolution. First, huge advances were made in the field of optical devices and systems during the 1990s. Optical amplifiers and WDM multiplexers are some of these key technologies at the origin of the explosion of the market of optical communications. Today, there is global consensus to admit that optical fibers are the most perennial medium for the next decades. First, the capacity and the range of an optical fiber may be extended to the Tbps range and to very long distance [7]. Second, several SMEs estimate that according to the increasing complexity of application softwares and to the increasing CPU speed of their computer equipment, the performance limits of their ADSL access systems should be noticeable in at most 5 years. Let us also recall that ADSL performance may be limited because of the electrical characteristics proper to each subscriber line. According to the FSAN objectives, BPON systems should be used by the carriers in the short or medium term to deliver multimedia services to both business and residential users.

12.2.3.2 BPON Access Systems Configuration

Since 2000, the FSAN members have introduced the concept BPON. A first motivation for this changing of acronyms is commercial: A BPON corresponds to an optical access system using ATM as an underlying layer but that is also able to provide to the application layers frame-oriented (e.g., frame relay and Ethernet) or circuit-switch–oriented services. A second motivation refers to an evolution of the technical characteristics of APON systems. The various aspects of BPON have been standardized by the ITU-T on the basis of numerous contributions of the members of the FSAN initiative. The G.983.1 recommendation [8] specifies the physical layer characteristics of a BPON system. The G.983.2 recommendation [9] specifies the management interface between the head and the leaves of the optical tree. The most recent recommendation

G.983.3 [10] dating from 2001 specifies that an additional wavelength band may be used on the system, for instance, to broadcast a video channel. Several additional optical carriers may eventually be used to increase the capacity of the system thanks to WDM multiplexers implemented within the OLT and the ONUs.

Weak points of APON systems are their lack of survivability in the case of fiber cut. The worse (and fortunately rare) scenario corresponds to a fiber cut between the OLT and the first splitting point. In that case, knowing that about 10 end users are connected to each ONU and that up to 64 ONUs are connected to the OLT, more than 600 customers are disconnected from the network. The *mean time to repair* (MTTR) could in that case fluctuate significantly from one carrier to an other, from a few days to a few weeks. A first objective of BPON systems is then to prevent such situations by considering protection and restoration techniques for PON systems. For that purpose, four different possible protection configurations have been investigated within the FSAN in order to protect and to restore the optical tree in case of fiber cut [11]. Only two of these configurations illustrated in Figure 12.6 have finally been selected in the ITU-T G.983.1 appendix IV recommendation.

Figure 12.6(a) depicts the first of these two protection alternatives. Protection is obtained by means of a duplication of the optical fiber between the OLT and the first splitting point and a duplication of the OLT itself. This portion of the optical tree concentrates all upstream and downstream traffic and is the most important of the global BPON infrastructure. One defines a primary OLT (LT-1) and a primary fiber link as illustrated in Figure 12.6(a). In the absence of a fiber cut or of OLT failure, the primary link and OLT are working whereas the secondary link and OLT (LT-2) are not activated. If the working fiber section is cut or if the working OLT fails, the OLT activates automatically switching from primary to secondary OLT. The primary and secondary links are not in the same conduit but are physically separated in order to reduce the probability of simultaneous cut on both of them.

Figure 12.6(b) illustrates the second protection configuration recommended by the ITU-T. Protection is obtained by means of a full duplication of the optical tree including the optical modems at the head-end and at the leaves. Both the primary and a secondary interfaces within the OLT and the ONUs are working in the absence of any fiber cut or interface failure. The information bits generated by the end users at each ONU are then duplicated and transmitted on the two parallel trees. At the OLT, only one version of the two copies of this same information is forwarded to the ATM switch. The same type of duplication is applied to downstream traffic at the OLT, a single copy of this traffic being exploited at the ONUs. In case of fiber cut or of interface failure, switching is thus carried out at the OLT in a very short delay. For this second protection configuration, a BPON may accommodate simultaneously mixed

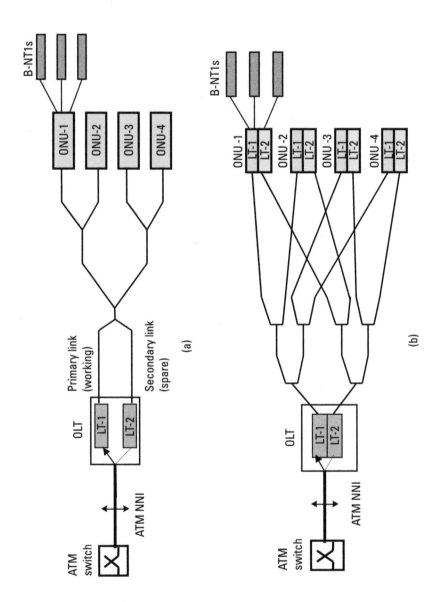

Figure 12.6 Two BPON protection configurations: (a) partial and (b) total.

situations with protected and nonprotected ONUs. A nonprotected ONU uses a single LT and a single optical fiber link up to the first combining point.

For the two BPON protection configurations recommended by the ITU-T, an upper-bounded switching time of 50 ms is specified. This delay already adopted for SONET/SDH equipment has been determined to prevent any noticeable interruption during a phone call by the two parties when restoration occurs.

12.2.3.3 Extended Services over BPON Access Systems

Confidentiality is carried out in APON systems by means of an encryption tool at the physical layer. Such tools being relatively expensive, BPON specifications replace it by a "churning" function that consists in permuting at the OLT and at reordering at the ONUs downstream ATM cells. The problem of confidentiality exists with downstream traffic, which is received by all the ONUs. This problem does not exist with upstream traffic. A churning key specifying the rules of these permutations are exchanged at most every second between the OLT and each ONU. The churning function is activated and deactivated at the setup and release of each ATM virtual path. Unlike APON systems, BPON systems use WDM for upstream and downstream traffic. The OLT and the ONUs are replaced in that case by extended versions called E-OLT and E-ONUs, the letter "E" standing for enhancement. Whereas a single carrier in the [1,530 nm, 1,560 nm] band[10] is specified for downstream traffic in APON systems, a BPON system considers a narrowed version of this bandwidth called *basic band* and limited to 20 nm ([1,480 nm, 1,500 nm]) for transporting downstream traffic. An additional bandwidth referred to as the *enhancement band* is defined either for the provision of *dense WDM* (DWDM) services or for the delivery of video services. These two bandwidth configurations are described in Figure 12.7.

As illustrated in Figure 12.7(a), the enhancement band used for the provision of very high-speed LLs by means of DWDM bidirectional channels corresponds to the [1,539 nm, 1,565 nm] band. Thanks to the transparency inherent in the all-optical nature of optical trees, these optical channels may be used for the transport of traffic with different bit rates and various data formats. Thus, a DWDM LL may be used, for instance, to transport asynchronous Ethernet frames, another one for the transport of ATM connections and a third one for the transport of frame relay frames. Figure 12.7(b) depicts the enhancement band used for video services delivery. In that case, the [1,530 nm, 1,560 nm] band is used only for downstream traffic. Again, thanks to all-optical transparency, both analog or digital video channels may be delivered to the end users via this channel.

---

10. This band is referred as the C-band. Two other bands can be used in optical fibers: the S-band around 1,470 nm and the L-band around 1,580 nm.

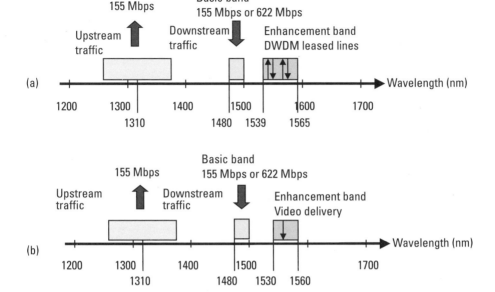

Figure 12.7 The two implementations of the BPON enhancement band: (a) for leased line services and (b) for video delivery services.

12.2.3.4 BPON Deployment and Evolution

Five major operators have deployed BPON access systems: NTT, BellSouth, British Telecom, France Telecom, and SBC. These operators either offer commercial services or plan to do so on their BPON infrastructure. NTT considers BPON access to be the most cost-effective method to provide ATM LLs services for business users in a FTTB approach. This operator has been very active for several years in the deployment of FTTH technologies with more than 2000 G.983-compatible APON or BPON access systems in Japan. Facing the commercial success of the offered service to the professionals, NTT expects the amount of installed systems to increase during the next few years, essentially for the benefit of enterprises.

These five operators have published a document summarizing the technical specifications adopted for their BPON access systems [12]. The five operators have all adopted UNI interfaces between terminal equipment and B-NT1s (see Figure 12.2) enabling either ATM25 at 25 Mbps, DS-3 at 45 Mbps and STM-1 at 155 Mbps. All of them intend to provide a new 10/100BASE-T interface. Three *ATM transfer capabilities* (ATCs) are taken into account in their systems: DBR, SBR, and GFR are commonly implemented.[11] All the

---

11. We have seen in the previous chapters that DBR and SBR ITU-T ATCs correspond to the CBR and VBR ATM Forum traffic classes, respectively.

operational systems operate with a symmetrical 155.52-Mbps bit rate. ATM cells conformance is estimated by means of *usage parameter control* (UPC) at the OLT for NTT, BellSouth, British Telecom, and SBC, whereas France Telecom implements UPC within the B-NT1s.

12.3 Ethernet Versus ATM in the Access

As mentioned in Section 12.1, two major evolutions have been observable for a few years in the field of access networks: first, the emergence of optical technologies, which was described in the first half of the chapter, and second, the progressive replacement of ATM packetization by Ethernet framing in the access networks.

12.3.1 Why EFM?

12.3.1.1 Ethernet Supremacy on the LAN Market

Section 9.2 discussed the motivations that have driven ILECs, service providers, and equipment manufacturers to adopt ATM in the access at the end of the 1980s. Ten years later, the supremacy of ATM in the access began to be discussed by some CLECs and service providers. Although ATM interface cards are now available on the market for PCs or work stations, such equipment remains relatively expensive in comparison to Ethernet 10BASE-T interface cards. Ethernet interface cards are currently largely predominant on the market of the enterprises. Indeed, successive versions of Ethernet LANs (IEEE 802.3 at 10 Mbps, IEEE 802.3u at 100 Mbps, IEEE 802.3z at 1 Gbps, and IEEE 802.3ae at 10 Gbps) represent today between 80% and 90% of LAN interfaces in the world. Within enterprises, as well as among residential users, Ethernet interfaces are quite familiar because of their simplicity of utilization.

12.3.1.2 ATM and Moore's Law

Although the ITU-T G.dmt or G.lite versions of the ADSL standard consider multiple protocol architectures in the access, only the versions based on an ATM transmission have been implemented by the vendors. The benefits inherent to ATM traffic management[12] are not in fact really exploited on ADSL links. The use of ATM in ADSL modems is essentially justified by its efficiency for multiplexing variable bit rate sources on the same physical channel. Because of the increasing speed of the hardware during these last 20 years (Moore's law specifies that processors speed doubles every 18 months), certain ATM chip designers estimate that segmentation of variable-size IP packets into 48-byte

---

12. Traffic management means intelligent routing depending on instantaneous network load and end users' QoS requirements.

blocks is not any longer necessary to manage such multirate multiplexing. In fact, a real debate occurs between the pros and the cons of ATM versus Ethernet in the access. Some vendors estimate that it will take several years before Ethernet reaches the same maturity [13].

12.3.1.3 From SONET/SDH Rings to IEEE 802.17

Several arguments may justify the emergence of Ethernet framing[13] in access networks. The objective of the EFM approach is to offer higher bit rates at lower costs to the end users. In fact, a similar approach is also considered for the metropolitan area. MANs located between access networks and core networks are traditionally based on SONET/SDH rings. The largest enterprises are in general directly connected to one of the ADMs of these rings by means of a digital LL or of an HDSL access system. SONET/SDH rings are historically managed by ILECs. They were designed 15 years ago for the transport of voice traffic and fixed bit-rate LLs. Bandwidth allocation on SONET/SDH rings is based on a granularity that lacks of flexibility. Indeed, a very limited range of bit rates are proposed to the customers: DS-0 at 64 Kbps, DS-1 at 1.5/2 Mbps (T1/E1), DS-3 at 45/34 Mbps (T3/E3). This rigidity imposes in general the customers to pay for excess unused bandwidth. On the other hand, the setting of a circuit through the SONET/SDH rings is much too long compared to the duration of the transmission of a few IP packets. In general, protection in SONET/SDH rings assumes a 50% unused capacity of the medium. For all these reasons, the IEEE has been promoting for a few years the emergence of a new generation of MANs known as the IEEE 802.17 standard [14]. The basic idea of this standard is to associate the efficiency of SONET/SDH protection and restoration techniques with a dynamic MAC protocol. The aim of this MAC protocol is to provide in real time bandwidth on-demand to asynchronous packet flows through the dual ring with a thin granularity. With granularities of a few kilobits per second being possible, the expression "fractional Ethernet" is sometimes used to characterize such a transport service. Interactive real-time services like voice or video may be offered on IEEE 802.17 MANs, thanks to specific QoS management tools, described in Section 12.3.3. The IEEE 802.17 standard was aimed to be finalized by March 2003.

12.3.1.4 EFMA

EFMA [14] and the IEEE currently promote the emergence of new standards for Ethernet-compatible access networks. In November 2002, the IEEE *LAN/MAN Standards Committee* (LMSC) formed the 802.3ah EFM task force

---

13. It is important to note that today most of the advanced versions of Ethernet LAN (at 100 Mbps or at higher speeds) are collision-free, thanks to the use of packet switches instead of a MAC protocol.

with the objective to develop the technical standards for various Ethernet-compatible access techniques. In parallel, the EFMA composed of several tens of companies (essentially equipment manufacturers) was created in December 2001. Many individual members of EFMA also contribute to the IEEE task force. It seems today that the roles of each of these two institutions are complementary. The technical aspects in terms of research and development are investigated within the IEEE task force whereas EFMA is more involved in industrial and commercial aspects. Thus, the EFMA discusses the various technical proposals of the IEEE task force in order to find a consensus among manufacturers and operators. Such a consensus is required to achieve interoperability of the future equipment of the market, especially with IEEE 802.17 rings. Three main variants are considered for EFM access systems:

- *EFM copper* (EFMC) based on point-to-point twisted copper pairs;
- *EFM fiber* (EFMF) based on point-to-point optical fibers;
- *EFM PON* (EFMP) based on point-to-multipoint optical fibers.

The members of the EFMA estimate that these three variants, or any combination between several of them, should satisfy the end users' service demands in the next few years.

12.3.2 EFM

12.3.2.1 EFM over Copper

Figure 12.8 illustrates the typical configuration of a EFMC access network. EFMC modems used at each end of the subscriber lines enable to transport Ethernet frames with a throughput of at least 10 Mbps over a range of at least 750m (2,500 ft). For that purpose, an adaptation of VDSL technology known as *Ethernet over VDSL* (EoVDSL) is under study.[14] According to the most efficient modulation techniques such as DMT modulation,[15] one knows that a 10-Mbps rate requires at least a 2-MHz band on the medium. At such frequencies, NEXT and FEXT represented in Figure 12.8 are very penalizing within binder cables. We have learned from FDM-ADSL that a simple way to circumvent this difficulty is to separate in the frequency domain upstream channels from downstream channels. Two FDM variants are under study for EoVDSL: [138 kHz, 3.75 MHz] and [3.75 MHz, 5.2 MHz] for upstream and downstream traffic, respectively, or [5.2 MHz, 8.5 MHz] and [8.5 MHz, 12 MHz] for upstream and downstream traffic, respectively. At the EO, a traffic concentrator (a gigabit

---

14. VDSL modems were discussed in Chapter 7.

15. DMT modulation is used in ITU-T ADSL modems.

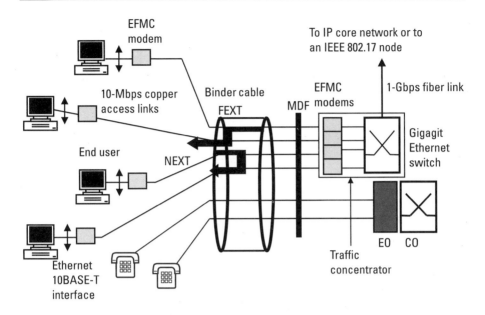

Figure 12.8 EFMC network configuration.

Ethernet switch) forwards Ethernet frames received from the various end users either toward the nearest IP router or toward an access node of an IEEE 802.17 metropolitan area network. In order to be able to serve end users located further than 750m (2,500 ft) from the EO, some operators consider using the DLC technique (see Section 2.6 and Figure 2.13). In that case, EFMC modems at the CO side are located in the RT.

Another approach of EFMC known as *Ethernet 100BASE-CU* is also under study. This approach is inspired by the SDMT modulation technique developed for VDSL access (see Section 7.3.6.2). The Ethernet 100BASE-CU technique aims at a reduction of NEXT within binder cables by scheduling alternate data transmissions in both directions. The basic idea of 100BASE-CU is that pending Ethernet frames within a customer premises are aggregated in bursts before being transmitted at 100 Mbps onto the copper medium. In order to prevent NEXT, the transmission of the bursts generated by the various EFMC modems transmissions are scheduled in a way that, ideally, a single modem generates upstream traffic at a time. This technique is then not efficient in terms of bandwidth utilization.

Figure 12.9 illustrates the protocol stacks used at the different nodes of an EFMC access network. EFMC access systems do not need any MAC protocol because of the clear separation between upstream and downstream channels. One assumes in Figure 12.9 that Ethernet links are also adopted beyond the

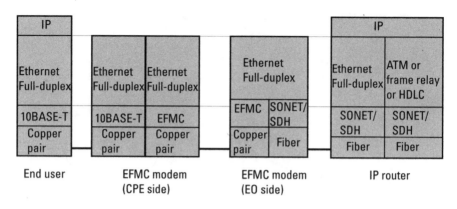

Figure 12.9 Protocol stacks involved in an AFMC access.

access network as far as to the nearest IP router. This is, for instance, the case if this IP router is directly connected to a node of an IEEE 802.17 MAN. EFMC modems operate in full-duplex (i.e., upstream and downstream frames are transported simultaneously on the subscriber line without contention). The terminal equipment at the customer premises also operates in full duplex, with two copper pairs being used between the terminal and the input port of the user's EFMC modem in order to prevent any collision.

12.3.2.2 EFM over Fiber

Like EFMC, EFMF is based on a point-to-point architecture. Whereas EFMC access rates are in the 10-Mbps range, EFMF access systems operate at speeds between 100 Mbps and 1 Gbps for a range up to 10 km. Unlike EFMC, which reuses installed copper pairs in the local loop, EFMF needs the setup of single-mode optical fibers between each customer premise and the EO. Figure 12.10 depicts the configuration of an EFMF access network.

An EFMF access modem is designed to serve simultaneously several end users. As it is described in the figure, three types of configurations are considered for the customer premises. New multiunit residential areas or multitenant buildings may share the same EFMF modem. This assumes a local cabling between each individual piece of end equipment and the EFMF modem. The specification of such a cabling system is out of the scope of EFM specifications. The technical solutions for home networking presented in Section 8.3 may be applied to the EFMF environment. Each EFMF modem is directly connected to a traffic concentrator located close to the CO by means of two contra-directional optical fibers. Like in the case of EFMC, the traffic concentrator is typically an Ethernet switch, in this case most likely at gigabit speeds. Multiple offices of an enterprise may also share a common EFMF modem. In this case,

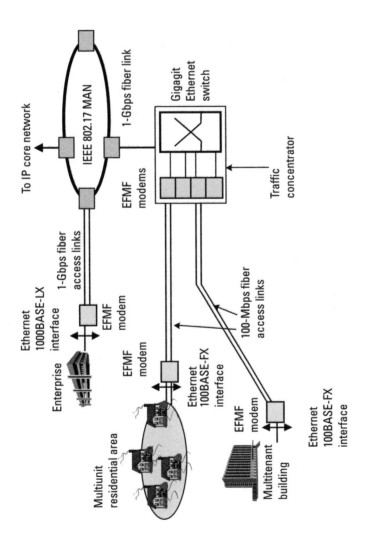

Figure 12.10 EFMF network configuration.

the enterprise gets a better service by being directly connected to an aggregation node, typically a SONET/SDH ADM or an IEEE 802.17 access node, by means of a 1-Gbps optical fiber pair.[16] Two versions are considered for the physical link between an EFMF modem and the traffic concentrator. In terms of civil engineering, it makes no difference whether one installs one or two fibers in the access network, the inherent cost of civil engineering remaining the same. Meanwhile, the EFMF working group also proposes using a single fiber for each end user with two optical channels λ_{up} and λ_{down} (for instance, $\lambda_{down} = 1,550$ nm and $\lambda_{up} = 1,310$ nm) requiring distinct transceivers at each end of the optical link. Like EFMC access systems, EFMF access systems assume a full-duplex transmission between the user's EFMC modem and the traffic concentrator.

According to the adopted data rate for the EFMF access, two different physical interfaces inspired by existing standards are considered for the physical layer. In the case of 100-Mbps access, the unidirectional fibers are based on the Ethernet 100BASE-FX standard. One uses in that case the 4B/5B+NRZI line code that associates in a first step to each suite (a block) of 4 bits issued by the MAC layer a symbol of 5 bits. The objective of such a line code referred as block code is to facilitate clock recovery at the receiver by imposing a minimum rate of transitions "1" to "0" or "0" to "1." After the 4B/5B encoding, the NRZI encoder associates to each bit a voltage level either "+V" or "−V," a "1" inverting the polarity of the voltage of the previous bit, a "0" keeping the same polarity as the previous bit. The 4B/5B + NRZI line code is used since about 2 decades in *fiber distributed data interface* (FDDI) LANs. In the case of 1-Gbps access, unidirectional fibers based on the Ethernet 1000BASE-LX standard are considered. This gigabit Ethernet standard reuses *Fibre Channel* (FC)[17] specifications based on the 8B/10B line code. The 8B/10B is a block code that has been developed specifically for gigabit-per-second transmission. Unlike traditional block codes, the 8B/10B coder determines the symbol of 10 bits to associate to an input block of 8 bits, not only on the basis of the value of these 8 bits, but also on the basis of the symbols adopted for a certain amount of previous input blocks. The objective is again to facilitate clock recovery at the receiver.

In terms of transceivers, one of the main challenges of EFMF is to reuse the same laser diodes as in FDDI for the 100-Mbps version, or as in FC-0 and FC-1 Fibre Channel specifications for the 1-Gbps version. The major constraint imposed to these specifications is the larger range of access networks than LANs. The required gain in distance is supposed to be achieved by means of higher voltages applied to the ports of laser diodes. The higher this voltage, the larger the rate of generated photons per bit and the better SNR at the receiver. The

16. This remark confirms the fact that today, large enterprises are more connected at the MAN level than at the access network level.

17. Fibre Channel is a gigabit LAN developed to connect terminals to a supercomputer.

other side of the coin of this approach is a reduction of the life duration of optical devices (between 5 to 10 years). Typically, 100BASE-FX and 1000BASE-LX standards have been specified for a LAN environment with a range of about 5 km, whereas the EFMF standard targets at least a 10-km range.

As a reminder, Figure 12.11 recalls the main physical layer standards adopted for the 100-Mbps and for the 1-Gbps versions of Ethernet LAN networks. The half-duplex version of the MAC protocol takes into account the possibility of collisions on the medium, upstream and downstream traffic being transmitted on the same physical medium and in the same frequency band. In that case, contention resolution is carried out by means of the CSMA/CD protocol. Today, the most popular versions of 100-Mbps and 1-Gbps Ethernet are based on full-duplex MAC protocols. These protocols are collision-free, with different physical resources being used by upstream and downstream traffic. In fact, only the Ethernet frame format is reused in EFMF, not the CSMA/CD MAC protocol.

12.3.2.3 EFM over PON

We have underlined in Section 12.3.1.2 the main arguments for replacing ATM by Ethernet in the access and in the metro. As shown in Figure 12.12, an EFMP also known as an EPON is quite similar in terms of topology to an APON. Like in APON access systems, the optical modems located at the head-end and at the

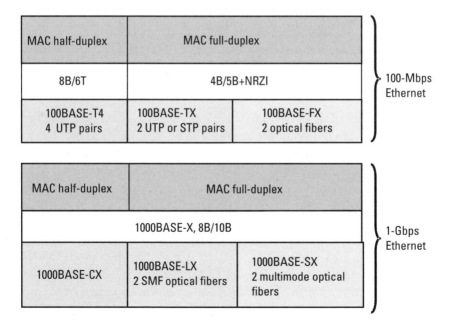

Figure 12.11 Line codes of 100-Mbps and 1-Gbps Ethernet.

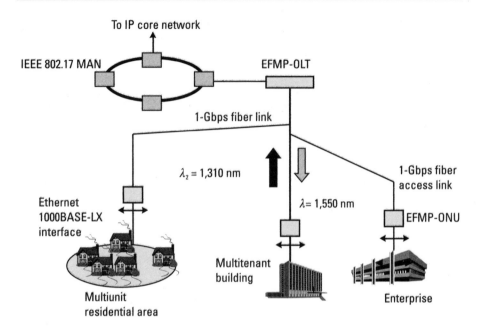

Figure 12.12 EFMP network configuration.

leaves of the optical tree are called EFMP-OLT and EFMP-ONUs, respectively. Distance ranging, power ranging, and clock ranging procedures quite similar to those described in APONs (see Section 12.2.2.2) are applied to EPONs. At least 64 EFMP-ONUs are connected to the same EFMP-OLT. In the worst case power budget, the range of an EFMP is fixed to 10 km. The same couple of wavelengths as in APON are considered in EFMP—that is, $\lambda_1 = 1,550$ nm and $\lambda_2 = 1,310$ nm.

One assumes in an EFMP a symmetrical 1.25-Gbps data rate, with future updating at 10 Gbps thanks to the emerging IEEE 802.3ae standard. The most advanced version of Ethernet known as IEEE 802.3ae operates at 10 Gbps. Only optical transmission using the 64B/66B line code is considered for this very high-speed version of Ethernet. Two versions of chip sets suited either to the LAN or WAN environment are specified at the physical layer. A *WAN interface sublayer* (WIS) is added to the 64B/66B line encoder in order to include a simplified SONET/SDH framing at the physical layer of the IEEE 802.3ae. In the upstream, the 1.25-Gbps EPON capacity is shared by the various ONUs. If the specification of the physical layer of EFMP access systems should be finalized in the very next months, the detailed aspects of the associated MAC protocol remain open to discussion within the IEEE 802.ah working group. Unlike the G.983 standardized APON MAC protocol, the EFMP MAC protocol specifies variable size MAC-PDUs for both the upstream and downstream. Both

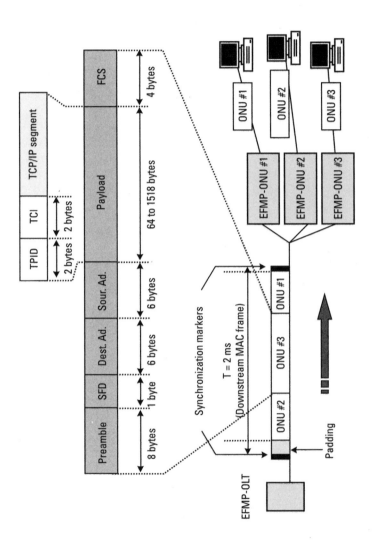

Figure 12.13 Downstream traffic format in an EFMP access system.

upstream and downstream traffic (variable size is TCP/IP packets) are encapsulated in Ethernet frames with a frame format recalled in Figure 12.13 (the meaning of the TPID and TCI fields is given in Section 12.3.3).

Downstream traffic access the medium at the OLT by means of a simple TDM policy. Every 2 ms, a 1-byte synchronization marker is sent by the EFMP-OLT. Between two synchronization markers (see Figure 12.13), several IEEE 802.3 frames addressed to various end users may be transported in a MAC frame. By commodity, we introduce the term *MAC frame* in order to distinguish the frames proper to the EPON system from the Ethernet frames generated by the end users. Nevertheless, this expression is not used in the standardization documents. At the EFMP-ONUs, the IEEE 802.3 frames are extracted from the MAC frames before being forwarded to the right ONUs (addressing rules are explained further).

The first version of the EPON MAC protocol is based on a TDMA access technique. Figure 12.14 illustrates the principle of this TDMA access applied to upstream traffic. Thanks to the distance and clock-ranging procedures, all the EFMP-ONUs are time-synchronized. Time is organized into a repetitive structure of N ONU-specific time slots, where N stands for the amount of EFMP-ONUs in the system. A set of N time slots represents a 2-ms upstream MAC frame. Time slot i is dedicated to EFMP-ONU i. During its time slot, an EFMP-ONU is allowed to send several variable-size IEEE 802.3 frames. The size of a time slot is at least equal to the maximum frame size. A time slot begins and ends with a header and a guard time, respectively. If the size of a pending IEEE 802.3 frame is such that the cumulated amount of bytes to be transmitted at a given EFMP-ONU in the next time slot exceeds the time-slot capacity, then the transmission of this IEEE 802.3 frame is delayed until the next available time slot. The remaining unused capacity of the current time slot is filled with padding bytes. In order to prevent spare capacity in upstream time slots, a reordering procedure of the pending IEEE 802.3 frames at each EFMP-ONU has been proposed. For instance, if an IEEE 802.3 frame is too long to be transported in the next upstream time slot, the considered EFMP-ONU looks among its pending IEEE 802.3 frames for one that could fit in the available space. This means that IEEE 802.3 frames buffered in an EFMP-ONU are not served according to a FIFO discipline.

The principle of fixed and periodic time slots dedicated cyclically to the N EFMP-ONUs is not efficient in the context of bursty traffic. This is the reason why a more dynamic bandwidth allocation by the EFMP-OLT to the EFMP-ONUs is preferable (upstream and downstream bandwidth is under the control of the head of the tree). As in APON systems, this dynamic access is based on a request/permit mechanism and on variable-size time slots (see Figure 12.15). Ideally, the requests R_1, R_2, ..., R_N sent by EFMP-ONU #1, #2, ..., #N are satisfied by permits P_1, P_2, ..., P_N in such a way that upstream channel utilization is

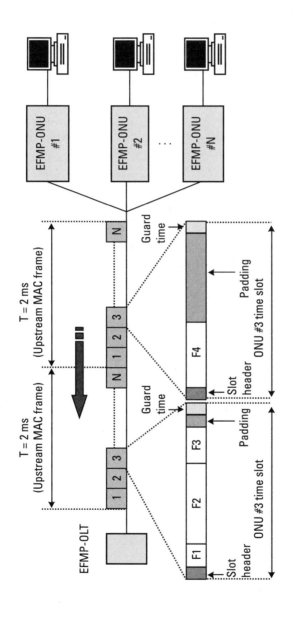

Figure 12.14 Upstream traffic format in an EFMP access system.

optimized. We have underlined in Section 12.2.2.4 dedicated to APON systems the trade-off necessary between bandwidth utilization efficiency and the satisfaction of real-time constraints of certain applications.

Several aspects of the EFMP MAC protocol have already been proposed to the IEEE for approval. One of these proposals called *interleaved polling with adaptive cycle time* (IPACT) optimizes the medium utilization ratio thanks to a request/permit mechanism based on three parameters: the ONU address, the length of the MAC-PDU generated by this ONU and the round-trip time between this ONU and the OLT [15]. Like with the APON-MAC protocol, one assumes with the IPACT protocol the transmission of a request for the transmission of bursts of Ethernet frames by the same ONU. Whereas permits are sent by batches[18] by the OLT in an APON PLOAM cell, the date of transmission of a permit by the EFMP-OLT to a given EFMP-ONU is fixed on the basis of the RTT inherent to this EFMP-ONU. The basic idea is that, as soon as it receives a permit, the EFMP-ONU sends its batch in a way that it arrives at the EFMP-OLT just after the previous batch received from another EFMP-ONU. This approach is illustrated by the scenario given in Figure 12.15.

Today, a consensus seems to be found on another MAC protocol called *Multipoint Control Protocol* (MPCP) [16]. The MPCP protocol uses MAC control frames for request/grant transmissions. MPCP assumes a permanent synchronization between the EFMP-OLT and the EFMP-ONUs. Like IPACT, MPCP schedules grants' transmission at the EFMP-OLT in order to compensate for RTT disparity between the various EFMP-ONUs. Unlike IPACT, which aims at a dynamic bandwidth allocation, MPCP assumes that this allocation is carried out by higher layers and is then vendor-dependent.

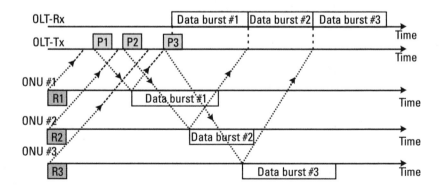

Figure 12.15 An example of scheduling in an EFMP access system.

---

18. Section 12.2.2.3 explained that these batches are sent in the PLOAM ATM cells of the
    G.983 frames.

Figure 12.11 shows that two types of media are possible for Ethernet, either shared or half-duplex (that is, with possible collisions) and full-duplex, which is collision-free. The EFMA has dedicated particular attention to the bridging function that can be operated by an EPON system. The fact that a duplex EPON system is collision-free means that two end users connected to different EFMP-ONUs cannot communicate with each other without a layer 3 routing function. The motivation of the EFMA is to implement an additional sublayer in APON systems below the MAC layer in order to emulate either a *shared medium* (SME) or a *point-to-point connection* (P2PE). The SME or the P2PE operation is carried out by means of a tagging procedure. These tags, called link IDs, are inserted between the preamble and the SFD field of the IEEE 802.3 frames by an SME or a P2PE MAC sublayer. The EFMP-OLT is in charge of the allocation of unique link-IDs to the EFMP-ONUs during an initialization procedure.

The P2PE mode is in fact the mode implicitly assumed in the previous sections, a downstream IEEE 802.3 frame being received logically by the MAC layer of a single EFMP-ONU [see Figure 12.16(a)]. Physically a downstream frame is received by all the ONUs. In the case of point-to-multipoint downstream communication like a video diffusion, the EFMP-OLT must duplicate the video packets as many times as the number of interested subscribers. The EFMP-OLT uses as many MAC ports as the number of EFMP-ONUs in the system. An upstream IEEE 802.3 frame cannot be forwarded directly by the EFMP-OLT to another EFMP-ONU belonging to the same EPON system. Such a forwarding requires a layer 3 routing carried out by the nearest IP router in the core network [see Figure 12.16(b)].

The SME mode requires a single MAC port at the EFMP-OLT. Downstream IEEE 802.3 frames are broadcast to the MAC layer of the various EFMP-ONUs [see Figure 12.16(c)]. The main motivation of this mode of operation is the possibility to forward directly an upstream IEEE 802.3 frame sent by a given EFMP-ONU to all the other EFMP-ONUs, like on an Ethernet bus, without requiring an IP router [see Figure 12.16(d)]. For that purpose, a bridging function is carried out at the EFMP-OLT. The SME mode is then better suited to broadcast services such as video distribution than the P2PE mode. The drawback of the SME mode is that if a return channel is considered (as in the case of VoD), a lot of bandwidth may be spared because of the bridging function at the EFMP-OLT.

In order to associate the benefits of these two modes, mixed configurations have been proposed by the EFMA (see Figure 12.17). If N stands for the amount of EFMP-ONUs, $N+1$ MAC ports are implemented within the EFMP-OLT, N for P2PE and 1 for SME. In each EFMP-ONU, two MAC ports are implemented for P2PE and SME, respectively.

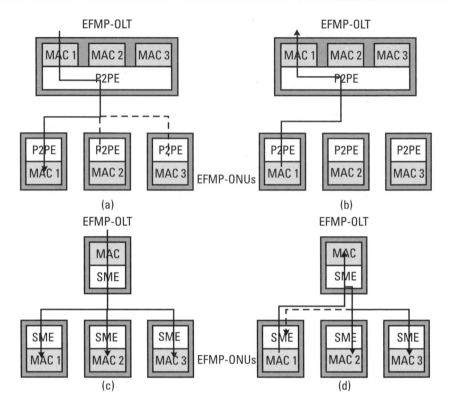

Figure 12.16 Illustration under the P2PE mode of (a) downstream and (b) upstream traffic and illustration under the SME mode of (c) downstream and (d) upstream traffic.

12.3.3 QoS in Ethernet-Based Networks

The current specifications of EFMP access systems need complementary work to provide QoS guarantees to real-time applications. The MAC layer specification of these systems is still open to discussion. Two approaches are possible in this matter. Either QoS will be included at the frame level in the MAC protocol itself (that is, the approach adopted for APON systems), or QoS will be managed by mechanisms implemented in the higher layers. This second approach seems to be the one adopted within IEEE 802.ah task force. Three alternatives to ATM are today available in order to provide QoS in Ethernet-based access systems:

- *VLAN:* Emerging standards such as IEEE 802.1p, IEEE 802.1q, IEEE 802.1s, and IEEE 802.1r facilitate QoS provisioning and traffic management in *virtual LANs* (VLANs). Traditionally, a VLAN is used in private network environment in order to isolate certain traffic categories

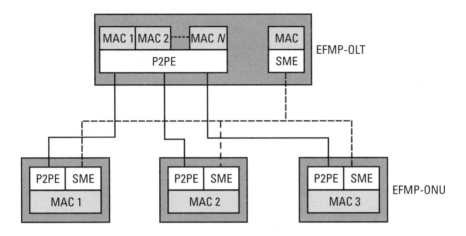

Figure 12.17 A mixed SME and P2PE configuration.

associated, for instance, with different group of users. Two motivations justify the use of VLAN in an enterprise: security and traffic engineering purposes. The VLAN concept relies on the IEEE 802.1q protocol, which consists of adding in the payload of an Ethernet frame (see Figure 12.13) two headers called *tag protocol identifier* (TPID) and *tag control information* (TCI). Each of these headers is 2 bytes long. The TPID field is set to a given value in the case of Ethernet frames. The main role of the TCI field is to identify the VLAN to which belongs the Ethernet frame. A *VLAN identifier* (VID) is coded on 12 bits. This means that at most 4,095 VLANs are possible. In a public network environment, this value is a strong scalability constraint. In order to enable enterprises to continue to manage their own VLANs via public Ethernet-based networks, a possible solution is today the VLAN stacking approach (in a VLAN, one creates another VLAN). This approach is in practice a limited response to the scalability problem.

- *IP-VPN:* Another approach for QoS provisioning in EFM networks can be based on IP-VPNs. This solution, depending partially on end user equipment, is more scalable than the VLAN approach. In the case of IP-VPN, there is no inherent addressing limitation. Thanks to the IPsec protocol, data encryption and end user authentication are carried out in IP-VPNs. We have referred in Section 12.2.3.3 to an encryption technique in EPON access systems. A carrier may decide to consider a single level of encryption, either at the EPON level or at the IP level. The latter approach seems more coherent because it is applied end-to-end. The principles of IP-VPN are described in Section 1.4.

- *MPLS:* The MPLS technique (see Section 1.3.2.3) consists of implementing a control plane in IP networks between layer 2 and layer 3. In MPLS clouds, layer 3 routing is replaced by a label swapping mechanism. In current implementations, label swapping is carried in layer 2 by ATM switches. The same principles of MPLS can be adapted to the case of a layer 2 made with Ethernet switches. MPLS is seen as a promising solution for resolving the scalability problem of VLANs in public Ethernet-based networks. The basic idea is to encapsulate in the access or at the MAN-level Ethernet frames into MPLS LSPs.

Current investigations in the field of OAM aim to establish in real-time point-to-point and point-to-multipoint connections. A great flexibility in bandwidth allocation is expected with a 64-Kbps increment from 10 Mbps up to 1 Gbps (fractional Ethernet concept).

12.4 Conclusion and Perspectives

From 1990 to 2000, several solutions based either on existing copper wires, coaxial cables, power line communications, radio access systems and optical fibers have been proposed for high bit rates in the local loop. Most of these proposals adopt the use of the ATM technique as a multiplexing layer. In terms of optical access networks, the ITU-T promotes BPON access systems as successors of APONs. BPONs introduce network resiliency in access networks and enlarge the range of offered services thanks to the concept of enhancement band. We have underlined the reasons why Ethernet-based techniques are considered as potential competitors to ATM in the access and in the metro. Since 2000, both the EFMA and the IEEE 802.ah working group are promoting three variants: EFMC, EFMF, and EFMP. Like VDSL, EFMC suffers from a limited range along the copper infrastructure. EFMP is more cost-effective than EFMF because of its point-to-multipoint configuration. EFMP access systems will be a direct concurrent of APON and BPON access systems. In comparison to APONs and BPONs, the P2PE and SME bridging modes of EFMPs open new perspectives in terms of services. The main benefit of EFM and IEEE 802.17 in the access and in the metro, respectively, is dynamic bandwidth allocation with a thin granularity. Nevertheless, end-to-end QoS provisioning in Ethernet-based networks remains an open problem due to the lack of scalability of VLANs. Ethernet VLANs over MPLS LSPs could be the solution for the next few years. The final version of the EFM standard was expected by the end of 2003.

References

[1] Stern, J. R., et al., "The Full-Service Access Network Requirements Specifications," *8th International Workshop on Optical Access Networks*, Atlanta, GA, March 2–5, 1997.

[2] Kramer, G., and G. Pesavento, "Ethernet Passive Optical Networks (EPON): Building a Next-Generation Optical Access Network," *IEEE Communications Magazine*, Vol. 40, Issue 2, February 2002.

[3] Angelopoulos, J., E. Fragoulopoulos, and E. N. Protonotarios, "Efficient Support of Best Effort Traffic in Passive Tree Local Loops," *SPIE Journal*, No. 2609, 1996, pp. 182–191.

[4] Gagnaire, M., and S. Stojanovski, "Stream Traffic Management over an ATM Passive Optical Network," *Computer Networks Journal*, Vol. 32, No. 5, April 2000, pp. 571–586.

[5] Stojanovski, S., M. Gagnaire, and R. Hoebeke, "Providing GFR Guarantees for TCP/IP Traffic over APON Access Systems," *IFIP Broadband Communications Networking 2000 Conference*, Paris, France, May 2000.

[6] Stojanovski, S., M. Gagnaire, and R. Hoebeke, "Support for the GFR Transfer Capability over PON Access Systems," *IEEE ATM Workshop*, Kochi City, Japan, May 1999.

[7] Gagnaire, M., "All-Optical Networks: Assets and Challenges," Special issue of the *Business Briefing Journal on Global Optical Communications, International Society for Optical Engineering (SPIE) and the Optical Internetworking Forum (OIF)*, World Markets Research Center, June 2001, pp. 18–22.

[8] ITU-T Recommendation G.983.1, Broadband Optical Access Systems Based on Passive Optical Networks (PON), 1998.

[9] ITU-T Recommendation G.983.2, ONT Management and Control Interface Specification for ATM-PON, 1999.

[10] ITU-T Recommendation G.983.3, Broadband Optical Access Systems with Increased Service Capability by Wavelength Allocation, 2001.

[11] Effeneberger, F. J., H. Ichibangase, and H. Yamashita, "Advances in Broadband Passive Optical Networking Technologies," *IEEE Communications Magazine*, December 2001, pp. 118–124.

[12] Ueda, H., et al., "Development Status and Common Technical Specifications for a BPON System," *IEEE Communications Magazine*, December 2001.

[13] Van de Voorde, I., et al., "Carrier-Grade Ethernet: Extending Ethernet into Next Generation Metro Networks," *Alcatel Telecommunications Review*, Third Trimester 2002.

[14] Cole, N., et al., "Global Optical Communications: An Analysis of the Global Optical Communications Industry and Perspective for the Future," *Resilient Packet Rings for Metro Networks*, World Market Series, Business Briefing publications, July 2001.

[15] Kramer, G., B. Mukherjee, and G. Pesavento, "IPACT: A Dynamic Protocol for an Ethernet PON (EPON)," *IEEE Communications Magazine*, Vol. 40, Issue 2, 2002, pp. 66–73.

[16] Gummalla, A., et al., "MPCP: A State of the Art," IEEE 802.ah working group, IEEE, http://ieee.org, 2002.

Selected Bibliography

Ethernet in the First Mile Alliance (EFMA), http://www.efmalliance.org.

Gagnaire, M., "From Asynchronous Transfer Mode to Ethernet in Optical Access Networks," Special issue of the *Business Briefing Journal on Global Optical Communications, International Society for Optical Engineering (SPIE) and the Optical Internetworking Forum (OIF), World Markets Research Center,* June 2002, pp. 36–42.

Acronyms

AAA   authentication, authorization, and accounting

AAL   ATM adaptation layer

ABR   available bit rate

ADM   add-drop multiplexer

ADPCM   adaptive delta pulse code modulation

ADSL   asymmetrical digital subscriber line

AF   assured forwarding

AGC   automatic gain control

AIU   air interface unit

AMI   alternate mark inversion

ANSI   American National Standard Institute

APON   ATM over passive optical network

ARQ   automatic repeat request

ATC   ATM transfer capability

ATU-C   ADSL transmission unit central office side

ATU-R   ADSL transmission unit remote side

AWG   American wire gauge

AWGN   additive white Gaussian noise

BAS   broadband access server

BBCP   broadband bearer control protocol

BCC   bearer channel connection

BER   bit error rate

BLES   broadband loop emulation service

B-NT   broadband network termination

BPF   bandpass filter

BPON   broadband passive optical network

BPSK   binary phase shift keying

CAC   call admission control

CAP   carrierless amplitude and phase

CATV   cable television

CBR   constant bit rate

CB-WFQ   class-based weighted fair queuing

CDMA   code division multiple access

CES   circuit emulation service

CHAP   Challenge-Handshake Authentication Protocol

CID   channel identifier

CLEC   competitive local exchange carrier

CM   cable modem

CMTS   cable modem termination system

CO   central office

COFDM   code orthogonal frequency division multiplexing

COT   central office terminal

CPA   clock phase alignment

CPE   customer premises equipment

CRC   cyclic redundancy checksum

CR-LDP   Constraint-based Routing Label Distribution Protocol

CSA   carrier serving area

CSMA/CA   carrier sense multiple access with collision avoidance

CTI   computer telephony integration

CVoDSL   channelized voice over DSL

DAVIC   Digital Audio Video Council

DBR   deterministic bit rate

DCA   dynamic channel allocation

DCT   discrete cosine transform

DDF   digital distribution frame

DECT   digital enhanced cordless telephone

DF   distribution frame

DFE   decision feedback equalizer

DFT   discrete Fourier transform

DHCP   Dynamic Host Configuration Protocol

DLC   digital loop carrier

DMT   discrete multitone

DQPSK   differential quadrature phase shift keying

DS-CDMA   direct sequence code division multiple access

DSCP   differentiated services code point

DSL   digital subscriber line

DSLAM   digital subscriber line access multiplexer

DSSS   direct sequence spread spectrum

DVB   digital video broadcasting

DWDM   dense wavelength division multiplexing

DWMT   discrete wavelet multitone

EC   echo cancellation

EDFA   Erbium-doped fiber amplifier

EF   expedited forwarding

EFM   Ethernet in the first mile

EFMA   Ethernet-in-the-First-Mile Alliance

EFMC   Ethernet in the first mile over copper

EFMF   Ethernet in the first mile over fiber

EMC   electromagnetic compatibility

EPON   Ethernet PON

EO   end office

EoVDSL   Ethernet over VDSL

EPG   electronic program guide

ETSI   European Telecommunication Standardization Institute

FC   Fibre Channel

FCC   Federal Communications Commission

FDD   frequency division duplexing

FDDI   fiber distributed data interface

FDL   full digital line

FDM   frequency division multiplexing

FDMA   frequency division multiple access

FEC   forward error correction

FEXT   far-end crosstalk

FFT   fast Fourier transform

FHSS   frequency hopping spread spectrum

FMT   filtered multitone

FP   fixed part

FSAN   full-service access network

FSK   frequency shift keying

FTTB   fiber to the building

FTTC   fiber to the curb

FTTH   fiber to the home

GFR   guaranteed frame rate

HAAP   high-altitude aerial platform

HDLC   high-level data link control

HDSL   high bit rate digital subscriber line

HFC   hybrid fiber coaxial

IAD   integrated access device

IAP   Internet access provider

ICI   interchannel interference

IDFT   inverse discrete Fourier transform

IDSL   integrated services digital network digital subscriber line

IEEE   Institute of Electrical and Electronics Engineers

IETF   Internet Engineering Task Force

IFFT   inverse fast Fourier transform

IGMP   Internet Group Management Protocol

ILEC   incumbent local exchange carrier

IP   Internet Protocol

ISDN   integrated services digital network

ISI   intersymbol interference

ISP   Internet service provider

ITU   International Telecommunication Union

L2CAP   Logical Link Control and Adaptation Protocol

L2F   layer 2 forwarding

L2TP   Layer 2 Tunneling Protocol

LAC   L2TP access concentrator

LAN   local area network

LCP   Link Control Protocol

LDP   Label Distribution Protocol

LES   loop emulation service

LEX   local exchange

LL   leased line

LLC   logical link control

LMDS   local multipoint distribution service

LMP   Link Manager Protocol

LNS   L2TP network server

LPF   lowpass filter

LSP   label-switched path

LT   line termination

MAC   medium access control

MAN   metropolitan area network

MBS   multiburst slot

MCM   multicarrier modulation

MCNS   multimedia cable network system

MCR   minimum cell rate

MDF   main distribution frame

MGC   media gateway controller

MPCP   Multipoint Control Protocol

MPEG   Moving Pictures Expert Group

MPEG2-TS   MPEG2 transport stream

MPLS   multiprotocol label switching

MTTR   mean time to repair

MVL   multiple virtual line

NAP   network access provider

NAS   network access server

NCP   Network Control Protocol

NEXT   near-end crosstalk

NGN   next generation network

NIC   network interface card

NID   network interface device

NIU   network interface unit

NRZ   nonreturn-to-zero

NSF   National Science Foundation

NSP   network service provider

NT   network termination

OAM   operation and maintenance

OFDM   orthogonal frequency division multiplexing

OLT   optical line termination

ONU   optical network unit

OSGI   open-services gateway initiative

P2PE   point-to-point emulation

PACS   public access communication system

PAM   pulse amplitude modulation

PAN   personal area network

PAP   Password Authentication Protocol

PBX   private branch exchange

PBRS   pseudobinary random sequence

PCM   pulse-coded modulation

PCR   peak cell rate

PDA   personal digital assistant

PDH   plesiochronous digital hierarchy

PDN   premises distribution network

PHS   personal handy-phone system

PLC   power line communication

PLL   phase-locked loop

PLOAM   physical layer operation and maintenance

PMD   physical medium–dependent

PON   passive optical access network

POP   point of presence

POTS   plain old telephone system

PP   portable part

PPP   Point-to-Point Protocol

PPTP   Point-to-Point Tunneling Protocol

PSD   power spectral density

PSK   phase shift keying

PSTN   public switched telephone network

PTA   PPP terminated aggregation architecture

PVC   permanent virtual circuit

PVP   permanent virtual path

QAM   quadrature amplitude and phase modulation

QoS   quality of service

QPSK   quadrature phase shift keying

RADSL   rate-adaptive digital subscriber line

RAM   remote access multiplexer

RBOC   regional Bell operating company

RED   random early discard

RFC   Request for Comment

RFI   radio frequency interference

RFP   radio fixed part

RR   repeater/regenerator

RS   Reed-Solomon

RSSI   radio signal strength intensity

RSVP   Resource Reservation Protocol

RT   remote terminal

RTCP   Real-Time Transport Control Protocol

RTP   Real-Time Transport Protocol

RTT   round-trip time

Rx   receiver

SABM   set asynchronous balanced mode

SAD   speech activity detector

SAN   storage area network

SBR statistical bit rate

SDMT synchronized discrete multitone

SDH synchronous digital hierarchy

SDP Service Discovery Protocol

SDSL single-pair digital subscriber line

SHDSL symmetric high bit rate digital subscriber line

SIP Session Initiation Protocol

SME shared-medium emulation

SNEXT self near-end crosstalk

SNMP simple network management protocol

SNR signal-to-noise ratio

SOHO small office home office

SONET synchronous optical network

SS7 signaling system number 7

STM synchronous time multiplexing

SVC switched virtual circuit

TA terminal adaptor

TC-F transmission convergence with fast buffer

TC-I transmission convergence with interleaved buffer

TCM trellis-coded modulation

TCP Transmission Control Protocol

TC-PAM trellis-coded pulse amplitude modulation

TDD time division duplexing

TDMA time division multiple access

TEQ time equalization

TE-RSVP traffic engineering resource reservation protocol

Tx transmitter

UA unnumbered acknowledgment

UAWG universal ADSL working group

UBR   unspecified bit rate

UDP   User Datagram Protocol

UMTS   Universal Mobile Telecommunication System

UNI   user network interface

UPC   usage parameter control

UTP   unshielded twisted pair

VBR   variable bit rate

VDSL   very high-speed digital subscriber line

VoD   video on demand

VoDSL   voice over DSL

VoIP   Voice over Internet Protocol

VoMBN   voice over multiservice broadband network

VPN   virtual private network

VTOA   voice traffic over ATM

WACS   wireless access communication system

WAN   wide area network

WDM   wavelength division multiplexing

WFQ   weighted fair queuing

WiFi   wireless fidelity

WITL   wireless in the loop

WRS   wireless relay station

WRED   weighted random early detection

About the Author

Maurice Gagnaire is a professor in the Computer Science and Networks Department of ENST in Paris, France. After starting his career in 1982 as a physicist in optical communications, he worked as an associate professor on the design and performance evaluation of MAC protocols for high-speed LANs and MANs from 1985 to 1995. Since 1995, he has progressively reoriented his teaching and research activities toward two distinct topics: all-optical WDM network planning and traffic management in broadband access systems. Most of his research activities are funded by industrial contracts. He is the author or coauthor of about 70 technical papers in IEEE or IFIP conferences and journals. He has been a coguest editor of two special issues of the *International Journal of Computer and Telecommunications Networking* and of the *Proceedings of the IEEE*, respectively. He is the author of a book on xDSL access systems, *Boucles d'acces Haut debit,* and coauthor of a book on ATM networks, *Resaux haut debit: ATM et reseaux locaux,* both in French (Dunod, 2001 and 1998, respectively). He has also contributed chapters to *Performance Evaluation and Applications of ATM Networks,* edited by Demetres Kouvatsos (Kluwer, 2000) and *IP over WDM,* edited by Sudhir Dixit (Addison-Wesley, 2003). From 2000 to 2002, he was involved in the ITEA-BRIC European research program dealing with video distribution over IP networks and ADSL access systems. He regularly serves as an expert for the French Ministry of Industry. In addition, he has served as an expert in the field of access networks for the Flemish government of Belgium and in the field of optical networks for the National Science Foundation of the United States. He graduated from the Institut National des Telecommunications in Evry, France. He obtained his Ph.D. from ENST in Paris, France. He is on the program committees of several IFIP or IEEE conferences.

Index

Recent Titles in the Artech House Telecommunications Library

Vinton G. Cerf, Senior Series Editor

For further information on these and other Artech House titles, including previously considered out-of-print books now available through our In-Print-Forever® (IPF®) program, contact:

Artech House
685 Canton Street
Norwood, MA 02062
Phone: 781-769-9750
Fax: 781-769-6334
e-mail: artech@artechhouse.com

Artech House
46 Gillingham Street
London SW1V 1AH UK
Phone: +44 (0)20 7596-8750
Fax: +44 (0)20 7630-0166
e-mail: artech-uk@artechhouse.com

Find us on the World Wide Web at:
www.artechhouse.com